机器学习及其应用2021

张敏灵　胡清华　李宇峰　主编

清华大学出版社
北　京

内 容 简 介

本书为 MLA 2019—2020 的部分专家以综述形式介绍机器学习研究进展的专著，内容涉及监督学习、深度学习、强化学习、对抗学习、贝叶斯学习的基本理论和方法；同时介绍了机器学习在计算机视觉、自然语言处理、城市计算、语音信号处理、模式识别中的应用。

本书可供计算机与人工智能、自动化及其他相关专业的高校和科研院所师生及工程技术人员阅读参考。

图书在版编目(CIP)数据

机器学习及其应用. 2021 / 张敏灵，胡清华，李宇峰主编.—北京：清华大学出版社，2021.10
(2022.5重印)
ISBN 978-7-302-59095-8

Ⅰ. ①机…　Ⅱ. ①张…　②胡…　③李…　Ⅲ. ①机器学习　Ⅳ. TP181

中国版本图书馆 CIP 数据核字(2021)第 182106 号

责任编辑： 陈朝晖
封面设计： 傅瑞学
责任校对： 王淑云
责任印制： 沈　露

出版发行： 清华大学出版社
网　　址： http://www.tup.com.cn, http://www.wqbook.com
地　　址： 北京清华大学学研大厦 A 座　　**邮　　编：** 100084
社 总 机： 010-83470000　　**邮　　购：** 010-62786544
投稿与读者服务： 010-62776969, c-service@tup.tsinghua.edu.cn
质量反馈： 010-62772015, zhiliang@tup.tsinghua.edu.cn
印 装 者： 三河市龙大印装有限公司
经　　销： 全国新华书店
开　　本： 185mm×230mm　**印　张：** 20.25　**插页：** 4　**字　数：** 415 千字
版　　次： 2021 年 10 月第 1 版　**印　次：** 2022 年 5 月第 2 次印刷
定　　价： 99.00 元

产品编号：093693-01

评审委员会

名誉主任委员：张效祥

主 任 委 员：唐泽圣

副主任委员：陆汝钤

委　　　员：（以姓氏笔画为序）

王　珊　吕　建　李晓明

林惠民　罗军舟　郑纬民

施伯乐　焦金生　谭铁牛

序

第一台电子计算机诞生于20世纪40年代。到目前为止，计算机的发展已远远超出了其创始者的想象。计算机的处理能力越来越强，应用面越来越广，应用领域也从单纯的科学计算渗透到社会生活的方方面面：从工业、国防、医疗、教育、娱乐直至人们的日常生活，计算机的影响可谓无处不在。

计算机之所以能取得上述地位并成为全球最具活力的产业，原因在于其高速的计算能力、庞大的存储能力以及友好灵活的用户界面。而这些新技术及其应用有赖于研究人员多年不懈的努力。学术研究是应用研究的基础，也是技术发展的动力。

自1992年起，清华大学出版社与广西科学技术出版社为促进我国计算机科学技术与产业的发展，推动计算机科技著作的出版，设立了“计算机学术著作出版基金”，并将资助出版的著作列为中国计算机学会的学术著作丛书。时至今日，本套丛书已出版学术专著近50种，产生了很好的社会影响，有的专著具有很高的学术水平，有的则奠定了一类学术研究的基础。中国计算机学会一直将学术著作的出版作为学会的一项主要工作。本届理事会将秉承这一传统，继续大力支持本套丛书的出版，鼓励科技工作者写出更多的优秀学术著作，多出好书，多出精品，为提高我国的知识创新和技术创新能力，促进计算机科学技术的发展和进步作出更大的贡献。

中国计算机学会

2002年6月14日

序　言

2002 年秋天，由王珏教授策划和组织，复旦大学智能信息处理开放实验室（即现在的上海市智能信息处理重点实验室）举办了一次“机器学习及其应用”研讨会。该研讨会属于实验室的“智能信息处理系列研讨会”之一。十余位学者在综述机器学习各个分支的发展的同时报告了他们自己的成果。鉴于研讨会取得了非常好的效果，而机器学习领域又是如此之广阔，有那么多重要的问题还没有涉及或还没有深入，2004 年秋天王珏教授又和周志华教授联合发起并组织第二届“机器学习及其应用”研讨会，仍由复旦大学的实验室举办。这次研讨会又取得了非常好的效果，并且参加的学者比上次更多，报告的内容也更丰富。根据与会者的意见，决定把报告及相关内容编成一本书出版，以便与广大的国内学者共享研讨会的成果。

机器学习是人工智能研究的核心课题之一，不但有深刻的理论内蕴，也是现代社会中人们获取和处理知识的重要技术来源。它的活力久盛不衰，并且日呈燎原之势。对此，国内已经有多种定期和不定期的学术活动。本书的出版反映了机器学习界一种新型的“华山论剑”：小范围、全视角、更专业、更深入，可与大、中型机器学习会议互相补充。值得赞扬的是，它没有任何学派和门户之见，无论是强调基础的“气宗”，还是注重技术的“剑宗”，都能在这里畅所欲言，自由交流。我很高兴地获悉：第三届“机器学习及其应用”研讨会已经于 2005 年 11 月由周志华教授和王珏教授主持在南京大学成功举行。并且以后还将有第四届、第五届…… 作为一直跟踪这项活动并从中获得许多教益的一个学习者，我真希望它发展成这个领域的一个品牌，希望机器学习的优秀成果不断地由这里飞出，飞向全世界。

值得一提的是王珏教授有一篇颇具特色的综述文章为本书开道。长期以来，许多有识之士为国内学术界缺少热烈的争鸣风气而不安。因为没有争鸣就没有学术繁荣。细心的读者可以看出，这篇综述的观点并非都是传统观点的翻版，并且很可能不是所有的同行都认同的。作者深刻反思了机器学习这门学科诞生以来走过的道路，对一些被行内人士几乎认作定论的观点摆出了自己的不同看法。其目的不是想推出一段惊世骇俗的宏论，而是为了寻求真理、辨明是非。在这个意义上，王珏教授也可算是一位“独孤求败”。如果有人能用充分的论据指出其中可能存在的瑕疵，他也许会比听到一片鼓掌之声更感到宽慰。

随着本书的出版，中国计算机学会丛书知识科学系列也正式挂牌了。在衷心庆贺这个系列诞生的同时，我想重复过去说过的一段话：“二十多年来，知识工程主要是一门实验性科学。知识处理的大量理论性问题尚待解决。我们认为对知识的研究应该是一门具有坚实理论基础的科学，应该把知识工程的概念上升为知识科学。知识科学的进步将从根本上回答在知识工程中遇到过，但是没有很好解决的一系列重大问题。”本系列为有关领域的学者提供了一个宽松的论坛。衷心感谢王珏、周志华、周傲英三位编者把这本精彩的文集贡献给知识科学系列的首发式。我相信今后机器学习著作仍将是这个系列的一个常客。据悉，第四届机器学习研讨会将于今秋在南京大学举行，届时各种观点又将有进一步的发展和碰撞。欲知争鸣烽火如何再燃，独孤如何锐意求败，且看本系列下回分解。

陆汝钤

2006 年 1 月

前　言

经过近 20 年的发展，“机器学习及其应用”研讨会已成为国内机器学习界的著名学术品牌，其历史大致可以分为四个阶段。“机器学习及其应用”研讨会的创始阶段是2002—2004 年，它起源于 2002 年陆汝钤院士在复旦大学智能信息处理实验室发起组织的“智能信息处理系列研讨会”，被列为当年支持的研讨会之一。是年 11 月，第一届“机器学习及其应用”研讨会邀请了 10 余位专家闭门研讨，在复旦大学成功举办。2004 年 11 月周志华教授与王珏研究员在复旦大学主持举办了第二届“机器学习及其应用”研讨会。此次研讨会除邀请报告人外，还吸引了闻讯自发与会旁听的学者和研究生 100 余人。自此，研讨会的模式从闭门的学术讨论变成了开放的学术交流。

“机器学习及其应用”研讨会的起步阶段是 2005—2010 年。由于当时机器学习受到的关注和支持较少，组织者和主办单位需承担较多压力，2005 年研讨会移师南京大学计算机软件新技术国家重点实验室，周志华教授和王珏研究员主持举办了第三届研讨会，吸引了来自全国近 10 个省市的 250 余人旁听。此次研讨会确定了会议不征文、不收费、报告人由组织者邀请，以及“学术至上、其余从简”的办会宗旨，同时确定了研讨会举办的时间，如无特殊，则为当年 11 月份的第一个周末。此后，组织者争取到兄弟单位给予协助，2006 年、2007 年分别由南京航空航天大学信息科学与技术学院、南京师范大学数学与计算机学院协办了第四届和第五届研讨会，均吸引了来自全国 10 余个省市的约 300 人旁听；2008 年举办的第六届研讨会，适逢南京大学计算机学科建立 50 周年，与会人数达到了 380 余人；2009 年和 2010 年在南京大学分别举办了第七届和第八届研讨会，均有 400 余人旁听。这一时期为国内机器学习界的“垦荒”阶段，众多学者为研讨会作出了重要贡献。研讨会在国内机器学习领域乃至整个计算机领域逐渐产生了重要的影响，学界美名为“南京会议”。

“机器学习及其应用”研讨会的发展阶段是 2011—2016 年。随着国内科研条件的改善，机器学习逐渐获得更多关注和支持。为进一步推动机器学习在国内的发展，从 2011 年起，研讨会的举办地开始走出南京。2011 年和 2012 年由张长水教授和杨强教授主持，清华大学自动化系、智能科学与系统国家重点实验室、清华大学信息科学与技术国家实验室（筹）举办了第九届和第十届研讨会，两次会议均有 500 多人参加。2013 年、2014 年由张军平教授和高新波教授主持，分别在复旦大学计算机科学技术学院和上海市智能信息处理实验室举办了第十一届研讨会，在西安电子科技大学举办了第十二届研讨会，这两次会议分别有 600 多人和 800 多人参加。2015 年和 2016 年，研讨会再次回到南京大学举办，南京航空航天大学协办。此时国内机器学习研究和应用已经发展到一个新的高峰，这两次会议均有约 1400 人参加。至此，研讨会已经成为备受国内机器学习及相关领域研究人员关注的盛会。

自 2017 年起，“机器学习及其应用”研讨会进入定型阶段，采用“一年外地、一年南京”的模式举行。2017 年，第十五届研讨会由北京交通大学主办，北京大学协办。2018 年，第十六届研讨会在南京大学举行。这两届研讨会的参会人数都超过了预期，由此采取了预注册的方式，两次会议参会人数分别超过 1500 人与 2000 人。2019 年，第十七届研讨会由天津大学主办，参会人数超过了 2200 人。2020 年，第十八届研讨会在南京大学举行，受疫情影响采用线上与线下结合的形式进行。综合疫情防控及场地因素，此次研讨会线下参会人数约 700 人，线上参会人数约 4.5 万人。

研讨会自发起之日起，清华大学出版社敏锐地了解到这一学术动态。研讨会主要组织者与清华大学出版社达成共识，每两年对研讨会上专家交流的部分技术内容，经过整理结集成书，以飨读者。十余年来，陆续出版了《机器学习及其应用》《机器学习及其应用 2007》《机器学习及其应用 2009》《机器学习及其应用 2011》《机器学习及其应用 2013》《机器学习及其应用 2015》《机器学习及其应用 2017》《机器学习及其应用 2019》等系列图书。本书是以上系列图书的延续。

本书是对第十七届和第十八届中国“机器学习及其应用”研讨会交流内容的部分总结，共邀请了与会的 14 位专家就其研究领域撰文，以综述的形式探讨了机器学习不同分支及相关领域的研究进展。全书共分 14 章，涉及深度学习、监督学习、因果学习、鲁棒优化、强化学习、对抗学习等，以及机器学习在视觉大数据、自然语言处理、脑影像分析等方面的应用。

王立威教授对深度学习中的优化方法进行了研究，从理论上证明了使用梯度下降方法优化过的参数化 ResNet 可以收敛至全局最优值，并提出了一种对足够宽的神经网络具有二阶收敛保障的高效优化算法。

吴建鑫教授介绍了基于结构化剪枝的深度神经网络压缩技术，包括结构化剪枝的基本方式以及三种深度模型结构化剪枝方法，并对结构化剪枝的应用和未来发展进行了总结展望。

秦涛博士讨论了基于深度神经网络的高效语音合成问题，介绍了作者在模型推断、训练数据、模型参数三方面效率问题上的研究工作。

刘成林研究员讨论了面向开放世界的分类器学习问题，包括面向开放集的分类决策规则、分类器设计与学习以及作者提出的一种开放集卷积原型网络。

耿新教授介绍了一种新型监督学习技术——标记增强，从理论解释、增强方法、应用背景三个方面对标记增强技术做了系统总结。

张坤博士讨论了因果关系研究的两个主要方向——因果推断与因果性学习，并从经典方法、隐变量场景以及非稳态/异质数据场景三个方面介绍了相关研究进展。

苏文藻教授介绍了基于Wasserstein距离的分布鲁棒优化模型及其在结构风险最小化和对抗训练问题中的应用，并讨论了该方向有待研究的问题。

俞扬教授对基于环境模型的强化学习进行了总结，包括环境模型学习相关背景、基于对抗生成的环境模型学习以及在两种具体推荐任务中的应用。

韩亚洪教授分析了基于迁移和基于决策的深度神经网络黑盒攻击方法的缺陷，介绍了作者提出的基于迁移的黑盒攻击方法以及基于决策的黑盒攻击方法。

易津锋博士从对抗攻击、对抗防御以及模型鲁棒性评估三个方面介绍了对抗机器学习领域的最新进展，并基于产业界应用需求展望了该领域未来研究方向。

华先胜博士围绕基于城市视觉大数据的交通预测与调度具体任务，介绍了感知推理层、预测层以及干预层三个层面上的算法研究进展与实际应用示例。

黄萱菁教授梳理了基于深度学习的命名实体识别现有工作，并介绍了作者在未登陆词表示等四个方面所做的系列研究工作。

邱锡鹏教授梳理了自然语言表示学习的主要架构和预训练模型，并对预训练模型的未来研究趋势进行了总结展望。

张道强教授介绍了基于机器学习的脑解码方法最新进展，并对面向脑影像分析的脑解码未来研究方向进行了展望。

本书概括了国内机器学习及其应用的最新研究进展，可供计算机、自动化、信息处理及其他相关专业的研究人员、教师、研究生和工程技术人员参考，也可作为人工智能、机器学习课程的辅助内容，希望对有志于从事机器学习研究的人员有所帮助。

张敏灵　胡清华　李宇峰

2021年6月22日

目　录

深度学习中的优化方法

王立威　杨运昌

（北京大学信息科学技术学院，北京 100871）

1　引言

深度学习在许多领域取得了突破性的进展，然而目前对于深度学习的训练过程仍然缺乏理论上的认识。一个重要的问题是为什么即使目标函数是高度非凸的，使用简单的一阶优化算法（如梯度下降法）仍然能够找到一个全局最优点。解决这个问题将会帮助我们更深刻地从理论上理解深度学习的训练过程，设计更好的算法。

深度学习中的训练过程常常会耗费大量的时间和计算资源。为加速深度学习的训练过程，学者提出了许多新算法和新结构，但是这些方法在理论上往往只能给出一阶收敛的保障，或是不能适用于特定的网络结构，如神经网络图。本文从理论和算法两个角度对深度学习的训练过程进行了深入的研究。从理论上证明了使用梯度下降方法(GD)优化过参数化的深度残差神经网络(ResNet)时，该算法可以保证收敛到全局最优值。在算法上，提出了一种对足够宽的神经网络有二阶收敛保障并且能高效地应用到实际问题的优化算法：Gram-Gauss-Newton(GGN)方法。在实验中发现，在RSNABone Age和AFAD-LITE任务上，GGN 的收敛速度比其他算法要快得多，并且能收敛到更低的训练损失。提出的方法可以从时间和计算资源两方面加速神经网络的训练，从而实现更快的训练、更少的能耗，进而向设计更高效的深度学习方法迈出重要一步。

2　梯度下降方法的全局收敛性

目前深度神经网络的一个问题是对于任意给定的标签，使用随机初始化的一阶优化算法（例如，梯度下降法）能够将训练误差降为 0[1]。大部分学者认为过参数化是导致这一现象的主要原因。之前有人证明了宽度少于输入向量长度的神经网络不能逼近所有的函数[2]。在实际中使用的网络结构也是高度过参数化的。另外一个问题是，一般认为层数越深的网络训练起来会更加困难。为了解决这一困难，有人提出了深度残差神经网络(ResNet)[3]。这一网络结构使随机初始化的一阶算法能够训练更深层的网络。可以从理论上证明，深层线性网络这种残差结构避免了梯度在 0 的一个较大邻域内消失[4]。但是对于带有非线性激活函数的网络结构，使用残差结构带来的好处还缺少理论方面的理解。

本文对上述两个问题给出了理论上的回答。对于数据集大小为 n、神经网络层数为 H、宽度为 m 的情况，在假设损失函数是平方损失函数且激活函数是 Lipschitz 并且连续的情况下，证明了：①对于全连接神经网络，当 $m=\Omega\left(\operatorname{poly}(n)2^{O(H)}\right)$ 时，随机初始化的梯度下降法可以以线性收敛速度收敛到零训练损失。②对于残差神经网络，只要 $m=\Omega\left(\operatorname{poly}(n,H)\right)$，随机初始化的梯度下降法可以以线性收敛速度收敛到零训练损失。这一结果说明了残差神经网络相比于全连接神经网络的优势。③用同样的技术分析了残差卷积神经网络。只要 $m=\operatorname{poly}(n,p,H)$，随机初始化的梯度下降法可以以线性收敛速度收敛到零训练损失（其中 p 是 patch 的个数）。

2.1 问题描述

考虑平方损失函数：

$$\min_{\theta} L(\theta)=\frac{1}{2}\sum_{i=1}^{n}[f(\theta,\boldsymbol{x}_i)-y_i]^2$$

其中，$\{\boldsymbol{x}_i\}_{i=1}^n$ 是训练数据；$\{y_i\}_{i=1}^n$ 是标签；θ 是需要优化的参数；f 是预测函数。在我们的问题中，f 是神经网络。考虑如下三种结构。

（1）多层全连接网络

设 $\boldsymbol{x}\in R^d$ 为输入，$\boldsymbol{W}^{(1)}\in R^{m\times d}$ 为第一层权重矩阵，$\boldsymbol{W}^{(h)}\in R^{m\times m}$ 为第 h 层权重矩阵，$\boldsymbol{a}\in R^m$ 为输出层，$\sigma(\cdot)$ 为激活函数。递归定义预测函数为

$$\boldsymbol{x}^{(0)}=\boldsymbol{x}$$

$$\boldsymbol{x}^{(h)}=\sqrt{\frac{c_\sigma}{m}}\sigma\left(\boldsymbol{W}^{(h)}\boldsymbol{x}^{(h-1)}\right),\quad 1\leqslant h\leqslant H$$

$$f(\boldsymbol{x},\theta)=\boldsymbol{a}^{\mathrm{T}}\boldsymbol{x}^{(H)}$$

其中，$c_\sigma=\left\{E_{x\sim N(\mathbf{0},\mathbf{1})}\left[\sigma(x)^2\right]\right\}^{-1}$ 是归一化因子。

（2）残差神经网络

残差神经网络可以递归定义为

$$\boldsymbol{x}^{(1)}=\sqrt{\frac{c_\sigma}{m}}\sigma\left(\boldsymbol{W}^{(1)}\boldsymbol{x}\right)$$

$$\boldsymbol{x}^{(h)}=\boldsymbol{x}^{(h-1)}+\frac{c_{\text{res}}}{H\sqrt{m}}\sigma\left(\boldsymbol{W}^{(h)}\boldsymbol{x}^{(h-1)}\right), 2\leqslant h\leqslant H$$

$$f_{\text{res}}\left(\boldsymbol{x},\theta\right)=\boldsymbol{a}^{\mathrm{T}}\boldsymbol{x}^{(H)}$$

其中，$0<c_{\text{res}}<1$是一个小常数。

（3）卷积残差网络

最后考虑卷积残差网络结构。同样地，用递归的方式定义网络结构。令$\boldsymbol{x}^{(0)}\in R^{d_0\times p}$为输入，其中$d_0$为输入通道数，$p$为像素数。对于任意的$h$，令通道数$d_h=m$，像素数为$p$。当给定$\boldsymbol{x}^{(h-1)}\in R^{d_{h-1}\times p}$时，定义$\phi_h(\cdot)$为将$\boldsymbol{x}^{(h-1)}$分割为$p$个部分的操作函数。每个部分的大小为$qd_{h-1}$，即$\phi_h\left(\boldsymbol{x}^{(h-1)}\right)\in R^{qd_{h-1}\times p}$。例如，当$q=3$时，有

$$\phi_h\left(\boldsymbol{x}^{(h-1)}\right)=\begin{pmatrix}\left(\boldsymbol{x}_{1,0:2}^{(h-1)}\right)^{\mathrm{T}} & \cdots & \left(\boldsymbol{x}_{1,p-1:p+1}^{(h-1)}\right)^{\mathrm{T}}\\ \vdots & \ddots & \vdots\\ \left(\boldsymbol{x}_{d_{h-1},0:2}^{(h-1)}\right)^{\mathrm{T}} & \cdots & \left(\boldsymbol{X}_{d_{h-1},p-1:p+1}^{(h-1)}\right)^{\mathrm{T}}\end{pmatrix}$$

其中，$\boldsymbol{x}_{:,0}^{(h-1)}=\boldsymbol{x}_{:,p+1}^{(h-1)}=\boldsymbol{0}$。这一函数有如下性质：

$$\|\boldsymbol{x}^{(h-1)}\|_F\leqslant\|\phi_h\left(\boldsymbol{x}^{(h-1)}\right)\|_F\leqslant\sqrt{q}\,\|\boldsymbol{x}^{(h-1)}\|_F$$

令$\boldsymbol{W}^{(h)}\in R^{d_h\times qd_{h-1}}$，递归定义如下：

$$\boldsymbol{x}^{(1)}=\sqrt{\frac{c_\sigma}{m}}\sigma\left(\boldsymbol{W}^{(1)}\phi_1\left(\boldsymbol{x}\right)\right)\in R^{m\times p}$$

$$\boldsymbol{x}^{(h)}=\boldsymbol{x}^{(h-1)}+\frac{c_{\text{res}}}{H\sqrt{m}}\sigma\left[\boldsymbol{W}^{(h)}\phi_h\left(\boldsymbol{x}^{(h-1)}\right)\right]\in R^{m\times p}$$

其中，$0<c_{\text{res}}<1$是一个小常数。对$\boldsymbol{a}\in R^{m\times p}$，输出定义为

$$f_{\text{cnn}}\left(\boldsymbol{x},\theta\right)=\langle\boldsymbol{a},\boldsymbol{x}^{(H)}\rangle$$

为了学习神经网络，考虑使用随机初始化的梯度下降方法来寻找损失函数的全局极小值点。具体来说，采用如下的随机初始化策略：对每一层h，每个权重都是从一个标

准高斯分布中采样，即 $\boldsymbol{W}_{ij}^{(h)} \sim N(0,1)$。输出层 $\boldsymbol{a}$ 中的每个元素也是从标准高斯分布中采样。使用梯度下降法训练每一层的权重。对 $k=1,2,\cdots$ 和 $\boldsymbol{h} \in [H]$，有

$$\boldsymbol{W}^{(h)}(k)=\boldsymbol{W}^{(h)}(k-1)-\eta \frac{\partial L[\theta(k-1)]}{\partial \boldsymbol{W}^{(h)}(k-1)}$$

$$\boldsymbol{a}(k)=\boldsymbol{a}(k-1)-\eta \frac{\partial L[\theta(k-1)]}{\partial \boldsymbol{a}(k-1)}$$

其中，$\eta>0$ 是步长。

2.2 梯度下降的收敛性质

首先分析全连接神经网络。对于步长为常数的情况，梯度下降会以线性的速度收敛到全局最优点。递归定义 Gram 矩阵如下：

$$\boldsymbol{K}_{ij}^{(0)}=\langle \boldsymbol{x}_i, \boldsymbol{x}_j \rangle$$

$$\boldsymbol{A}_{ij}^{(h)}=\begin{pmatrix} \boldsymbol{K}_{ii}^{(h-1)} & \boldsymbol{K}_{ij}^{(h-1)} \\ \boldsymbol{K}_{ji}^{(h-1)} & \boldsymbol{K}_{jj}^{(h-1)} \end{pmatrix}$$

$$\boldsymbol{K}_{ij}^{(h)}=c_\sigma E_{(u,v)^{\mathrm{T}} \sim N\left(\mathbf{0}, \boldsymbol{A}_{ij}^{(h)}\right)}\left[\sigma(u)\sigma(v)\right]$$

$$\boldsymbol{K}_{ij}^{(H)}=c_\sigma \boldsymbol{K}_{ij}^{(H-1)} E_{(u,v)^{\mathrm{T}} \sim N\left(\mathbf{0}, \boldsymbol{A}_{ij}^{(H-1)}\right)}\left[\sigma'(u)\sigma'(v)\right]$$

梯度下降的收敛性质与 Gram 矩阵的最小特征值 $\lambda_{\min}\left(\boldsymbol{K}^{(H)}\right)$ 密切相关。下面介绍用梯度下降法训练全连接网络的收敛性定理。

定理 1 梯度下降训练全连接神经网络的收敛速度。假设对任意的 $i \in [n], \| \boldsymbol{x}_i \|_2 = 1$, $|\boldsymbol{y}_i| = \boldsymbol{O}(1)$，并且每层的隐藏节点个数为

$$m=\Omega\left(2^{O(H)} \max\left\{\frac{n^4}{\lambda_{\min}^4\left(\boldsymbol{K}^{(H)}\right)}, \frac{n}{\delta}, \frac{n^2 \log\left(\frac{Hn}{\delta}\right)}{\lambda_{\min}^2\left(\boldsymbol{K}^{(H)}\right)}\right\}\right)$$

对于梯度下降法，当步长取为

$$\eta=O\left(\frac{\lambda_{\min}\left(\boldsymbol{K}^{(H)}\right)}{n^2 2^{O(H)}}\right)$$

时，则以至少$1-\delta$的概率，对$k=1,2,\cdots$每次迭代的损失函数满足：

$$L(\theta(k))\leqslant\left[1-\frac{\eta\lambda_{\min}\left(\boldsymbol{K}^{(H)}\right)}{2}\right]^{k}L(\theta(0))$$

这个定理表明，如果m足够大，并且将步长调至合适值，则梯度下降法会以线性速度收敛到全局最优值。这个定理的主要假设是需要一个足够大的网络，该网络的宽度m取决于n,H和$1/\lambda_{\min}\left(\boldsymbol{K}^{(H)}\right)$。对$n$和$1/\lambda_{\min}\left(\boldsymbol{K}^{(H)}\right)$的依赖都是多项式级别的，但是对$H$的依赖是指数级的。

接下来分析残差神经网络。同样，定义 Gram 矩阵如下：

$$\boldsymbol{K}_{ij}^{(0)}=\langle\boldsymbol{x}_i,\boldsymbol{x}_j\rangle$$

$$\boldsymbol{A}_{ij}^{(h)}=\begin{pmatrix}\boldsymbol{K}_{ii}^{(h-1)} & \boldsymbol{K}_{ij}^{(h-1)}\\ \boldsymbol{K}_{ji}^{(h-1)} & \boldsymbol{K}_{jj}^{(h-1)}\end{pmatrix}$$

$$\boldsymbol{b}_i^{(1)}=\sqrt{c_\sigma}E_{u\sim N\left(\mathbb{0},\boldsymbol{K}_{ii}^{(0)}\right)}\left[\sigma(u)\right]$$

$$\boldsymbol{K}_{ij}^{(1)}=c_\sigma E_{(u,v)^{\mathrm{T}}\sim N\left(\boldsymbol{0},\boldsymbol{A}_{ij}^{(1)}\right)}\left[\sigma(u)\sigma(v)\right]$$

$$\boldsymbol{K}_{ij}^{(h)}=\boldsymbol{K}_{ij}^{(h-1)}+E_{(u,v)^{\mathrm{T}}\sim N\left(\boldsymbol{0},\boldsymbol{A}_{ij}^{(h)}\right)}\left[\frac{c_{\text{res}}\boldsymbol{b}_i^{(h-1)}\sigma(u)}{H}+\frac{c_{\text{res}}\boldsymbol{b}_j^{(h-1)}\sigma(v)}{H}+\frac{c_{\text{res}}^2\sigma(u)\sigma(v)}{H^2}\right]$$

$$\boldsymbol{b}_i^{(h)}=\boldsymbol{b}_i^{(h-1)}+\frac{c_{\text{res}}}{H}E_{u\sim N\left(\mathbb{0},\boldsymbol{K}_{ii}^{(h-1)}\right)}\left[\sigma(u)\right]$$

$$\boldsymbol{K}_{ij}^{(H)}=\frac{c_{\text{res}}^2}{H^2}\boldsymbol{K}_{ij}^{(H-1)}E_{(u,v)^{\mathrm{T}}\sim N\left(\boldsymbol{0},\boldsymbol{A}_{ij}^{(H-1)}\right)}\left[\sigma'(u)\sigma'(v)\right]$$

与全连接相比，残差神经网络的 Gram 矩阵取决于序列$\left\{\boldsymbol{b}^{(h)}\right\}_{h=1}^{H-1}$。这是由于残差神经网络包含了跳跃连接。可以证明，只要训练数据是非退化的，那么$\lambda_{\min}\left(\boldsymbol{K}^{(H)}\right)$就是严格为正的，并且不指数依赖于$H$。接下来介绍关于残差神经网络的收敛定理。

定理 2 残差神经网络的梯度下降法收敛速度定理。假设对任意的$i\in[n],\|\boldsymbol{x}_i\|_2=1$，$|\boldsymbol{y}_i|=O(1)$，并且每层的隐藏节点个数为

$$m=\Omega\left(\max\left\{\frac{n^4}{\lambda_{\min}^4\left(\boldsymbol{K}^{(H)}\right)H^6},\frac{n^2}{\lambda_{\min}^2\left(\boldsymbol{K}^{(H)}\right)H^2},\frac{n}{\delta},\frac{n^2\log\left(\frac{\boldsymbol{Hn}}{\delta}\right)}{\lambda_{\min}^2\left(\boldsymbol{K}^{(H)}\right)}\right\}\right)$$

对于梯度下降法，当步长取为 $\eta=O\left(\frac{\lambda_{\min}\left(\boldsymbol{K}^{(H)}\right)H^2}{n^2}\right)$ 时，则以至少 $1-\delta$ 的概率，对 $k=1,2,\cdots$，每次迭代的损失函数满足:

$$L(\theta(k))\leqslant\left[1-\frac{\eta\lambda_{\min}\left(\boldsymbol{K}^{(H)}\right)}{2}\right]^k L(\theta(0))$$

与全连接网络相比，这个定理中的神经元个数和收敛速度都是多项式依赖于 n 和 H 的。这里不包含指数依赖的主要原因是残差网络的跳跃连接使整个网络结构在初始化阶段和训练阶段更加稳定。

对 $\boldsymbol{m}$ 的要求包括四项，前两项表明在训练中 Gram 矩阵始终保持稳定，第三项用来保证在初始化阶段每一层的输出都是近似正则化的，第四项用来约束 Gram 矩阵在初始阶段的扰动大小。

接下来分析残差卷积神经网络。首先定义该网络结构的 Gram 矩阵。对任意的 $(i,j)\in[n]\times[n],(l,r)\in[p]\times[p]$ 及 $h=2,3,\cdots,H-1$ 有

$$\boldsymbol{K}_{ij}^{(0)}=\phi_1\left(\boldsymbol{x}_i\right)^{\mathrm{T}}\phi_1\left(\boldsymbol{x}_j\right)\in R^{p\times p}$$

$$\boldsymbol{A}_{ij}^{(h)}=\begin{pmatrix}\boldsymbol{K}_{ii}^{(h-1)} & \boldsymbol{K}_{ij}^{(h-1)}\\ \boldsymbol{K}_{ji}^{(h-1)} & \boldsymbol{K}_{jj}^{(h-1)}\end{pmatrix}$$

$$\boldsymbol{K}_{ij}^{(1)}=c_\sigma E_{(u,v)^{\mathrm{T}}\sim N\left(\boldsymbol{0},\boldsymbol{A}_{ij}^{(1)}\right)}\left[\sigma(u)\sigma(v)\right]$$

$$\boldsymbol{b}_i^{(1)}=\sqrt{c_\sigma}E_{u\sim N\left(\boldsymbol{0},\boldsymbol{K}_{ii}^{(0)}\right)}\left[\sigma(u)\right]$$

$$\boldsymbol{H}_{ij}^{(h)}=\boldsymbol{K}_{ij}^{(h-1)}+E_{(u,v)^{\mathrm{T}}\sim N\left(\boldsymbol{0},\boldsymbol{A}_{ij}^{(h-1)}\right)}\left[\frac{c_{\text{res}}\boldsymbol{b}_i^{(h-1)\mathrm{T}}\sigma(u)}{H}+\frac{c_{\text{res}}\boldsymbol{b}_j^{(h-1)\mathrm{T}}\sigma(v)}{H}+\frac{c_{\text{res}}^2\sigma(u)^{\mathrm{T}}\sigma(v)}{H^2}\right]$$

$$\boldsymbol{K}_{ij,lr}^{(h)}=\operatorname{tr}\left(\boldsymbol{H}_{ij,\boldsymbol{D}_l^{(h)}\boldsymbol{D}_r^{(h)}}\right)$$

$$\boldsymbol{b}_i^{(h)}=\boldsymbol{b}_i^{(h-1)}+\frac{c_{\text{res}}}{H}E_{\boldsymbol{u}\sim N\left(\boldsymbol{0},\boldsymbol{K}_{ii}^{(h-1)}\right)}\left[\sigma(\boldsymbol{u})\right]$$

$$\boldsymbol{M}_{ij,lr}^{(H)} = \boldsymbol{K}_{ij,lr}^{(H-1)} E_{(\boldsymbol{u},\boldsymbol{v})\sim N\left(\boldsymbol{0},\boldsymbol{A}_{ij}^{(H-1)}\right)}\left[\sigma'(u_l)\sigma'(v_r)\right]$$

$$\boldsymbol{K}_{ij}^{(H)} = \operatorname{tr}\left(\boldsymbol{M}_{ij}^{(H)}\right)$$

其中，$\boldsymbol{u}$ 和 $\boldsymbol{v}$ 都是随机向量，$D_l^{(h)} = \left\{s : \boldsymbol{x}_{:,s}^{(h-1)} \in \text{the } l\text{th patch}\right\}$；$\boldsymbol{K}_{ij}^{(h)}$ 是 $p \times p$ 维矩阵；$\boldsymbol{K}_{ij,lr}$ 表示 (l,r) 元素。下面介绍残差卷积网络的收敛定理。

定理 3 残差卷积神经网络的梯度下降法收敛速度定理。假设对任意的 $i \in [n]$，$\|\boldsymbol{x}_i\|_F = 1, |\boldsymbol{y}_i| = O(1)$，并且每层的隐藏节点个数为

$$m = \Omega\left(\max\left\{\frac{n^4}{\lambda_0^4 H^6}, \frac{n^2}{\lambda_0^2 H^2}, \frac{n}{\delta}, \frac{n^2 \log\left(\frac{\boldsymbol{Hn}}{\delta}\right)}{\lambda_0^2}\right\}\operatorname{poly}(p)\right)$$

对于梯度下降法，当步长取为 $\eta = O\left(\frac{\lambda_0 H^2}{n^2 \operatorname{poly}(p)}\right)$ 时，则以至少 $1-\delta$ 的概率，对 $k = 1, 2, \cdots$，每次迭代的损失函数满足：

$$L(\theta(k)) \leqslant \left[1 - \frac{\eta\lambda_{\min}\left(\boldsymbol{K}^{(H)}\right)}{2}\right]^k L(\theta(0))$$

定理 3 与残差神经网络的收敛定理相似。每层所需的神经元个数仅仅多项式取决于数据个数和深度，并且步长也是多项式级别的大小。唯一的不同是这里多了对于 p 的多项式依赖。

3 Gram-Gauss-Newton 方法

虽然深度神经网络的优化是高度非凸的，但即使在具有随机标签的数据集上，随机梯度下降法（SGD）等简单算法也能实现接近零的训练损失。为了从理论上理解这一现象，最近有一系列的工作考虑了基于神经正切核（neural tangent kernel，NTK）思想的过参数化神经网络的优化[5]。粗略地讲，NTK 的思想是对网络的输出在参数的局部区域进行线性逼近，从而得到一个核特征图，即输出关于参数的梯度。然而，对于足够宽的神经网络，GD 或 SGD 的动态优化相当于使用 GD 或 SGD 来解决 NTK 核的回归问题。那么，一个自然的问题就出现了，能否在每一步通过直接求解核回归来获得加速？

本节对这个问题给出了肯定的答案，并揭示了 NTK 回归与 Gauss-Newton 方法[6]之间的联系。本节提出了一种 Gauss-Newton 方法，称为 Gram-Gauss-Newton(GGN)方法，用于优化具有平方损失函数的神经网络。GGN 不做梯度下降，而是在优化的每一步中求解对 NTK 的核回归。按照这个思路，证明了在与前人工作类似的过参数化条件下，full-batchGGN 可以优化网络，与梯度下降的线性速率相比，具有二次收敛速率。进一步将 GGN 与 mini-batch 处理方案相结合。尽管传统的观点认为 mini-batch 会在二阶方法中引入偏梯度，并可能导致优化的发散，但还是给出了 mini-batch 版本的 GGN 的线性收敛结果。进一步指出 GGN 隐含 Gauss-Newton 方法的重构，Gauss-Newton 方法是一种经典的二阶算法，常用于解决有平方损耗的非线性回归问题。

本节不仅从理论上证明了 GGN 的有效性，而且从经验上验证了其在实际应用中的潜力。首先对标量回归任务进行了实验。在这种情况下，与 SGD 等一阶方法相比，mini-batch GGN 在每个纪元的计算开销很小。这与大多数二阶方法形成了鲜明的对比，后者计算二阶信息的成本很高。证明了在这两个实际应用中使用具有标准宽度的实用神经网络(如 ResNet-32)，我们提出的 GGN 算法可以比 SGD、Adam 和 K-FAC 等几种基准算法收敛得更快，性能更好。

3.1 过参数化神经网络的 Gram-Gauss-Newton 方法

对于足够宽的网络，使用梯度下降法求解回归问题与使用梯度下降法在每一步求解 NTK 核回归具有相似的动态性。然而，也可以使用核最小二乘回归的显式公式来求解每一步的 NTK 核回归问题。通过显式求解 NTK 核回归而不是使用梯度下降法，可以预期优化的速度会加快。我们提出的 Gram-Gauss-Newton(GGN)方法使用每个时间步的 Gram 矩阵 $\boldsymbol{G}_t$ 直接求解 NTK 核回归。NTK 在时间 t 的特征图可以表示为 $x \to \nabla_w f(w_t, x)$，RKHS 中的线性参数为 $w - w_t$，目标为 $\boldsymbol{f}(w, x_i) - \boldsymbol{f}(w_t, x_i)$。因此，对于 $\boldsymbol{J}_{t,S}(w - w_t) = (\boldsymbol{f}(w, x_i) - \boldsymbol{f}(w_t, x_i))$ 关于 $(w - w_t)$ 的核(无脊)回归问题，有这样的更新方法：

$$w_{t+1} = w_t - \boldsymbol{J}_{t,S}^{\mathrm{T}} \boldsymbol{G}_{t,S}^{-1} (\boldsymbol{f}_{t,S} - \boldsymbol{y}_S)$$

其中，$\boldsymbol{J}_{t,S}$ 是在训练数据集 S 上计算出的 t 时刻的特征矩阵，它等价于 Jacobian 矩阵；$\boldsymbol{f}_{t,S}$ 和 $\boldsymbol{y}_S$ 分别是神经网络的矢量化输出和 S 上对应的目标；$\boldsymbol{G}_{t,S} = \boldsymbol{J}_{t,S} \boldsymbol{J}_{t,S}^{\mathrm{T}}$ 是 NTK 在 S 上的 Gram 矩阵。

我们注意到，现有的 Gauss-Newton 方法中没有一个工作利用这一点来加速对过参数化模型的优化。相反，受 NTK 回归的启发，GGN 自然地给出了这一更新规则，避免了 Gauss-Newton 矩阵的计算。

学习算法的设计不仅要考虑优化，还要考虑泛化能力。已有研究表明，使用 mini-batch 代替 full-batch 来计算导数对于学习的模型是否具有良好的泛化能力至关重要[7-9]。因此，提出了一个 mini-batch 版本的 GGN。更新规则如下：

$$w_{t+1} = w_t - \boldsymbol{J}_{t,B_t}^{\mathrm{T}} \boldsymbol{G}_{t,B_t}^{-1} (\boldsymbol{f}_{t,B_t} - \boldsymbol{y}_{B_t})$$

其中，B_t 是迭代 t 时使用的 mini-batch；$\boldsymbol{J}_{t,B_t}$ 和 $\boldsymbol{G}_{t,B_t}$ 分别是使用 B_t 的数据计算出的 Jacobian 和 Gram 矩阵；$\boldsymbol{f}_{t,B_t}, \boldsymbol{y}_{B_t}$ 分别是 B_t 上的矢量化输出和对应的目标。当使用典型的 batch 大小时，$\boldsymbol{G}_{t,B_t} = \boldsymbol{J}_{t,B_t} \boldsymbol{J}_{t,B_t}^{\mathrm{T}}$ 是一个很小的矩阵。我们的更新规则只需要计算 Gram 矩阵 $\boldsymbol{G}_{t,B_t}$ 及其逆矩阵。需要注意的是，$\boldsymbol{G}_{t,B_t}$ 的大小等于 batch 的大小，通常非常小，所以这也大大降低了计算成本。

利用核脊回归的思想(也可以看作是 Gauss-Newton 方法的 Levenberg-Marquardt 扩展)引入 GGN 的以下变体：

$$w_{t+1} = w_t - \boldsymbol{J}_{t,B_t}^{\mathrm{T}} (\lambda \boldsymbol{G}_{t,B_t} + \alpha I)^{-1}$$

其中，$\lambda > 0$ 是控制学习过程的另一个超参数。我们的算法如下：

1. 输入：训练集 S，超参数 λ 和 α
2. 初始化网络参数 w_0，设置 $t = 0$
3. 对每次迭代：
4. 在数据集中取一个 mini-batch B_t
5. 计算 Jacobian 矩阵 $\boldsymbol{J}_{t,B_t}$
6. 计算 Gram 矩阵 $\boldsymbol{G}_{t,B_t} = \boldsymbol{J}_{t,B_t} \boldsymbol{J}_{t,B_t}^{\mathrm{T}}$
7. 更新参数

$$w_{t+1} = w_t - \boldsymbol{J}_{t,B_t}^{\mathrm{T}} \boldsymbol{G}_{t,B_t}^{-1} \left(\boldsymbol{f}_{t,B_t} - \boldsymbol{y}_{B_t} \right)$$

8. $t = t + 1$
9. 结束

3.2 GGN 在过参数化神经网络上的收敛性

接下来证明，对于足够宽的两层神经网络，有：

（1）full-batch 版本的 GGN 以二次收敛速度收敛；

（2）mini-batch 版本的 GGN 以线性速度收敛。

正如通过 NTK 的视角所解释的那样，结果是这样一个事实的结论：对于足够宽的神经网络，如果按照合适的概率分布对权重进行初始化，那么在高概率的情况下，在包含初始化点和全局最优的邻域内，网络关于参数的输出会接近一个线性函数（但网络的输入是非线性的）[10-13]。虽然实际中使用的神经网络远没有这么宽，但这仍然激励了我们设计 GGN 算法。下面介绍设置和结果。

使用以下结构的两层网络：

$$f(w,x)=\frac{1}{\sqrt{M}}a^{\mathrm{T}}\sigma\left(W^{\mathrm{T}}x\right)=\frac{1}{\sqrt{M}}\sum_{r=1}^{M}a_r\sigma(w_r x)$$

其中，$x\in\mathbb{R}^d$ 为输入；M 为网络宽度；$W=\left(w_1^{\mathrm{T}},\cdots,w_M^{\mathrm{T}}\right)^{\mathrm{T}}$；$\sigma(\cdot)$ 为激活函数。W 的每个元素都用标准高斯分布初始化 $w_r\sim\mathcal{N}(0,I_d)$，并且 a 的每个元素都是从 $\{\pm 1\}$ 上的均匀分布初始化的。

为了证明的清晰性，只对参数 W 进行网络训练，并假设激活函数 $\sigma(\cdot)$ 是 ℓ-Lipschitz 和 β-smooth 的，而 ℓ,β 被认为是 $O(1)$ 常数。关键的发现是，在这种初始化下，Gram 矩阵 $\boldsymbol{G}$ 有一个渐近极限，在温和的条件下（如输入数据不是退化的等），它是一个正定矩阵。

$$\boldsymbol{K}\left(x_i,x_j\right)=\mathbb{E}_{w\sim\mathcal{N}(0,I)}[x_i^{\mathrm{T}}x_j\sigma'\left(wx_i\right)\sigma'\left(wx_j\right)]$$

假设矩阵 $\boldsymbol{K}$ 是正定矩阵，并将其最小特征值表示为 $\lambda_0=\lambda_{\min}\left(\boldsymbol{K}\right)>0$。下面介绍 full-batch GGN 定理。

定理 4　过参数化神经网络上 full-batch GGN 的二次收敛。如果假设成立，假设数据的规模是 $\|\boldsymbol{x}_i\|_2=O(1),|\boldsymbol{y}_i|=O(1),i\in\{1,\cdots,n\}$，网络宽度 $M=\Omega\left(\max\left(\frac{n^4}{\lambda_0^4},\frac{n^2 d\log\left(\frac{16n}{\delta}\right)}{\lambda_0^2}\right)\right)$，且为随机初始化，那么 GGN 的 full-batch 版本的更新规则以 $1-\delta$ 的概率满足：

（1）每次迭代时的 Gram 矩阵 $\boldsymbol{G}_{t,S}$ 是可逆的；

（2）损失函数以下列方式趋近于

$$\|\boldsymbol{f}_{t+1}-\boldsymbol{y}\|_2\leqslant\frac{C}{\sqrt{M}}\|\boldsymbol{f}_t-\boldsymbol{y}\|_2^2$$

对于一些与 M 无关的 C，这是一个二阶收敛。

注意，将 λ 设为 1 对定理中的加速收敛很重要。而结合对 Jacobian 的更精细的分析，将 Jacobian 关于参数扰动是这个想法稳定的形式化，能够取得比线性收敛更好的结果，其中还涉及学习率超参数。我们的理论表明，在过参数化的设置下或者 NTK 保持稳定的情况下，将学习率设置为接近 1 是加快收敛的好选择，这一点在实验中得到了验证。

对于 mini-batch 版本的 GGN，通过对其 NTK 极限的分析，该算法实质上是在 mini-batch 引起的子空间上做串行子空间修正，所以 mini-batch GGN 与用于求解线性方程系统的 Gauss-Siedel 方法类似。与 full-batch 情况类似，GGN 采取的是“子空间上的核回归问题”的精确解，这比只在子空间上做梯度步长优化要快。此外注意到，现有的 SGD 在过参数化网络上的收敛结果通常采用这样的思路：当步长大小被一个与平稳性相关的量所约束时，SGD 可以简化为 GD。然而，我们的分析采取了与 GD 的分析不同的方式，因此并不依赖于小步长。

在下文中，将 $\boldsymbol{G}_0 \in \mathbb{R}^{n\times n}$ 表示为初始 Gram 矩阵。令 $n=bk$，其中 b 为 batch 大小，k 为 batch 的序号，并令 $\boldsymbol{G}_{0,ij} := \boldsymbol{G}_0((i-1)b+1:ib,(j-1)b+1:jb)$ 为 $\boldsymbol{G}_0$ 的第 $(i,\ j)$ 个 $b\times b$ 块。定义迭代矩阵 $\boldsymbol{A}=\boldsymbol{L}^{\mathrm{T}}(\boldsymbol{D}-\boldsymbol{L})^{-1}\in\mathbb{R}^{n\times n}$，其中 $\boldsymbol{D}$ 和 $\boldsymbol{L}$ 分别代表 $\boldsymbol{G}_0$ 的块对角线部分和块下三角部分。下面将证明 mini-batch GGN 的收敛性与 $\boldsymbol{A}$ 的谱半径有很大关系，为了简化证明，假设矩阵 $\boldsymbol{A}$ 是可对角化的。选择 $\boldsymbol{A}$ 的任意对角化为 $\boldsymbol{A}=P^{-1}QP$，并表示 $\mu := \|P\|_2\|P^{-1}\|_2$。下面介绍 mini-batch GGN 的定理。

定理 5　mini-batch GGN 对过参数化神经网络的收敛性。假设假设 1 成立且数据的规模是 $\|\boldsymbol{x}_i\|_2=O(1),|\boldsymbol{y}_i|=\boldsymbol{O}(1),i\in\{1,\cdots,n\}$，$\mu$ 定义如上。使用 GGN 的 mini-batch 版本，batch $\boldsymbol{B}_t$ 按照循环序列被选择，batch 大小固定为 b，每个 epoch 更新 $k=\dfrac{n}{b}$。如果网络宽度 $M=\max\left(\Omega\left(\dfrac{\mu^2 n^{18}}{\lambda_0^{16}}\right),\Omega\left(\dfrac{n^2\mathrm{d}\log\left(\dfrac{16n}{\delta}\right)}{\lambda_0^2}\right)\right)$，在随机初始化下，下列以 $1-\delta$ 的概率成立。

（1）每次迭代时的 Gram 矩阵 $\boldsymbol{G}_{t,B_t}$ 是可逆的；

（2）在 T 个 epoch 后损失收敛为零，有

$$\|\boldsymbol{f}_{Tk}-\boldsymbol{y}\|_2 \leqslant \mu\sqrt{n}\left[1-\Omega\left(\frac{\lambda_0^2}{n^2}\right)\right]^{\mathrm{T}}$$

4　实验

在理论结果的激励下，还测试了 GGN 在网络宽度没有定理要求的较温和条件下的实际任务上的性能，并将我们提出的 GGN 算法与实际应用中的几种基准算法进行了比

较。特别是，主要研究了两个标量回归任务——RSNABone Age[14]和 AFAD-LITE[15]。

在介绍结果之前，首先简单分析一下 GGN 对标量回归问题的计算复杂性。对于每一次迭代，GGN 的计算主要有两个步骤。

（1）正向，然后反向推进，计算 Jacobian 矩阵 $\boldsymbol{J}$。

（2）计算更新 $\boldsymbol{J}^{\mathrm{T}}\left(\lambda G+\alpha I\right)^{-1}(f-y)$。

对于步骤（1），我们注意到，虽然在常见的优化算法如 SGD 中没有明确计算 Jacobian 矩阵，但它的计算只需要对自动微分框架进行一些修改，这一步引入的计算开销可以相当小。对于步骤(2)中的计算，由于 Jacobian 的大小为 $b\times m$，其中 b 为 batch 大小，m 为参数量，因此只需要 $O(b^2m+b^3)$ 来计算矩阵的乘法和求逆。在实际设置下，步骤(2)中的矩阵计算要比步骤（1）中的反推快得多。因此，对于标量回归任务，与 SGD 相比，GGN 每次迭代只引入了比较小的计算开销。

4.1 实验设置

按照标准设置训练神经网络，在这里列出一些关键的实验内容。

（1）基准

将我们的 GGN 算法与两个一阶基准进行比较，即 SGD with momentum[16]和 Adam[17]，以及 KFAC[16]，后者是一种流行的二阶优化算法。

（2）模型

研究 ResNet-32[3]架构，为了适应不同的优化方法，有三种变体：带有 batch 归一化的 ResNet-32（简称 ResNetBN）、带有 group 归一化的 ResNet-32（简称 ResNetGN）和无归一化的 ResNet-32（简称 ResNet）。对于基准，将 SGD 和 Adam 应用于 ResNetBN 和 ResNetGN，只将 KFAC 应用于 ResNet。就我们的 GGN 算法而言，batch 归一化与我们之前的假设不一致，即回归函数的形式为 $f(w,x)$，它只取决于 w 和单一的输入数据 x。出于这个原因，GGN 算法并不直接适用于 ResNetBN，只在 ResNetGN 和 ResNet 上测试我们提出的算法。

4.2 实验结果

（1）收敛性

RSNA Bone Age 和 AFAD-LITE 任务的不同优化算法的训练损失曲线如图 1 所示。图 1 给出了 SGD 和 Adam 在 ResNetBN 的效果，其速度始终快于 ResNetG。在这两个任务上，我们提出的方法的收敛速度比 SGD 快得多。Adam 和 KFAC 在一开始可能比 GGN 优化得更快，然而 GGN 很快就追上了，并在最后收敛到更低的训练损失。

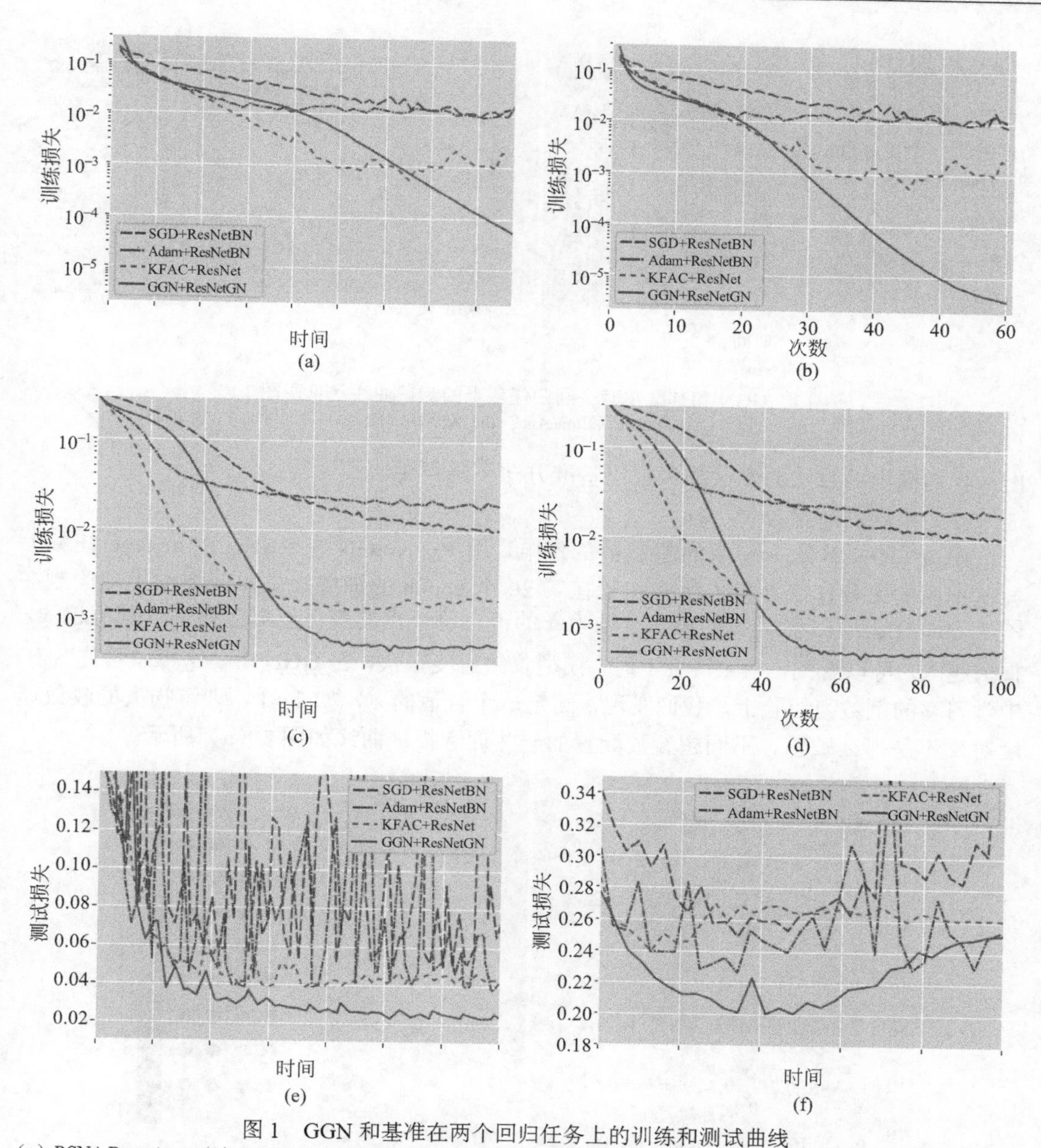

图 1　GGN 和基准在两个回归任务上的训练和测试曲线

（a）RSNA Bone Age；（b）RSNA Bone Age；（c）AFAD-LITE；（d）AFAD-LITE；（e）RSNA Bone Age；（f）AFAD-LITE

（2）泛化性能

通过两个数据集上的测试损耗曲线来评估泛化性能。从图 2（a）和图 2（b）可以看

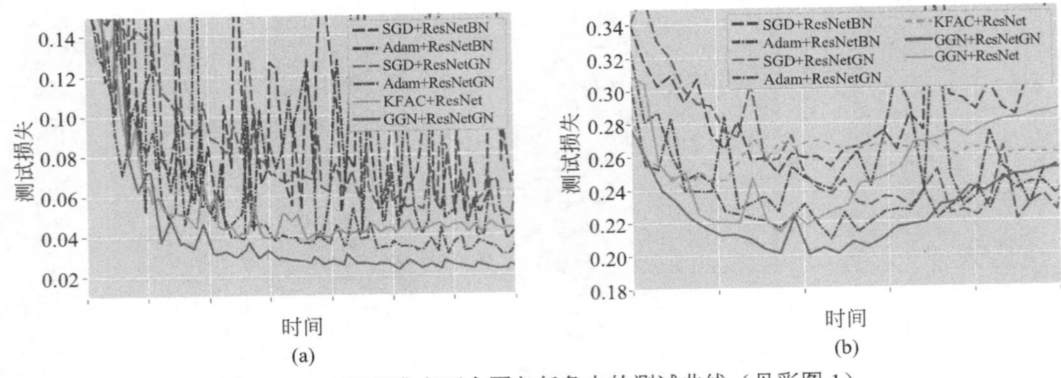

图 2　GGN 和基准在两个回归任务上的测试曲线（见彩图 1）

（a）RSNA Bone Age；（b）AFAD-LITE

出，我们提出的算法的测试损失也比基准方法下降得更快。

（3）不同的超参数

研究 GGN 算法中使用的超参数的影响。用 ResNetGN 在 RSNA Bone Age 任务上尝试不同的 λ 和 α，并报告所有实验在第 20 个 epoch 的训练损失。所有结果都绘制在图 3（b）中,图中 x 轴为 λ 的值，y 轴为 α 的值，每个点的灰色值对应损失，颜色越浅，损失越大。可以看到，当 λ 接近 1 时，模型收敛速度较快，在 GGN 中，α 可以看作 SGD 中学习率的倒数。经验上，我们发现，给定一个合适的 λ，如 $\lambda=1$，训练损失的收敛速度对 α 不是那么敏感。不同超参数配置的一些训练损失曲线如图 3（a）所示。

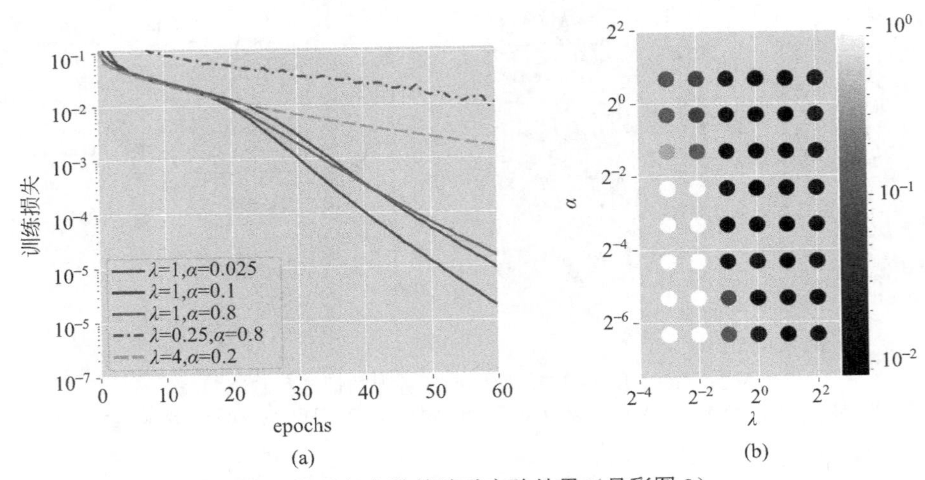

图 3　关于超参数的消融实验结果（见彩图 2）

（a）不同超参数配置的训练损失曲线；（b）所有实验在第 20 个 epoch 的训练损失

5 小结

本文从 NTK 和 Gauss-Newton 方法之间的联系中得到启发，提出了一种利用过参数化神经网络优化具有平方损失的神经网络的 GGN 方法，并保证了对过参数化神经网络的加速收敛。在标量回归任务上，证明了 GGN 算法在每次迭代时的计算开销比 SGD 小。在两个任务上的实验结果表明，GGN 与标准网络架构的其他一阶和二阶基准相比，具有很好的优势。

提出了 GGN——一种在过参数化环境下具有可证明的快速收敛保证的神经网络优化方法。该方法可以从时间和计算资源两方面加速神经网络的训练，从而实现更快的训练和更少的能耗。本文提出的理论还可以为深度学习算法提供更多的启示，从而有可能在理论保证的前提下得到更好的深度学习算法。

参考文献

[1] ZHANG C, BENGIO S, HARDT M, et al. Understanding deep learning requires rethinking generalization[Z]. arXiv preprint arXiv:1611.03530, 2016.

[2] LU Z, PU H, WANG F, et al. The expressive power of neural networks: A view from the width[Z]. arXiv preprint arXiv:1709.02540, 2017.

[3] HE K, ZHANG X, REN S, et al. Deep residual learning for image recognition[C]//Proceedings of the IEEE conference on computer vision and pattern recognition. 2016: 770-778.

[4] HARDT M, MA T. Identity matters in deep learning[Z]. arXiv preprint arXiv:1611.04231, 2016.

[5] JACOT A, GABRIEL F, HONGLER C. Neural tangent kernel: Convergence and generalization in neural networks[Z]. arXiv preprint arXiv:1806.07572, 2018.

[6] GOLUB G. Numerical methods for solving linear least squares problems[J]. NumerischeMathematik, 1965, 7(3): 206-216.

[7] HARDT M, RECHT B, SINGER Y. Train faster, generalize better: Stability of stochastic gradient descent[C]// International Conference on Machine Learning. PMLR. 2016: 1225-1234.

[8] KESKAR N S, MUDIGERE D, NOCEDAL J, et al. On large-batch training for deep learning: Generalization gap and sharp minima[Z]. arXiv preprint arXiv:1609.04836, 2016.

[9] MASTERS D, LUSCHI C. Revisiting small batch training for deep neural networks[Z]. arXiv preprint arXiv:1804.07612, 2018.

[10] DU S, LEE J, LI H, et al. Gradient descent finds global minima of deep neural networks[C]// International Conference on Machine Learning. PMLR, 2019: 1675-1685.

[11] DU S S, ZHAI X, POCZOS B, et al. Gradient descent provably optimizes over-parameterized neural networks[Z]. arXiv preprint arXiv:1810.02054, 2018.

[12] ALLEN-ZHU Z, LI Y, LIANG Y. Learning and generalization in overparameterized neural networks, going beyond two layers[Z]. arXiv preprint arXiv:1811.04918, 2018.

[13] ZOU D, CAO Y, ZHOU D, et al. Stochastic gradient descent optimizes over-parameterized deep relu networks. arxiv e-prints, art[Z]. arXiv preprint arXiv:1811.08888, 2018.

[14] HALABI S S, PREVEDELLO L M, KALPATHY-CRAMER J, et al. The RSNA pediatric bone age machine learning challenge[J]. Radiology, 2019, 290(2): 498-503.

[15] NIU Z, ZHOU M, WANG L, et al. Ordinal regression with multiple output cnn for age estimation[C]// Proceedings of the IEEE conference on computer vision and pattern recognition. 2016: 4920-4928.

[16] MARTENS J, GROSSE R. Optimizing neural networks with kronecker-factored approximate curvature[C]// International conference on machine learning. PMLR. 2015: 2408-2417.

[17] KINGMA D P, BA J. Adam: A method for stochastic optimization[Z]. arXiv preprint arXiv:1412.6980, 2014.

[18] QIAN N. On the momentum term in gradient descent learning algorithms[J]. Neural networks, 1999, 12(1): 145-151.

结构化剪枝综述

吴建鑫　王环宇　张永顺

（计算机软件新技术国家重点实验室（南京大学），南京 210023）

1　引言

卷积神经网络（convolutional neural networks）特别是深度神经网络（deep neural networks）在计算机视觉任务上发挥着愈发重要的作用。在一个神经网络模型中，通常包含卷积层、汇合层、全连接层、非线形层等基本结构，通过这些基本结构的堆叠，最终形成我们常用的深度神经网络。早在 1998 年，LeCun 等人使用少数几个基本结构组成 5 层的 LeNet-5 网络[1]，并在 MNIST 数据集上得到了 98.9%的分类精度，但此时的深度神经网络还相对简单，并且只能用于简单的任务；在 2012 年的 ImageNet 图像分类竞赛中，AlexNet[2]将深度提高到了 8 层，并且达到了远超传统方法的结果；此后，VGG 团队提出的 VGG-Net[3]进一步加深了网络，使网络最高达到了 19 层。虽然增加网络的深度能够带来性能的提升，但也不能无限制地增加网络深度，随着网络的加深，梯度消失会愈发严重，并且模型会变得愈发难以训练。因此在 2016 年，He 等人提出 ResNet[4]，在模型中加入残差结构，并一举将网络的深度提高到 152 层。至此，随着深度学习研究的逐步推进，神经网络可以变得更宽、更深、更复杂，与此同时带来更好的表示能力和性能表现。

当一些研究者将模型变得更大、更深时，另一些研究者则考虑在保持模型精度的同时使模型变得更小、更快，其中一类重要的方法为模型压缩。模型压缩大致上可以分为四类：模型量化、模型剪枝、低秩近似和知识蒸馏。

通常来说，我们用 32 位浮点数来保存模型，模型量化主要考虑用更小位数来保存模型参数，通常使用的有 16 位浮点数和 8 位整数，其参数量和计算量都会相应地随着存储位数而成倍降低；更有甚者，将模型量化成二值网络[5]、三元权重[6]或者同或网络[7]。例如，经过简单量化之后的 MobileNetV1[8]仅仅只有 4～5 MB，能够轻松部署在各种移动平台上。

模型剪枝[9-12]主要分为结构化剪枝和非结构化剪枝，非结构化剪枝去除不重要的神经元，相应地，被剪除的神经元和其他神经元之间的连接在计算时会被忽略。由于剪枝后的模型通常很稀疏，并且破坏了原有模型的结构，所以这类方法被称为非结构化剪枝。非结构化剪枝能极大降低模型的参数量和理论计算量，但是现有硬件架构的计算方式无法对其进行加速，所以在实际运行速度上得不到提升，需要设计特定的硬件才可能加速。与非结构化剪枝相对应的是结构化剪枝，结构化剪枝通常以滤波器或者整个网络层为基本单位进行剪枝。一个滤波器被剪枝，那么其前一个特征图和下一个特征图都会发生相应的变化，但是模型的结构却没有被破坏，仍然能够通过 GPU 或其他硬件来加速，因此这类方法被称为结构化剪枝。

低秩近似[13-15]将一个较大的卷积运算或者全连接运算替换成多个低维的运算。常用的低秩近似方法有 CP 分解法[13]、Tucker 分解[14]和奇异值分解[15]。例如，一个 $M \times N$ 的全连接操作若能近似分解为 $M \times d$ 和 $d \times N$（其中 $d \ll M,N$），那么这一层全连接操作的计算量和参数量将被极大地缩减。

知识蒸馏（knowledge distillation）[16]通过使用一个足够冗余的教师模型，将其知识“传授”给紧凑的学生模型。在训练时同时使用教师模型的软标签和真实标记的硬标签来共同训练学生模型，从而使学生模型达到接近教师模型的性能，也因此能够降低达到目标精度所需的计算量和模型大小。

上述模型压缩方法能配合使用，一个模型经过结构化剪枝之后，由于其结构没有发生重要变化，所以能紧接着进行低秩近似以减少参数量和计算量，最后再通过参数量化进一步减少参数量并加速。近些年来，随着物联网的发展，企业将深度学习模型部署在嵌入式设备的需求在快速增长，而嵌入式设备计算能力有限，并且由于成本原因希望部署的模型尽可能地小。模型压缩的意义在于保证精度的同时尽可能减少计算量和参数量，因此对于嵌入式设备的部署有切实的价值。模型压缩包含很多内容，本文将主要关注剪枝算法中的结构化剪枝，首先介绍结构化剪枝的一些基本方式，然后介绍一些经典的和最新的结构化剪枝算法，最后对结构化剪枝的应用和未来发展进行总结和展望。

2 剪枝方式介绍

在结构化剪枝中，最基本的方式是滤波器剪枝（也称为通道剪枝），本节将首先介绍滤波器剪枝的内容，并以此为基础介绍三种常见的模型剪枝方式。

FLOPs（floating-point operations）是浮点计算量的简称[17]，通常使用 FLOPs 来表示模型的计算复杂度。将一个输入通道数为 C_{in}，输出通道数为 C_{out} 的卷积层简化记为 $[C_{\text{out}}, C_{\text{in}}, K]$，表示这个卷积层里有 C_{out} 个 $C_{\text{in}} \times K \times K$ 的卷积单元，分别和输入特征进行卷积操作。这里将一次乘加（multiply and accumulate）算作两个浮点运算，所以对于输入为 $H \times W \times C_{\text{in}}$ 的特征，经过这一层卷积的浮点运算量为 $\text{FLOPs} = 2HWC_{\text{in}}K^2C_{\text{out}} + HWC_{\text{out}}$，其中 HWC_{out} 表示偏置加法带来的计算量。参数量即为所含参数的数量，一个卷积层的参数量 $\text{Params} = K^2C_{\text{in}}C_{\text{out}} + C_{\text{out}}$，等号右边的两项分别表示卷积核和偏置的参数量。我们使用一个两层的卷积网络来展示滤波器剪枝的细节，为简化模型，均使用卷积核为 3、步长为 1 的普通卷积，并且不考虑卷积中的偏置操作。在我们的简化模型中，每层卷积的浮点计算量为 $18HWC_{\text{in}}C_{\text{out}}$，参数量为 $9C_{\text{in}}C_{\text{out}}$。

如图 1 所示，以这个简化的两层卷积网络为例来对剪枝过程进行分析。假定第一层卷积的输入和输出通道数分别为 C_1 和 C_2，第二层卷积的输入和输出通道数分别为 C_2 和 C_3。那么一个高度为 H、宽度为 W、通道数为 C_1 的输入特征 $F_1 \in \mathbb{R}^{H \times W \times C_1}$ 经过第一层卷积将得到中间特征 $F_2 \in \mathbb{R}^{H \times W \times C_2}$，然后再经过第二层卷积操作得到输出特征 $F_3 \in \mathbb{R}^{H \times W \times C_3}$。这个双层卷积网络的参数量为 $9C_2(C_1 + C_3)$，计算量为 $18HWC_2(C_1 + C_3)$。对这个简单的双层卷积模型进行剪枝，将中间层的特征 F_2 由 C_2 个通道减少到 C_2' 个通道，那么相应地，第一个卷积层中产生 $C_2 - C_2'$ 个对应通道特征的卷积单元变得不再需要，所以新的卷积层可以表示为 $[C_2', C_1, 3]$；第二个卷积层中和相应通道特征进行卷积的参数也变得不再需要，所以新的卷积层可以表示为 $[C_3, C_2', 3]$。因此，第一个卷积的输出通道数量和第二个卷积的输入通道数量都由 C_2 相应减少到 C_2'，至此一个通道剪枝过程结束。

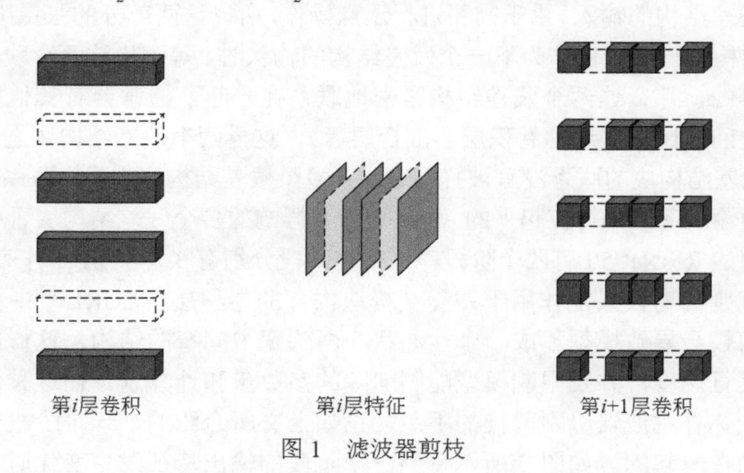

图 1　滤波器剪枝

剪枝之后的双层卷积网络的参数量变为 $9C_2'(C_1+C_3)$，计算量变为 $18HWC_2'(C_1+C_3)$。当 C_2' 减小到 C_2 的一半时，这个简单双层网络的参数量和计算量都将减半。

2.1 滤波器级别剪枝

在上述的例子中，相邻的两个卷积层紧紧联系在一起，第一个卷积层输出维度的变化将引起第二个卷积层的输入维度发生相应的变化。这样的卷积操作可以由第一个卷积层和第二个卷积层的组合一直进行到倒数第二个卷积层和最后一个卷积层的组合。

由于卷积层是用于视觉任务的神经网络的基础，所以对于几乎所有的网络结构，都可以使用滤波器级别剪枝（filter-level pruning）来减少参数量和计算量。大多数剪枝算法也是基于滤波器级别剪枝来精简模型。

滤波器级别剪枝的核心在于减少一个中间特征的数量，其前一个和后一个卷积层需要发生相应的变化。

2.2 阶段级别的剪枝

现在使用的卷积神经网络大多都参考了 ResNet[4]中的残差结构。残差结构的具体形式不唯一，在 ResNet-50 中的残差结构由三层卷积和一个跨层连接构成；在 MobileNetV2[18]中，采用逐点卷积和逐层卷积来减少参数量和计算量，也有一个跨层连接。除此之外也有很多其他的形式，但是不管具体的构建形式如何，残差结构都可以表示为

$$y=f(x)+x$$

即残差结构的输出由两部分逐点相加得到：一部分是残差结构内最后一层卷积的输出，另一部分是残差结构的输入。由于需要进行逐点操作，所以这两部分的张量维度必须一致。当一个残差结构的输出直接作为下一个残差结构的输入时，等于说这两个残差结构的输出通道数是相等的，那么这两个残差结构紧密地联系在一起。通常会有其他操作夹在两个残差结构块之间（比如普通的卷积层、汇合层等），这样两个残差结构块之间的联系便被打破了。在残差结构块的联系没有被打破之前，多个残差结构块将紧密联系在一起，这些残差结构块的输出通道是对应起来的。将这样相互影响的多个残差结构块看成一个“阶段”（stage），常用的 ResNet-50 有四个阶段，这四个阶段分别有 3，4，6，3 个残差结构块。

滤波器级别的剪枝只能作用于残差结构块内部的卷积层，CURL[19]中指出只进行滤波器级别的剪枝会导致模型形成一个沙漏状、两头宽中间窄的结构，这样的结构会影响模型的实际运行速度，并且限制模型能够剪除的参数量和计算量。在这种情况下，阶段级别的剪枝能弥补滤波器级别剪枝的不足。正如上文所介绍的，一个阶段中的残差结构块是紧密联系在一起的，如图 2 所示，当一个阶段的输出特征发生变化时（一些特征被

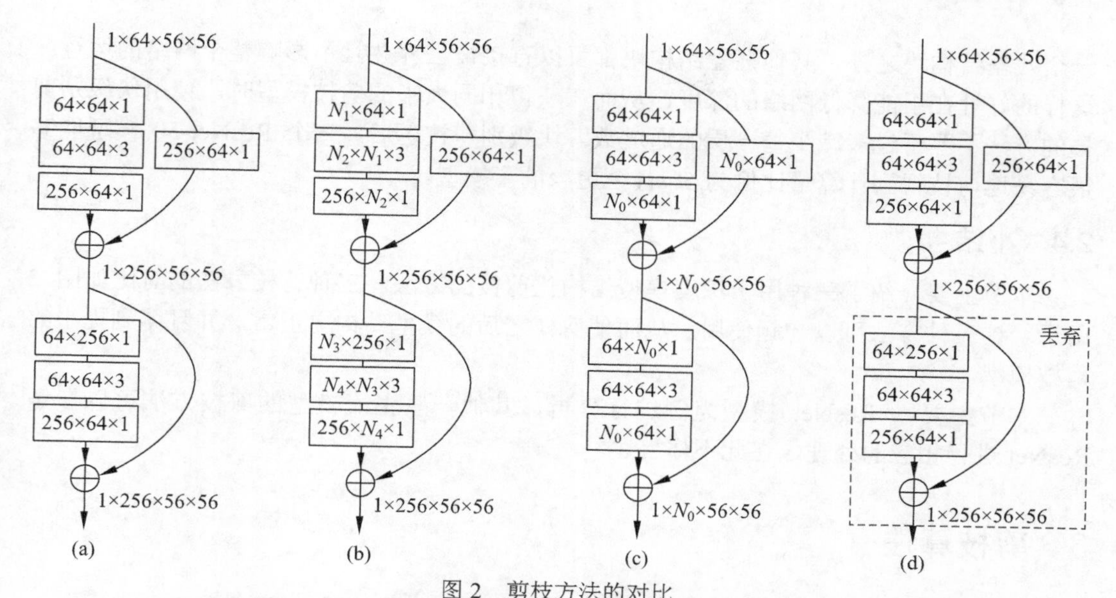

图 2　剪枝方法的对比

（a）原始模型；（b）滤波器级别剪枝；（c）阶段级别剪枝；（d）块级别剪枝

抛弃），其对应的每个残差结构的输入特征和输出特征都要发生相应的变化，所以整个阶段中，每个残差结构的第一个卷积层的输入通道数及最后一个卷积层的输出通道数都要发生相同的变化。由于这样的影响只限定在当前的阶段，不会影响之前和之后的阶段，因此称这个剪枝过程为阶段级别的剪枝（stage-level pruning）。

阶段级别的剪枝加上滤波器级别的剪枝能够使网络的形状更均匀，以避免出现沙漏状的网络结构。此外，阶段级别的剪枝能够剪除更多的网络参数，这给网络进一步压缩提供了支持。

2.3　块级别的剪枝

当我们想得到比较小的剪枝模型（计算量是原模型的10%甚至更少）时，需要大量缩减各层特征的通道数，这样模型的宽度会大幅减少，而深度不会发生改变，最终将形成一个特别窄但是仍然比较深的模型。此时，模型的深度会限制模型速度，并且一味减少模型的宽度可能会导致在某些层出现极端窄的情况。此时块级别的剪枝（block-level pruning）能帮助解决这些困难。

块级别的剪枝是直接丢弃某些残差结构块，由于残差结构的数学形式可以表达为 $y=f(x)+x$，丢弃残差结构后等于这一层变为 $y=x$。以 ResNet 为例，其每个阶段的第一个残差结构通常会降低特征图的分辨率并提高特征图的通道数，所以除了每个阶段第一个残差结构块之外，其他残差结构块都可以直接被丢弃并且不影响整个网络的运行。这样的好处在于能降低网络的深度，从而在获得相同大小的剪枝模型时，使用块级别剪枝的方法不至于过多减少每一层的通道数。块级别剪枝之后，一个 ResNet-50 模型能够很容易地剪枝得到计算量比例为 5%甚至更少的子模型。

2.4 小结

本节主要介绍了三种用于深度模型结构化剪枝的方法。三种剪枝方法的对比如图 2 所示。这三种方法可以一起使用，从而使剪枝之后的模型结构更正常，并且得到更小的剪枝模型。

本节主要以 ResNet 模型为例进行分析，其他一些主流模型的剪枝方法可以参考 ResNet 进行相应的处理，在此不作赘述。

3 剪枝算法

3.1 剪枝算法的运行流程及评价方式

剪枝算法的目的在于减少原模型的参数及计算量，同时尽可能保证得到的子模型表达能力相比原模型来说损失较少。本节介绍当前主要的几种剪枝流程及其典型的代表方法，同时在剪枝实验上，由于本文只针对基于卷积神经网络的图像分类领域模型剪枝，所以在实验方面只介绍此领域主要的评价框架，包括数据集、常用模型及评估准则。

3.1.1 剪枝的具体流程

剪枝是减少模型参数量和计算量的经典方法。随着深度学习的兴起及卷积神经网络在图像分类领域的大量应用，各种各样的剪枝方法不断涌现出来。虽然剪枝方法的种类很多，但是其核心思想还是对神经网络的结构进行剪枝，目前剪枝算法的总体流程大同小异，可以归结为三类：标准剪枝、基于子模型采样的剪枝和基于搜索的剪枝，如图 3 所示。

标准剪枝主要包含三个部分：训练、剪枝、微调，如图 3（a）所示。标准剪枝流程的详细说明如下：

（1）训练。在剪枝流程中，训练部分只需进行一次即可，训练的目的是为剪枝算法获取在特定任务上训练好的原始模型。

（2）剪枝。剪枝最重要的环节是对网络结构进行重要性评估，而这一重要性评估的环节也是各种剪枝算法最主要的区别之一。评估的模型结构主要包含滤波器、块等结构。

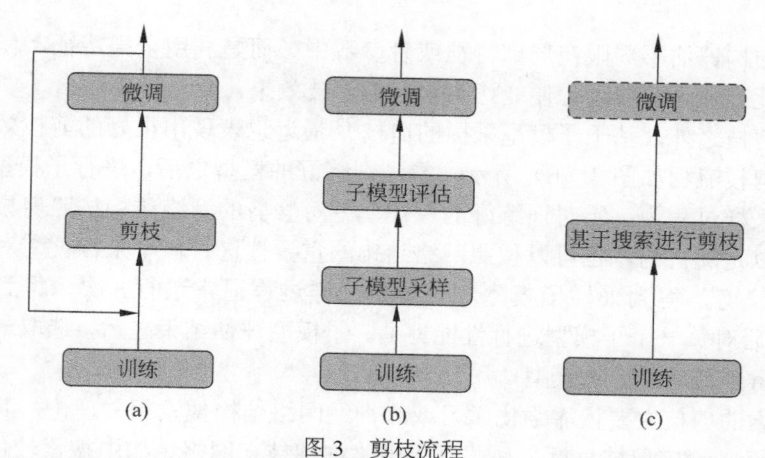

图3 剪枝流程

（a）标准剪枝；（b）基于子模型采样的剪枝；（c）基于搜索的剪枝

对网络结构的重要性评估可以分为网络参数驱动的评估及数据驱动的评估两类方法。

- 基于网络参数驱动的方法利用模型本身的参数信息来衡量模型结构的重要性，如参数的 l_1 正则化或者 l_2 正则化[9-12]，该类方法评估过程不依赖输入数据。
- 基于数据驱动的方法通过利用训练数据来对网络结构的重要性作出评估，如通过统计滤波器输出结果经过激活层之后 0 值的个数来评价该滤波器的重要性[20]。

在 3.2 节中，会对参数驱动及数据驱动两类评估方法做更详细的介绍。在网络结构评估完成之后，只需根据网络结构的评估结果按照第 2 节“剪枝方式介绍”中的模型剪枝方式，修剪掉不重要的网络结构即可。

（3）微调。微调是恢复被剪枝操作影响的模型表达能力的必要步骤。结构化模型剪枝会对原始模型结构进行调整，因此剪枝后的模型参数虽然保留了原始的模型参数，但是由于模型结构的改变，剪枝后模型的表达能力会受到一定程度的影响。微调过程通过将剪枝后的子模型在训练集进行微调训练，能够恢复子模型的表达能力。

（4）再剪枝。再剪枝过程将微调之后的子模型再送到剪枝模块中，再次进行模型结构评估和剪枝过程。通过再剪枝过程，使每次剪枝都在性能更优的模型上进行，不断阶段性地优化剪枝模型，直到模型能够满足剪枝目标需求。

标准剪枝流程是目前剪枝算法的主要流程[10,20-26]，同时在标准剪枝的基础上，有些相关工作对标准剪枝过程进行改进[17,19]。文献[17]将剪枝过程集成到模型微调中，不再区分微调和剪枝两部分，并提出一个新的可训练的网络层用于剪枝过程，该网络层生成二进制码，二进制码中的 0 值对应的网络结构将被剪掉。文献[19]通过计算原始模型和去掉对应网络结构的子模型之间的 KL 散度来衡量每一个网络结构的重要性，这种计算方式使网络结构评估不局限在局部特征或者参数中，而是利用全局特征，使评估结果更为精确，因此无须再剪枝过程便能达到很好的剪枝效果。

除标准剪枝之外，基于子模型采样的剪枝[27]最近也表现出很好的剪枝效果。基于子模型采样的剪枝流程如图 3（b）所示，得到训练好的模型之后，进行子模型采样过程。一次子模型采样过程为：①对训练好的原模型中可修剪的网络结构按照剪枝目标进行采样，采样可以是随机的，也可以按照网络结构的重要性进行概率采样；②对采样出的网络结构进行修剪，得到采样子模型。子模型采样过程通常进行 n 次，得到 n 个子模型（$n \geqslant 1$），之后对每一个子模型进行性能评估。子模型评估结束之后，选取最优的子模型进行微调以得到最后的剪枝模型。

基于搜索的剪枝主要依靠强化学习或者神经网络结构搜索相关理论，其主要流程如图 3（c）所示。给定剪枝目标之后，基于搜索的剪枝在网络结构中搜索较优的子结构，这个搜索过程往往伴随着网络参数的学习过程，因此一些基于搜索的剪枝算法在剪枝结束后无须再进行微调。在 3.3 节“基于搜索的剪枝算法”中，我们会对其进一步做详细的介绍和探讨。

3.1.2　评价剪枝效果使用的数据集、模型和评估准则

图像分类领域的剪枝实验常用的数据集包括 CIFAR-10[28]，ILSVRC-2012[29]。

- CIFAR-10 数据集包括 10 个类别，其训练集和验证集分别包含 50 000 张和 10 000 张 RGB 彩色图像。CIFAR-10 数据集类别平衡，即每个类别包含 5000 张训练图像和 1000 张验证图像，每张图像的分辨率为 32×32。
- ILSVRC-2012 是 ImageNet2012 竞赛数据集，其总类别数量为 1000。ILSVRC-2012 的训练集包含 1 281 167 张训练图像，每个类别的训练图像数量从 732 到 1300 不等。同时其验证集包含 50 000 张图像，验证集类别平衡，每一个类别包含 50 张验证图像。

在图像分类领域卷积神经网络的众多种类中，VGG-Net[3]和 ResNet[4]应用广泛。由于移动端设备部署的需求越来越强烈，MobileNet 系列[8,18]也逐渐成为剪枝实验的主要网络结构之一。

在实际实验中，VGG-Net 通常采用 VGG-16（16 层网络结构的 VGG-Net）作为实验网络结构，而 MobileNet 通常采用 MobileNetV1[8]和 MobileNetV2[18]作为 MobileNet 在剪枝实验中的网络结构。同时对于 ResNet，CIFAR-10 实验上通常采用 ResNet-52（针对 CIFAR-10 数据集设计的 52 层网络结构的 ResNet），而 ILSVRC-2012 数据集上通常采用 ResNet-50（50 层网络结构的 ResNet）。

通常从三个方面评价神经网络剪枝算法的优劣：准确率及准确率变化、模型大小变化、网络前向时间变化。在实际实验中，这三种指标都需要统计，以综合体现剪枝算法的效果。

- 准确率及准确率变化。在测试剪枝算法的实验结果时，需要在数据集的验证集上统计剪枝子模型的 Top-1 验证集准确率及 Top-5 验证集准确率，作为剪枝子网络的性能指标。同时，需要记录剪枝子模型准确率相对于原始模型的准确率变化（通常是负数），准确率变化大表明该子模型相比原模型准确率下降明显。
- 模型大小变化。剪枝实验的一个重要目的是减少原始模型的参数量，因此需要统计剪枝子模型的模型参数量相比于原始模型的变化量。该模型参数变化量作为模型大小变化的指标，变化量越大，表示模型大小压缩得越多。
- 网络前向时间变化。对神经网络进行剪枝的最终目的在于保证子模型准确率的前提下，减少子模型的部署需求及其运行时间，而运行时间仅靠模型大小甚至 FLOPs 大小来衡量都是不准确的，因为有些模型参数量、FLOPs 虽然小，但是其结构不适用于硬件架构，所以其实际运行时间可能并未提升。模型前向时间变化可以由模型在硬件上的前向时间来体现，FLOPs 可供参考。

3.2 剪枝重要性评估准则

剪枝过程通过定义重要性指标来剪除相对不重要的参数。剪枝的指标往往是启发式的，根据评估指标时利用的信息，基于重要性的模型剪枝算法评估准则可分为数据驱动和参数驱动两类。数据驱动和参数驱动的评估准则部分方法可见表 1。

3.2.1 参数驱动的评估准则

基于参数驱动的评估准则的网络结构评估无须依赖输入数据，最常用的方式为 l_1 正则化和 l_2 正则化。

在文献[10]中，对每一个卷积层的滤波器计算其权重绝对值之和（l_1 正则化）作为其重要性分数，之后根据每一个滤波器的重要性分数排序，修剪掉重要性较低的滤波器。Networks[38]limming[11]将 l_1 正则化加在损失函数中，用以约束批归一化层（batch normalization）[38]的参数 γ 稀疏程度，之后通过参数 γ 来评估滤波器的重要性，选取不重要的滤波器进行剪枝。SFP[30]中利用滤波器权重的 l_2 正则化来衡量每一个滤波器的重要性，在迭代剪枝的过程中将 l_2 正则化值较小的滤波器权重赋值为 0，网络剪枝结束时将网络中权重为 0 的滤波器修剪掉来获得最终的剪枝模型。文献[39]从变分贝叶斯方法的角度说明批归一化层中参数 γ 可作为滤波器的评估指标，该方法也将参数 γ 加入损失函数中予以更新，最后修剪掉较小 γ 值对应的滤波器。

表 1　结构化剪枝中的网络结构重要性评估准则

	具体方法	评价准则简介
参数驱动	文献[10]和文献[11]	l_1 正则化
	文献[30]	l_2 正则化
	文献[3]和文献[11]	批归一化层[38]中的 γ 参数
	文献[24]	几何中位数
	文献[31]	滤波器的谱聚类
数据驱动	文献[20]	激活输出值中 0 的激活值数量
	文献[26]	特征层的秩分解
	文献[32]	对特征图进行子空间聚类
	文献[23]和文献[33]	网络特征重建误差
	文献[34]	利用主成分分析来确定滤波器重要性
	文献[35]	网络结构重要值
	文献[36]	根据网络梯度进行剪枝
	文献[22]	利用剪枝前后网络特征熵值
	文献[19]	子模型和原模型之间的特征 KL 散度
	文献[27]	子模型进行 AdaptiveBN[27]操作后的准确率
	文献[37]	为网络每一层通过训练选择不同的评价准则

FPGM[24]从几何中位数的角度进行滤波器重要性评估。针对每一个卷积层包含的滤波器，FPGM 先计算其几何中位数，认为如果某个滤波器接近几何中位数，可以认为这个滤波器的信息跟其他滤波器重合，于是可以去掉这个滤波器而不对网络产生大的影响。SCSP[32]利用谱聚类的思想，先对各层滤波器进行谱聚类，聚类之后根据滤波器的贡献进行排序，贡献小的滤波器会被剪枝。

基于参数驱动的评估准则部分代表方法总结在表 1 中，由于其评估过程不需要输入数据，只依赖模型本身的参数，因此对计算资源要求较少。但基于参数驱动的方法通常需要预先设定一个参数剪枝阈值来决定要修剪掉的网络结构，这个阈值通常会根据网络结构变化而变化，需要按照实际任务进行调整，给剪枝工作带来了一定的时间损耗。

3.2.2　数据驱动的评估准则

基于数据驱动的评估准则进行网络结构重要性评估时需要利用输入数据，通常在特征层面、梯度层面、网络输出结果等方面进行分析。

APoZ[20]为数据驱动的经典方法之一，其将滤波器经过激活层后输出的特征层当中数值为 0 所占百分比作为该滤波器重要性的评估依据，值为 0 所占百分比越多，则该滤波器重要性越低。HRank[26]发现单个滤波器生成的特征层的平均秩总是变化不大的，同时证明了秩越低的特征层对精度的贡献越小，因此 HRank 通过修剪掉这些生成低秩特征层的滤波器达到剪枝目的。文献[32]通过实验发现特征层之间具有线性关系，因此利用子空间聚类的思想对滤波器产生的特征层进行聚类来消除滤波器的冗余，达到剪枝目的。

除了对特征层进行直接利用之外，文献[23]和文献[33]通过特征重建的思想来达到网络剪枝目的。ThiNet[23]根据下一层的特征输出来修剪当前层的滤波器，其主要思想是通过重构下一层的输入特征，即用下一层输入的子集代替原来的输入得到与原来尽可能类似的输出，这样子集以外的输入就可以去掉，同时其对应的前一层的滤波器也就可以修剪掉。文献[33]中也采用了类似的特征层重建思想，通过迭代算法进行逐层剪枝，将网络每一层的特征重建问题转换为 LASSO 回归和最小二乘法重建误差，通过特征重建过程修剪掉对特征层重建贡献较小的滤波器。PFA[34]将主成分分析（PCA）用于特征分析中，对于每一层滤波器输出的特征层，先将特征层进行池化输出为向量，而后对各个特征向量进行 PCA 分析，去掉信息含量少的特征层对应的滤波器来达到剪枝目的。

最近的一些工作[19,22,27,35-36]认为只单独考虑深度网络某一层或相邻两层滤波器的重要性并裁剪掉看似不重要的滤波器，可能会对后续层甚至更深层的响应输出产生影响，而且误差会逐层累积，因此需要从网络整体考虑网络结构的重要性。

SNIP[35]在全连接层前加入 Inf-FS 滤波，利用矩阵幂级数的性质有效计算各个特征相对于所有其他特征的重要性，而后将重要性分数逐层反推，得到整个网络所有层的重要性评估结果，最后根据评估结果对原有的模型进行剪枝。文献[36]将滤波器的剪枝过程看作一个优化问题，将代价函数反向传播以泰勒展开作为评估准则来决定滤波器的重要性，目的是使剪枝之后的子模型在训练集上的代价值和原始模型在训练集上的代价值接近，对应在梯度层面上即为滤波器的重要性可由滤波器内的每一个权重的梯度表示。文献[22]认为滤波器产生的特征之间熵越大，则该滤波器的重要性越高，对于每一个卷积层产生的特征，通过全局平均池化转换为特征向量，在训练子集上，对于该卷积层中每一个滤波器对应的特征向量集合，计算其熵作为对应滤波器的重要性指标。

文献[19]和文献[27]在模型层面上对网络结构重要性进行评估。CURL[19]对网络中每一个滤波器，计算去掉该滤波器之后的子网络最后一层输出的池化特征和原始网络最后一层输出的池化特征之间的 KL 散度，认为该滤波器越重要，则去掉该滤波器之后的网络输出特征和原始特征之间相差应该越大，即 KL 散度越大，这种计算方式利用全局特征进行重要性评估。EagleEye[27]直接将剪枝过程转化为对可剪枝的网络结构采样的过程，直接评估每一个采样出来的子网络的重要性来完成剪枝目标，而不直接评估每一个滤波器的重要性。文献[27]对原始网络中可剪枝的滤波器进行采样，将采样出的子网络除批归一化层之外的参数固定之后，经过训练子集调节批归一化层参数（AdaptiveBN），AdaptiveBN之后的子网络在训练子集上的Top-1准确率作为该子网络的重要性，通过实验展示了子网络经过AdaptiveBN之后的准确率和该子网络经过微调之后的模型准确率之间的正相关关系，但由于其采样过程是随机的，所以通常需要针对一个剪枝目标采样上千次子模型，这种子模型采样及评估过程计算量极大。

以上基于数据驱动的剪枝方法对网络中的每一层都应用了相同的评价准则，LFPC[38]认为不同的卷积层中特征表达意义不同，应该为其设置不同的评价准则来适应卷积层的多样性。LFPC 设计可以学习的评价准则组合，每一层对应一组评价准则组合系数，该系数为可学习的参数。

和基于参数驱动的评估准则相比，基于数据驱动的评估准则评估方式不限于网络本身的参数，虽然需要的计算资源和评估时间增加，但是其可利用的空间由于输入数据的参与而变得更多。同时，针对不同的场景输入数据能够使评估结果符合该应用场景下的数据特点，避免基于参数驱动的评估方法只局限于网络本身参数的缺点。通过表 1 中总结的代表方法可以看出，基于数据驱动的评估准则方法受剪枝领域欢迎的程度高于基于参数驱动的方法，近几年提出的评估准则多为数据驱动的方法。

3.3 基于搜索的剪枝方法

神经网络结构搜索（neural architecture search）近些年来在模型结构设计上得到了广泛的关注和研究。在传统的模式下，研究者基于经验手工设计神经网络的结构，这不仅需要巨大的人力成本，同时对于一些普通的研究者来说也是一个困难的任务。Zoph 和Le[40]首先提出使用基于强化学习的神经网络结构搜索来代替人工设计，但搜索过程需要耗费巨大的计算资源。随着 DARTS[41]的提出，基于梯度的搜索方法凭借其速度优势逐渐成为主流。此时，开始有研究者将结构化剪枝建模成一个基于原型网络的搜索过程，基于原型网络搜索得到的子模型就是剪枝后的小模型。本节将简要介绍一些基于搜索的剪枝方法。

AMC[42]早在 2018 年就使用强化学习方法来搜索剪枝结构，在 AMC 中，将每一层的输入特征维度、卷积核大小、步长等信息编码成一个状态，使用 DDPG[43]智能体来预测行动，使用分类错误和 FLOPs 作为奖赏。

ABCPruner[44]使用人工蜂群算法来寻找最优的剪枝结构。和常规的剪枝方法不同，ABCPruner 寻找每层的通道数量，而不是去选择相对重要的通道。具体来说，每层对应的通道数量组成一个可行解，所有可行解构成目标种群，直接使用子模型在数据集上的表现作为其适应度，使用人工蜂群算法来进行搜索。在 ILSVRC-2012 数据集上，ABCPruner 能够在减少 62.87% FLOPs 和 60%参数量的同时，使 ResNet-152 的精度保持不变。

MetaPruning[45]使用元学习来自动进行通道剪枝。MetaPruning 首先训练一个元网络（meta network），然后使用演化过程来搜索表现好的剪枝模型。MetaPruning 为每层搜索通道的数量，而不是具体的每一个通道，能极大缩小搜索空间；当减少 ResNet-50 模型 25% FLOPs 时，在 ILSVRC-2012 上只比原模型降低 0.4%的 Top-1 准确率。

Network adjustment[46]将模型准确率看作关于计算量 FLOPs 的一个方程，由此来计算每层的 FLOPs 利用率，最终根据这个利用率来自动调整每层的通道数。

基于演化算法的搜索虽然对搜索过程进行了优化，但仍然有巨大的搜索时间花销。DMCP[25]将剪枝建模成一个可微分的马尔可夫过程，然后直接对网络进行梯度下降进行优化。每一层通道的剪枝构造成一个单独的马尔可夫过程，通道数量的增加和减少对应马尔可夫状态的转移。在 ILSVRC-2012 数据集上，DMCP 在减少 ResNet-50 接近 46% FLOPs 的同时，能得到精度降低 0.4%的剪枝模型。

使用表 2 来展示不同剪枝算法的性能表现。为了方便比较，使用 ResNet-50 作为基础网络，展示不同算法在 ILSVRC-2012 数据集上的剪枝表现。由于不同算法使用的深度学习框架不同，所以模型的基准结果会不同，我们比较其相对基准结果的变化。

表 2 不同剪枝算法在 **ResNet-50** 上的性能表现

	具体方法	模型 FLOPs	剪枝百分比/%	子模型 Top-1 准确率/%	准确率变化/%
基于重要性的剪枝方法	SFP[30]	8.17	0	76.15	+0.00
		4.75	42	74.61	−1.54
	FPGM[24]	8.17	0	76.15	+0.00
		4.74	42	75.59	−0.56
		3.80	53	74.83	−1.32
	HRank[26]	8.17	0	76.15	+0.00
		4.60	44	74.98	−0.17
		3.10	62	71.98	−4.17
		1.96	76	69.10	−7.05
	ThiNet[23]	7.72	0	72.88	+0.00
		4.88	37	72.04	−0.84
		3.41	56	71.01	−1.87
		2.20	72	68.42	−4.46

续表

	具体方法	模型 FLOPs	剪枝百分比/%	子模型 Top-1 准确率/%	准确率变化/%
基于重要性的剪枝方法	Entropy[22]	7.72	0	72.88	+0.00
		7.16	7	73.56	+0.68
		6.38	17	72.89	+0.01
		5.04	35	70.84	−2.04
	CURL[19]	8.17	0	76.15	+0.00
		2.22	73	73.39	−2.76
	EagleEye[27]	8.17	0	76.60	+0.00
		6.00	27	77.10	+0.50
		4.00	51	76.40	−0.20
		2.00	76	74.20	−2.40
	LFPC[37]	8.17	0	76.15	+0.00
		3.20	61	74.46	−1.69
基于搜索的剪枝方法	ABCPruner[44]	8.17	0	76.01	+0.00
		5.12	37	74.84	−1.17
		2.60	68	72.58	−3.42
		1.88	77	70.29	−5.72
	MetaPruning[45]	8.17	0	76.60	+0.00
		6.00	27	76.20	−0.40
		4.00	51	75.40	−1.20
		2.00	76	73.40	−3.20
	TAS[47]	8.17	0	77.46	+0.00
		2.31	72	76.20	−1.26
	AutoSlim[48]	8.17	0	76.15	+0.00
		6.00	27	76.10	−0.15
		4.00	51	75.60	−0.55
		2.00	76	74.00	−2.15
	DMCP[25]	8.17	0	76.60	+0.00
		5.60	31	77.00	+0.40
		4.40	46	76.20	−0.40
		2.20	73	74.40	−2.20

3.4 小结

本节首先介绍了模型剪枝的基本流程，并且介绍了评估剪枝算法常用的数据集、模型和一些评估准则。由于剪枝算法基本上使用相同的数据集和模型进行实验说明验证，所以不同剪枝算法之间的比较十分方便和直接。还介绍了一些常见的基于重要性的结构化剪枝算法，并且按照参数驱动和数据驱动对其分别进行介绍。这类方法认为不同的通道具有不同的重要性，其核心区别在于重要性的定义方式不同。此外，还介绍了一些基于搜索的结构化剪枝算法。这些方法直接对各层通道数量进行搜索。总结起来，无论是哪种剪枝算法，剪枝之后的模型都可以看作原模型的一个子模型。这些子模型从原模型中继承部分参数，再辅以一定的微调训练，能达到和原模型相当甚至超过原模型的分类结果，与此同时用到更少的参数量和计算量。

4 讨论与展望

现在的结构化剪枝算法主要在图像分类任务上进行研究，但由于神经网络结构的通用性，在分类任务上证明了有效的剪枝算法能够方便地移植到其他视觉任务上。Ghosh 等人[49]在 YOLOv3 模型上运用剪枝算法进一步提升模型在目标检测任务上的运行速度；Hou 和 Kung[50]在图像超分辨率重建任务上使用剪枝算法得到更小的模型。也有一些研究者在特定任务上进行优化，Luo 和 Wu[19]在细粒度图像分类和小数据集上优化剪枝算法，比起直接运用大规模图像分类任务上的算法，优化过后的剪枝算法能够进一步带来精度的提升。

随着物联网的兴起和智能手机的发展，越来越多的移动设备搭载深度学习模型来提供更智能的服务。而这些移动设备的计算能力相对较低、存储空间相对不足，轻量级的模型变得愈发重要。结构化剪枝是一个重要并且有效的精简模型的方法，通常会和参数量化及低秩近似等其他方法共同使用来最大化压缩模型。非结构化剪枝破坏了模型原有的结构，虽然参数量和理论计算量都极大地降低了，但实际运行速度却并不乐观。和非结构化剪枝不同的是，结构化剪枝算法能在通用平台上运行，并且由于保持了原模型的结构，能在剪枝的基础上进行低秩近似，这是非结构化剪枝无法做到的。

然而结构化剪枝也存在一些问题，通常需要经过评估-选择-微调这三个步骤才能得到一个精简的模型，更有甚者会逐层进行选择和微调，获取一个精简模型需要花费极长的时间；此外，绝大多数研究都集中在图像分类的研究上，在目标检测、语义分割等更难的视觉任务上，成果还相对欠缺。

参考文献

[1] LECUN Y, BUTTOU L, BENGIO Y, et al. Gradient-based learning applied to document recognition[J]// Proceedings of the IEEE. 1998, 86(11): 2278-2324.

[2] KRIZHEVSKY A, SUTSKEVER I, HINTON G E. ImageNet classification with deep convolutional neural networks[C]//Proceedings of Advances in Neural Information Processing Systems 25 (NIPS). 2012: 1097-1105.

[3] SIMONYAN K, ZISSERMAN A. Very deep convolutional networks for large-scale image recognition[C]// Proceedings of the International Conference on Learning Representations (ICLR). 2015: 1-14.

[4] HE K M, ZHANG X Y, REN S Q, et al. Deep residual learning for image recognition[C]//Proceedings of the IEEE Conference on Computer Vision and Pattern Recognition (CVPR). 2016: 770-778.

[5] HUBARA I, COURBARIAUX M, SOUDRY D, et al. Binarized neural networks[C]//Proceedings of Advances in Neural Information Processing Systems 29 (NIPS). 2016: 4107-4115.

[6] LI F F, ZHANG B, LIU B. Ternary weight networks[Z]. arXiv preprint arXiv:1605.04711, 2016.

[7] RASTEGARI M, ORDONEZ V, REDMON J, et al. XNOR-Net: ImageNet classification using binary convolutional neural networks[C]//Proceedings of the European Conference on Computer Vision (ECCV). Springer, 2016, 9908: 525-542.

[8] HOWARD A G, ZHU M L, CHEN B, et al. Mobilenets: Efficient convolutional neural networks for mobile vision applications[Z]. arXiv preprint arXiv:1704.04861, 2017.

[9] LEBEDEV V, LEMPITSKY V. Fast convnets using groupwise brain damage[C]//Proceedings of the IEEE Conference on Computer Vision and Pattern Recognition (CVPR). 2016: 2554-2564.

[10] LI H, KADAV A, DURDANOVIC I, et al. Pruning filters for efficient convents[C]//Proceedings of The International Conference on Learning Representations (ICLR). 2017: 1-13.

[11] LIU Z, LI J G, SHEN Z Q, et al. Learning efficient convolutional networks through network slimming[C]// Proceedings of the IEEE International Conference on Computer Vision (CVPR). 2017: 2736-2744.

[12] WEN W, WU C P, WANG Y D, et al. Learning structured sparsity in deep neural networks[C]// Proceedings of Advances in Neural Information Processing Systems 29 (NIPS). 2016: 2074－2082.

[13] LEBEDEV V, GANIN Y, RAKHUBA M, et al. Speeding-up convolutional neural networks using fine-tuned CP-Decomposition[C]//Proceedings of the International Conference on Learning Representations (ICLR). 2015: 1-11.

[14] KIM Y-D, PARK E, YOO S, et al. Compression of deep convolutional neural networks for fast and low power mobile applications[C]//Proceedings of the International Conference on Learning Representations (ICLR). 2016: 1-16.

[15] ZHANG X Y, ZOU J H, HE K M, et al. Accelerating very deep convolutional networks for classification and detection[J]. IEEE Transactions on Pattern Analysis and Machine Intelligence (PAMI), 2016, 38(10): 1943-1955.

[16] HINTON G, VINYALS O, DEAN J. Distilling the knowledge in a neural network[C]//Proceedings of NIPS Deep Learning and Representation Learning Workshop, 2015.

[17] LUO J-H, WU J X. AutoPruner: An end-to-end trainable filter pruning method for efficient deep model inference[J]. Pattern Recognition (PR), 2020, 107.

[18] SANDLER M, HOWARD A, ZHU M L, et al. MobileNetV2: Inverted residuals and linear bottlenecks[C]// Proceedings of the IEEE Conference on Computer Vision and Pattern Recognition (CVPR). 2018: 4510-4520.

[19] LUO J-H, WU J X. Neural network pruning with residual-connections and limited-data[C]//Proceedings of the IEEE Conference on Computer Vision and Pattern Recognition (CVPR). 2020: 1458-1467.

[20] HU H Y, PENG R, TAI Y-W, et al. Network trimming: A data-driven neuron pruning approach towards efficient deep architectures[Z]. arXiv preprint arXiv:1607.03250, 2016.

[21] HAN S, MAO H Z, DALLY W J. Deep compression: Compressing deep neural networks with pruning, trained quantization and huffman coding[C]//Proceedings of the International Conference on Learning Representations (ICLR). 2016: 1-14.

[22] LUO J-H, WU J X. An entropy-based pruning method for CNN compression[Z]. arXiv preprint arXiv: 1706.05791, 2017.

[23] LUO J-H, WU J X, LIN W Y. ThiNet: A filter level pruning method for deep neural network compression[C]// Proceedings of the IEEE International Conference on Computer Vision (ICCV). 2017: 5058-5066.

[24] HE Y, LIU P, WANG Z W, et al. Filter pruning via geometric median for deep convolutional neural networks acceleration[C]//Proceedings of the IEEE Conference on Computer Vision and Pattern Recognition (CVPR). 2019: 4340-4349.

[25] GUO S P, WANG Y J, LI Q Q, et al. DMCP: Differentiable Markov channel pruning for neural networks[C]//Proceedings of the IEEE Conference on Computer Vision and Pattern Recognition (CVPR). 2020: 1539-1547.

[26] LIN M B, JI R R, WANG Y, et al. HRank: Filter pruning using high-rank feature map[C]//Proceedings of the IEEE Conference on Computer Vision and Pattern Recognition (CVPR). 2020: 1529-1538.

[27] LI B L, WU B, SU J, et al. EagleEye: Fast sub-net evaluation for efficient neural network pruning[C]// Proceedings of the European Conference on Computer Vision (ECCV). Springer, 2020, 12347: 639-654.

[28] KRIZHEVSKY A, HINTON G. Learning multiple layers of features from tiny images[R]. University of Toronto, 2009.

[29] RUSSAKOVSKY O, DENG J, SU H, et al. ImageNet large scale visual recognition challenge[J]. International Journal of Computer Vision, 2015, 115(3): 211-252.

[30] HE Y, KANG G L, DONG X Y, et al. Soft filter pruning for accelerating deep convolutional neural networks[C]//Proceedings of the International Joint Conference on Artificial Intelligence (IJCAI). 2018: 2234-2240.

[31] ZHUO H Y, QIAN X L, FU Y W, et al. SCSP: Specral clustering filter pruning with soft self-adaption manners[Z]. arXiv preprint arXiv:1806.05320, 2018.

[32] WANG D, ZHOU L, ZHANG X N, et al. Exploring linear relationship in feature map subspace for convnets compression[Z]. arXiv preprint arXiv:1803.05729, 2018.

[33] HE Y H, ZHANG X Y, SUN J. Channel pruning for accelerating very deep neural networks[C]// Proceedings of the IEEE International Conference on Computer Vision (ICCV). 2017: 1389-1397.

[34] SUAU X, ZAPPELLA L, PALAKKODE V, et al. Principal filter analysis for guided network compression[Z]. arXiv preprint arXiv:1807.10585, 2018.

[35] YU R C, LI A, CHEN C-F, et al. NISP: Pruning networks using neuron importance score propagation[C]// Proceedings of the IEEE Conference on Computer Vision and Pattern Recognition (CVPR). 2018: 9194-9203.

[36] MOLCHANOV P, TYREE S, KARRAS T, et al. Pruning convolutional neural networks for resource efficient inference[Z]. arXiv preprint arXiv:1611.06440, 2016.

[37] HE Y, DING Y H, LIU P, et al. Learning filter pruning criteria for deep convolutional neural networks acceleration[C]//Proceedings of the IEEE Conference on Computer Vision and Pattern Recognition (CVPR). 2020: 2006-2015.

[38] IOFFE S, SZEGEDY C. Batch normalization: Accelerating deep network training by reducing internal covariate shift[C]//Proceedings of the 32nd International Conference on Machine Learning (ICML). 2015: 448-456.

[39] ZHAO C L, NI B B, ZHANG J, et al. Variational convolutional neural network pruning[C]//Proceedings of the IEEE Conference on Computer Vision and Pattern Recognition (CVPR). 2019: 2775-2784.

[40] ZOPH B, LE Q V. Neural architecture search with reinforcement learning[C]//Proceedings of the International Conference on Learning Representations (ICLR). 2017: 1-16.

[41] LIU H X, SIMONYAN K, YANG Y M. DARTS: Differentiable architecture search[C]//Proceedings of the International Conference on Learning Representations (ICLR). 2019: 1-13.

[42] HE Y H, LIN J, LIU Z J, et al. AMC: AutoML for model compression and acceleration on mobile devices[C]//Proceedings of the European Conference on Computer Vision (ECCV). Springer, 2018, 11211: 815-832.

[43] LILLICRAP T P, HUNT J J, PRITZEL A, et al. Continuous control with deep reinforcement learning[Z]. arXiv preprint arXiv:1509.02971, 2015.

[44] LIN M B, JI R R, ZHANG Y X, et al. Channel pruning via automatic structure search[C]//Proceedings of the International Joint Conference on Artificial Intelligence (IJCAI). 2020: 673-679.

[45] LIU Z C, MU H Y, ZHANG X Y, et al. MetaPruning: Meta learning for automatic neural network channel pruning[C]//Proceedings of the IEEE International Conference on Computer Vision (ICCV). 2019: 3296-3305.

[46] CHEN Z S, NIU J W, XIE L X, et al. Network adjustment: Channel search guided by FLOPs utilization ratio[C]//Proceedings of the IEEE Conference on Computer Vision and Pattern Recognition (CVPR). 2020: 10658-10667.

[47] DONG X Y, YANG Y. Network pruning via transformable architecture search[C]//Proceedings of Advances in Neural Information Processing Systems 32 (NeurIPS). 2019: 760-771.

[48] YU J H, HUANG T. AutoSlim: Towards one-shot architecture search for channel numbers[Z]. arXiv preprint arXiv:1903.11728, 2019.

[49] GHOSH S, SRINIVASA S K K, Amon P, et al. Deep network pruning for object detection[C]// Proceedings of IEEE International Conference on Image Processing (ICIP). 2019: 3915-3919.

[50] HOU Z J, KUNG S-Y. Efficient image super resolution via channel discriminative deep neural network pruning[C]//Proceedings of IEEE International Conference on Acoustics, Speech and Signal Processing (ICASSP). 2020: 3647-3651.

Efficient Neural Speech Synthesis

Tao Qin

(Microsoft Research Asia, Beijing 100080)

1 Introduction

Speech synthesis, also known as text-to-speech [1] or TTS, which aims to produce intelligible and natural speech from text, covers many applications in human communication [2] and has been a key research problem in artificial intelligence, natural language and speech processing [3-5].

Speech synthesis technologies/systems have been evolved for multiple stages, from early formant synthesis [6-8], articulatory synthesis [9], concatenative approaches [10-13] to statistical parametric approaches [14-17] and today's deep learning based neural approaches [18-25].

In recent years, deep neural networks based systems have become the dominant approach for speech synthesis, such as Tacotron [23], Tacotron 2 [26], Deep Voice 3 [27] and the fully end-to-end ClariNet [28]. Those models usually first generate mel-spectrograms autoregressively from text input and then synthesize speech from the mel-spectrograms using vocoder such as Griffin-Lim [29], WaveNet [21], Parallel WaveNet [30] or WaveGlow [31]. Neural network based TTS has outperformed conventional concatenative and statistical parametric approaches [11,32] in terms of speech quality.

While neural approaches can synthesize high-quality speech, they are inefficient in multiple aspects:

- Inference inefficiency. Neural TTS models usually generate mel-spectrograms one by one in an autoregressive manner conditioned on previously generated mel-spectrograms. Since speech sequences are usually of hundreds or thousands of mel-spectrograms, autoregressive neural TTS models suffer from slow inference speed.

- Data inefficiency. The success of neural TTS is built on large scale of paired speech-text data, which is easy to collect for top languages but brings challenges for many languages that are scarce of paired speech-text data.
- Parameter inefficiency. To synthesize high-quality speech, neural TTS systems usually employ large neural networks with tens of millions of parameters, which block the applications in mobile and IoT devices due to their limited memory and power consumption.

To address those challenges and push forward the research and applications of TTS, we conduct a set of research to improve the inference efficiency, data efficiency and parameter efficiency for neural TTS. In this paper, we summarize our recent research on efficient TTS and cover following works:

- FastSpeech series [33-34] that improve inference (and training) efficiency of neural TTS. FastSpeech [34], which is a novel feed-forward network based on Transformer to generate mel-spectrograms in parallel, is the first non-autoregressive TTS model and achieves significant inference speedup: it accelerates mel-spectrogram generation by 270x and the end-to-end speech synthesis by 38x compared with its autoregressive counterpart. FastSpeech 2 [33] is further developed to speed up the training efficiency of FastSpeech by directly training the model with ground-truth target instead of the simplified output from a teacher and improve the quality of synthesized speech by FastSpeech by introducing more variation information of speech(e.g., pitch, energy and more accurate duration) as conditional inputs. In addition, FastSpeech 2s [33] is designed to directly generate speech waveform from text in parallel, enjoying the benefit of fully end-to-end inference and further improving inference efficiency of FastSpeech and FastSpeech 2.
- DualSpeech [35] and LRSpeech [36] that improve data efficiency for neural TTS. To handle languages with limited paired speech-text data, inspired by the duality between the TTS and automatic speech recognition (ASR) tasks [37], we propose to leverage ASR for TTS and jointly train an ASR and a TTS models together, using a few amount of paired speech-text data and extra unpaired data. Our method achieves 99.84% in terms of word level intelligible rate and 2.68 MOS for TTS, by leveraging only 200 paired speech and text data (about 20 minutes audio), together with extra

unpaired speech and text data. Taking one step further, we extend DualSpeech to extremely low-resource settings for TTS, ASR and build a system LRSpeech [36] for industrial deployment under two constraints: ① Extremely low data collection cost; ② High accuracy to satisfy the deployment requirement. LRSpeech achieves 98.08% intelligibility rate and 3.57 MOS score, satisfying the online deployment requirements.

- LightSpeech [38] and AdaSpeech series [39-40] that improve parameter efficiency for neural TTS. Deploying TTS in various end devices such as mobile phones or embedded devices requires extremely small memory usage and inference latency. While non-autoregressive TTS models such as FastSpeech and FastSpeech 2/2s are fast while making inference than autoregressive models, their model size and inference latency are still large for the deployment in resource-constrained devices. Our proposed LightSpeech [38] leverages neural architecture search (NAS) to automatically design more lightweight and efficient models based on FastSpeech. The TTS model discovered by ourLightSpeech achieves 15x model compression ratio and 6.5x inference speedup on CPU with on par voice quality. To support a large number of customers in custom voice, a specific TTS service in commercial speech platforms that aims to adapt a source TTS model to synthesize personal voice for a target speaker using few speech from her/him, the adaptation parameters need to be as few as possible for each target speaker to reduce memory usage while maintaining high voice quality. To improve parameter efficiency for this adaptation scenario and better trade off the number of adaptation parameters and voice quality, we build AdaSpeech [39], an adaptive TTS system for high-quality and efficient customization of new voices, in which we introduce conditional layer normalization in the mel-spectrogram decoder and fine-tune this newly introduced component in addition to speaker embedding for adaptation. Furthermore, considering that in many scenarios only untranscribed speech data is available for adaptation, we develop AdaSpeech 2 [40], an adaptive TTS system that only leverages untranscribed speech data for adaptation.

2 Inference Efficiency: FastSpeech Series

In this section, we introduce our recent works on improving computation efficiency of neural TTS: FastSpeech [34] that is super efficient in inference, FastSpeech 2 [33] that improves the quality and speeds up the training of FastSpeech, and FastSpeech 2s [33] that further improves inference efficiency over FastSpeech and FastSpeech 2.

2.1 FastSpeech

Previous neural TTS systems (before FastSpeech) generate mel-spectrograms autoregressively, and face several challenges:

- Inference is slow, given that the mel-spectrogram sequence is usually with a length of hundreds or thousands.
- Synthesized speech is not robust. Due to error propagation [41] and the wrong attention alignments between text and speech in the autoregressive generation, the problem of words skipping and repeating [27] happens here and there.
- Synthesized speech is lack of controllability. Previous autoregressive models generate mel-spectrograms one by one automatically, without explicitly leveraging the alignments between text and speech. Consequently, it is usually hard to directly control the voice speed and prosody in the autoregressive generation.

Considering the monotonous alignment between text and speech, to speed up mel-spectrogram generation, we build a novel model FastSpeech [34], which takes a text (phoneme) sequence as input and generates mel-spectrograms non-autoregressively. It adopts a feed-forward network based on the self-attention in Transformer [42] and 1D convolution [27,43-44]. Since a mel-spectrogram sequence is much longer than its corresponding phoneme sequence, in order to solve the problem of length mismatch between the two sequences, FastSpeech adopts a length regulator that up-samples the phoneme sequence according to the phoneme duration (i.e., the number of mel-spectrograms that each phoneme corresponds to) to match the length of the mel-spectrogram sequence. The regulator is built on a phoneme duration predictor, which predicts the duration of each phoneme.

FastSpeech can address the above-mentioned three challenges as follows:

- Through parallel mel-spectrogram generation, FastSpeech greatly speeds up the synthesis process.
- Phoneme duration predictor ensures hard alignments between a phoneme and its mel-spectrograms, which is very different from soft and automatic attention alignments in the autoregressive models. Thus, FastSpeech avoids the issues of error propagation and wrong attention alignments, consequently reducing the ratio of the skipped words and repeated words.

- The length regulator can easily adjust voice speed by lengthening or shortening the phoneme duration to determine the length of the generated mel-spectrograms, and control part of the prosody by adding breaks between adjacent phonemes.

Experiments on the LJSpeech dataset show that in terms of speech quality, FastSpeech nearly matches the autoregressive Transformer model. Furthermore, FastSpeech achieves 270x speedup on mel-spectrogram generation and 38x speedup on final speech synthesis compared with the autoregressive Transformer TTS model, almost eliminates the problem of word skipping and repeating, and can adjust voice speed smoothly.

The overall model architecture of FastSpeech is shown in Figure 1. We describe individual components of FastSpeech as follows.

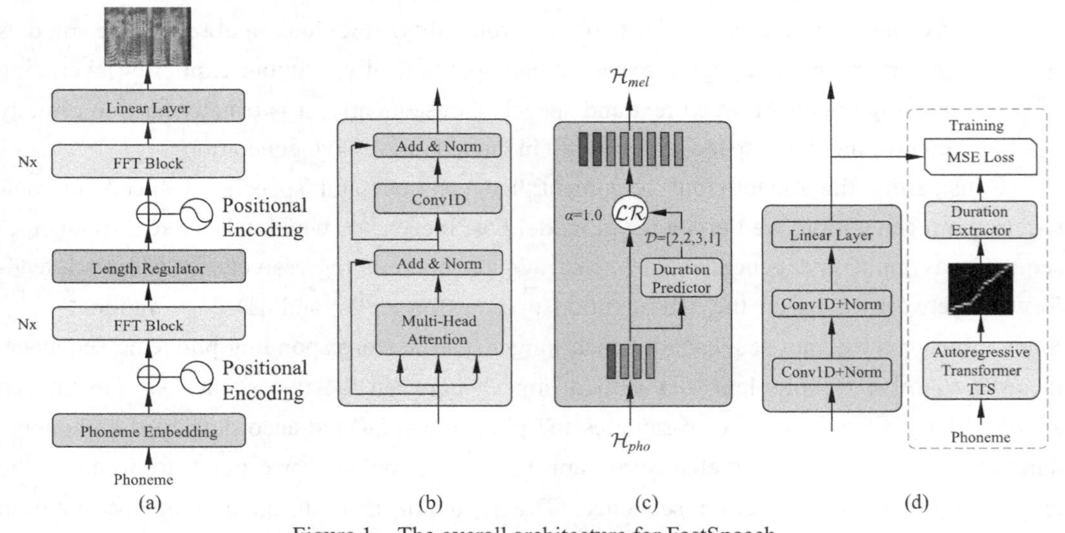

Figure 1　The overall architecture for FastSpeech

(a) The feed-forward Transformer; (b) The feed-forward Transformer block; (c) The length regulator; (d) The duration predictor MSE loss denotes the loss between predicted and extracted duration, and is only used in training.

2.1.1　Feed-Forward Transformer

FastSpeech adopts a feed-forward structure based on self-attention proposed in Transformer [42] and 1D convolution [27,43]. We call this structure as Feed-Forward Transformer (FFT), as shown in Figure 1(a). Feed-Forward Transformer stacks multiple FFT blocks for phoneme to mel-spectrogram transformation, with N blocks on the phoneme side, and N blocks on the mel-spectrogram side, with a length regulator in between to bridge the length gap between the phoneme and mel-spectrogram sequence.

Each FFT block consists of a self-attention network and 1D convolutional network, as shown in Figure 1(b). The self-attention network consists of a multi-head attention to extract the cross-position information. Different from the 2-layer densenetwork in Transformer, we use a 2-layer 1D convolutional network with ReLU activation. The motivation is that the adjacent hidden states are more closely related in the character/phoneme and mel-spectrogram sequence in speech tasks. Following Transformer [42], residual connections, layer normalization, and dropout are added after the self-attention network and 1D convolutional network respectively.

2.1.2 Length Regulator

The length regulator (Figure 1(c)) is used to address the length mismatch between the phoneme and spectrogram sequence in the Feed-Forward Transformer, as well as to control the voice speed and part of prosody. The length of a phoneme sequence is usually smaller than that of its mel-spectrogram sequence, and each phoneme corresponds to multiple mel-spectrograms. We refer to the length of the mel-spectrograms that corresponds to a phoneme as the phoneme duration. Based on the phoneme duration d, the length regulator expands the hidden states of the phoneme sequence d times, and then the total length of the hidden states equals the length of the mel-spectrograms. Denote the hidden states of the phoneme sequence as $\mathcal{H}_{\text{pho}} = [h_1, h_2, \cdots, h_n]$, where n is the length of the sequence. Denote the phoneme duration sequence as $\mathcal{D} = [d_1, d_2, \cdots, d_n]$, where $\sum_{i=1}^{n} d_i = m$ and m is the length of the mel-spectrogram sequence. We denote the length regulator $\mathcal{LR}$ as

$$\mathcal{H}_{\text{mel}} = \mathcal{LR}(\mathcal{H}_{\text{pho}}, \mathcal{D}, \alpha) \tag{1}$$

where α is a hyperparameter to determine the length of the expanded sequence $\mathcal{H}_{\text{mel}}$, thereby controlling the voice speed. For example, given $\mathcal{H}_{\text{pho}} = [h_1, h_2, h_3, h_4]$ and the corresponding phoneme duration sequence $\mathcal{D} = [2, 2, 3, 1]$, the expanded sequence $\mathcal{H}_{\text{mel}}$ based on Equation 1 becomes $[h_1, h_1, h_2, h_2, h_3, h_3, h_3, h_4]$ if $\alpha = 1$ (normal speed). When $\alpha = 1.3$ (slow speed) and 0.5 (fast speed), the duration sequences become $\mathcal{D}_{\alpha=1.3} = [2.6, 2.6, 3.9, 1.3] \approx [3, 3, 4, 1]$ and $\mathcal{D}_{\alpha=0.5} = [1, 1, 1.5, 0.5] \approx [1, 1, 2, 1]$, and the expanded sequences become $[h_1, h_1, h_1, h_2, h_2, h_2, h_3, h_3, h_3, h_3, h_4]$ and $[h_1, h_2, h_3, h_3, h_4]$ respectively. We can also control the break between words by adjusting the duration of the space characters in the sentence, so as to adjust part of prosody of synthesized speech.

2.1.3 Duration Predictor

Phoneme duration prediction is important for the length regulator. As shown in Figure 1(d), the duration predictor consists of a 2-layer 1D convolutional network with ReLU activation, each followed by a layer normalization and a dropout layer, and an extra linear layer to output a scalar, which is exactly the predicted phoneme duration. This module is stacked on top of the FFT blocks on the phoneme side and is jointly trained with the FastSpeech model to predict the length of mel-spectrograms for each phoneme with the mean square error (MSE) loss. We predict the length in the logarithmic domain, which makes them more Gaussian and easier to train. Note that the trained duration predictor is only used in inference, because we can directly use thephoneme duration extracted from an autoregressive teacher model in training.

In order to train the duration predictor, we extract the ground-truth phoneme duration from an autoregressive teacher TTS model, as shown in Figure 1(d). We describe the detailed steps as follows:

- We first train an autoregressive encoder-attention-decoder based Transformer TTS model following [45].
- For each training sequence pair, we extract the decoder-to-encoder attention alignments from the trained teacher model. There are multiple attention alignments due to the multiples heads in self-attention [42], and not all attention heads demonstrate the diagonal property (the phoneme and mel-spectrogram sequence are monotonously aligned). We propose a focus rate F to measure how an attention head is close to diagonal: $F = \frac{1}{S}\sum_{s=1}^{S} \max_{1 \leqslant t \leqslant T} a_{s,t}$, where S and t are the lengths of the ground-truth spectrograms and phonemes, and $a_{s,t}$ denote the element in the s-th row and t-th column of the attention matrix. We compute the focus rate for each head and choose the head with the largest F as the attention alignment.
- Finally, we extract the phoneme duration sequence $\mathcal{D} = [d_1, d_2, \ldots, d_n]$ according to the duration extractor $d_i = \sum_{s=1}^{S} [\text{argmax}_t\, a_{s,t} = i]$. That is, the duration of a phoneme is the number of mel-spectrograms attended to it according to the attention head selected in the above step.

2.2 FastSpeech 2

TTS is a one-to-many mapping problem [23,39,46], since multiple possible speech sequences can correspond to a text sequence due to variations in speech, such as pitch, duration, sound volume and prosody. In non-autoregressive TTS, the only input information is text which is not enough to fully predict the variance in speech. In this case, the model is prone to overfit to the variations of the target speech in the training set, resulting in poor generalization ability.

The training of FastSpeech relies on an autoregressive teacher model for duration prediction (to provide more information as input) and knowledge distillation (to simplify the data distribution in output), which can ease the one-to-many mapping problem (i.e., multiple speech variations correspond to the same text) in TTS. However, FastSpeech has several disadvantages: ① The teacher-student distillation pipeline in training is complicated and time-consuming; ② The duration extracted from the teacher model is not accurate enough, and the target mel-spectrograms distilled from teacher model suffer from information loss due to data simplification, both of which limit the voice quality.

To address the issues in FastSpeech and better solve the one-to-many mapping problem in non-autoregressive TTS, we build FastSpeech 2 [33], a TTS system that directly trains the model with ground-truth target instead of the simplified output from a teacher and introduces more variation information of speech (e.g., pitch, energy and more accurate duration) as conditional inputs. Experimental results show that FastSpeech 2 achieves a 3x training speedup over FastSpeech and outperform FastSpeech in voice quality. Furthermore, its voice quality even surpasses autoregressive models.

2.2.1 Architecture

The overall architecture of FastSpeech 2 is shown in Figure 2(a). The text encoder converts the phoneme embedding sequence into the phoneme hidden sequence, then the variance adaptor adds different variance information such as duration, pitch and energy into the hidden sequence, and finally the mel-spectrogram decoder converts the adapted hidden sequence into mel-spectrogram sequence in parallel. Similar to FastSpeech [34], we use the feed-forward Transformer block as the basic structure for the text encoder and mel-spectrogram decoder. Different from FastSpeech that relies on a teacher-student distillation pipeline and the phoneme duration from a teacher model, FastSpeech 2 introduces several improvements.

- We remove the teacher-student distillation pipeline, and directly use ground-truth mel-spectrograms as target for model training, which avoids the information loss in distilled mel-spectrograms and increases the upper bound of the voice quality.

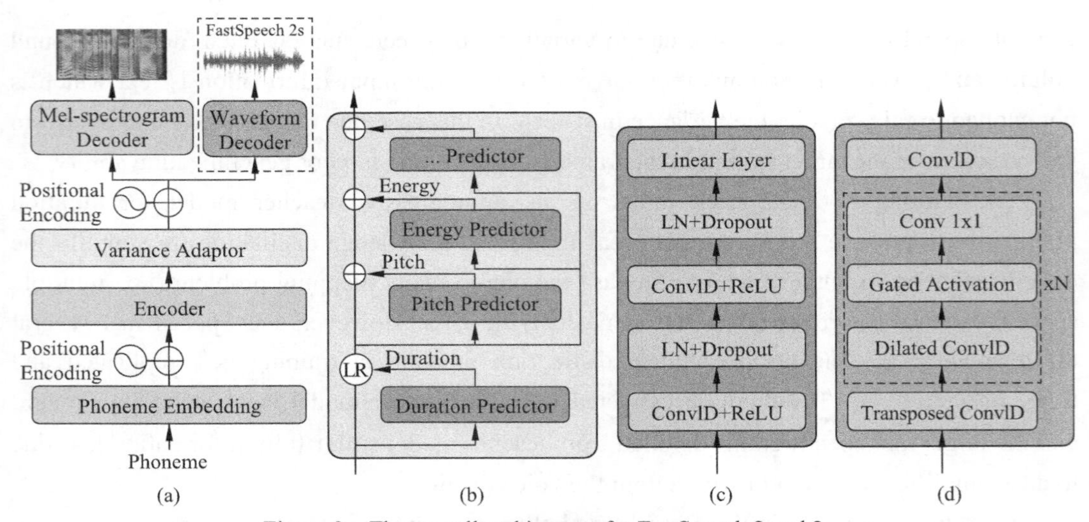

Figure 2 The overall architecture for FastSpeech 2 and 2s

(a) FastSpeech 2/2s; (b) Variance adaptor; (c) Duration/pitch/energy predictor; (d) Waveform decoder

LR in figure (b) denotes the length regulator proposed in FastSpeech, LN in figure (c) denotes layer normalization

- The variance adaptor consists of not only duration predictor but also pitch and energy predictors, where the duration predictor uses the phoneme duration obtained by forced alignment[47] as training target, which is more accurate than that extracted from the attention map of autoregressive teacher models; and the additional pitch and energy predictors can provide more variance information, which is important to ease the one-to-many mapping problem in TTS.

2.2.2 Variance Adaptor

The variance adaptor adds variance information (e.g., duration, pitch, energy, etc.) to the phoneme hidden sequence, and provides more information to predict variant speech to alleviate the one-to-many mapping problem in TTS. We consider several kinds of variance information: ① phoneme duration, which represents how long the speech voice sounds; ② pitch, which is a key feature to convey emotions and greatly affects the speech prosody; ③ energy, which indicates frame-level magnitude of mel-spectrograms and directly affects the volume and prosody of speech. More variance information can be added in the variance adaptor, such as emotion, style and speaker, and we leave it for future work. Correspondingly, the variance adaptor consists of a duration predictor (i.e., the length regulator, as used in FastSpeech), a pitch predictor, and an energy predictor, as shown in Figure 2(b).

In training, we take the ground-truth value of duration, pitch and energy extracted from the recordings as input into the hidden sequence to predict the target speech. At the same time, we use the ground-truth duration, pitch and energy as targets to train the duration, pitch and energy predictors, which are then used in inference to synthesize speech. As shown in Figure 2(c), the duration, pitch and energy predictors share similar model structure (but different model parameters), which consists of a 2-layer 1D-convolutional network with ReLU activation, each followed by layer normalization and a dropout layer, and an extra linear layer to project the hidden states into the output sequence. In the following paragraphs, we describe the details of the three predictors respectively.

(1) Duration Predictor. The duration predictor takes the phoneme hidden sequence as input and predicts the duration of each phoneme, which represents how many mel frames correspond to this phoneme, and is converted into logarithmic domain for ease of prediction. The duration predictor is optimized with mean square error (MSE) loss, taking the extracted duration as training target. Instead of extracting the phoneme duration using a pre-trained autoregressive TTS model in FastSpeech, we use Montreal forced alignment (MFA) [47] tool① to extract the phoneme duration, in order to improve the alignment accuracy and thus reduce the information gap between the model input and output.

(2) Pitch Predictor. Previous neural network based TTS systems with pitch prediction [48-49] often predict pitch contour directly. However, due to high variations of ground-truth pitch, the distribution of predicted pitch values is very different from ground-truth distribution. To better predict the variations in pitch contour, we use continuous wavelet transform (CWT) to decompose the continuous pitch series into pitch spectrogram [50-51] and take the pitch spectrogram as the training target for the pitch predictor which is optimized with MSE loss. In inference, the pitch predictor predicts the pitch spectrogram, which is further converted back into pitch contour using inverse continuous wavelet transform (iCWT). To take the pitch contour as input in both training and inference, we quantize pitch F_0 (ground-truth/predicted value for train/inference respectively) of each frame to 256 possible values in log-scale and further convert it into pitch embedding vector p and add it to the expanded hidden sequence.

(3) Energy Predictor. We compute L2-norm of the amplitude of each short-time Fourier

① MFA is an open-source system for speech-text alignment with good performance, which can be trained on paired text-audio corpus without any manual alignment annotations. We train MFA on our training set only without other external dataset. We will work on non-autoregressive TTS without external alignment models in the future.

transform (STFT) frame as the energy. Then we quantize energy of each frame to 256 possible values uniformly, encoded it into energy embedding e and add it to the expanded hidden sequence similarly to pitch. We use an energy predictor to predict the original values of energy instead of the quantized values and optimize the energy predictor with MSE loss①.

2.3 FastSpeech 2s

To further speed up speech synthesis, we extend FastSpeech 2 to FastSpeech 2s [33], which directly generates waveform from text, without cascaded mel-spectrogram generation (acoustic model) and waveform generation (vocoder). As shown in Figure 2(a), FastSpeech 2s generates waveform conditioning on intermediate hidden representations, which makes it more compact and efficient in inference by discarding mel-spectrogram decoder. We first discuss the challenges in non-autoregressive text-to-waveform generation, and then describe details of FastSpeech 2s, including model structure and training and inference processes.

There are several challenges for end-to-end text-to-waveform generation. First, since waveform contains more variance information (e.g., phase) than mel-spectrograms, the information gap between the input text and output waveform is larger than that in text-to-spectrogram generation. Second, it is difficult to train on the audio clip that corresponds to the full text sequence due to the extremely long waveform samples and limited GPU memory. As a result, we can only train on a short audio clip that corresponds to a partial text sequence which makes it hard for the model to capture the relationship among phonemes in different partial text sequences and thus harms the text feature extraction.

To tackle the above challenges, we make several designs in the waveform decoder. First, considering that the phase information is difficult to predict using a variance predictor [52], we introduce adversarial training in the waveform decoder to force it to implicitly recover the phase information [53]. Second, we leverage the mel-spectrogram decoder of FastSpeech 2, which is trained on the full text sequence to help on the text feature extraction. As shown in Figure 2(d), the waveform decoder is based on the structure of WaveNet [21] including non-causal convolutions and gated activation [54]. The waveform decoder takes a sliced hidden sequence corresponding to a short audio clip as input and upsamples it with transposed 1D-convolution to match the length of audio clip. The discriminator in the adversarial training adopts the same structure in

---

① We do not transform energy using CWT since energy is notas highly variable as pitch on LJSpeech dataset, and we do not observe gains when using it.

Parallel WaveGAN [53] which consists of ten layers of non-causal dilated 1-D convolutions with leaky ReLU activation function. The waveform decoder is optimized by the multi-resolution STFT loss and the LSGAN discriminator loss following Parallel WaveGAN. In inference, we discard the mel-spectrogram decoder and only use the waveform decoder to synthesize speech audio.

3 Data Efficiency: DualSpeech and LRSpeech

Building a high-quality neural TTS model requires large scale of paired speech-text data. A natural idea to handle the scenarios with limited paired data is to leverage unpaired speech and text data. Speech synthesis and recognition are naturally in dual form [37]: one maps from text to speech and the other maps from speech to text. It is obvious that structure duality can help speech synthesis (and recognition models) to learn from unpaired speech and text data. In this section, we introduce two TTS systems based on the idea of dual learning [37]: DualSpeech [35] for low-resource settings and LRSpeech [36] for extremely low-resource settings.

3.1 DualSpeech

DualSpeech [35] jointly trains a TTS model and an ASR model using only a few amount of paired speech and text data and extra unpaired data. It consists of the following components:

First, we leverage the idea of self-supervised learning for the unpaired speech and text data, to build the capability of the language understanding and modeling in both speech and text domain. Specifically, we use denoising auto-encoder to reconstruct the corrupt speech and text in an encoder-decoder framework, as shown in Figure 3(a) and Figure 3(b).

Second, we use the dual transformation, which is in spirit of dual learning [37,55], to develop the capability of transforming text to speech (TTS) and speech to text (ASR): ① The ASR model transforms the speech x into text $\hat{y}$, and then the TTS model leverages the transformed pair $(\hat{y}, x)$ for training, as shown in Figure 3(c); ② The TTS model transforms the text y into speech $\hat{x}$, and then the ASR model leverages the transformed pair $(\hat{x}, y)$ for training, as shown in Figure 3(d). Dual transformation iterates between TTS and ASR, and boosts the accuracy of the two tasks gradually.

Third, considering the speech and text sequence are usually longer than other sequence to sequence learning tasks such as neural machine translation, they will suffer more from error propagation [41], which refers to the problem that the right part of the generated sequence is

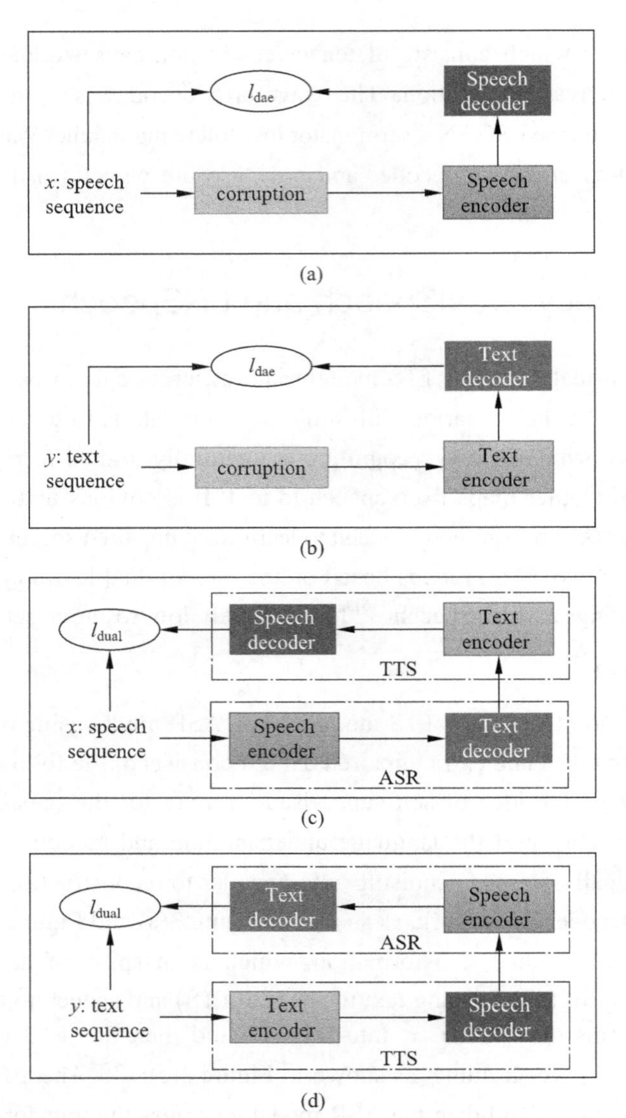

Figure 3 Training flow of low-resource TTS and ASR in DualSpeech [35]

(a) Denoising auto-encoding for speech; (b) Denoising auto-encoding for text; (c) Dual reconstruction for speech; (d) Dual reconstruction for text

Note that the corruption operation only be used in denoising auto-encoding. l_{dae} stands for the training loss of denoising auto-encoding, and l_{dual} stands for the training loss of dual learning/reconstruction

usually worse than the left part, especially in zero- or low-resource setting due to the lack of supervised data. Therefore, based on denoising auto-encoder and dual transformation, we further leverage bidirectional sequence modeling for both text and speech to alleviate the error propagation problem①.

Fourth, we design a unified model structure based on Transformer [42] that can take speech or text as input or output, in order to incorporate the above components for TTS and ASR together.

We conduct experiments on the LJSpeech dataset by leveraging only 200 paired speech and text data and extra unpaired data. The results show that DualSpeech generates intelligible voice with a word level intelligible rate of 99.84%, compared with nearly 0 intelligible rate if training on only 200 paired data. Furthermore, it achieves 2.68 MOS for TTS and 11.7% PER for ASR, outperforming the baseline model trained on only 200 paired data.

3.2 LRSpeech

Although DualSpeech[35] produces reasonably good artificial speech for low-resource settings by leveraging unpaired speech and text data through dual learning, the quality of synthesized speech still does not meet the requirement of commercial speech services. We extend DualSpeech to extremely low-resource settings for speech synthesis (and speech recognition) and target at industrial deployment under two constraints: ① Extremely low data collection cost; ② High accuracy to satisfy the deployment requirement.

Industrial neural speech synthesis and recognition systems typically use several kinds of data for model training.

- Speech synthesis needs high-quality single-speaker (the target speaker) recordings that are collected in professional recording studio. To improve the pronunciation accuracy, speech synthesis also requires a pronunciation lexicon to convert the character sequence into phoneme sequence as the model input (e.g., "speech" is converted into "s p iy ch"), which is called as grapheme-to-phoneme conversion [56]. In addition, speech synthesis models use text normalization rules to convert the irregular word into the normalized type that is easier to pronounce (e.g., "Dec 6th" is converted into "December sixth").

① We train the models to generate speech and text sequence in both left-to-right and right-to-left directions. During dual transformation, for example, we use TTS model to transform the text y to the speech $\boldsymbol{x}$ (left-to-right) and $\boldsymbol{x}$ (right-to-left). Then the ASR model leverages $(\boldsymbol{x}, y)$ and $(\boldsymbol{x}, y)$ for training, where $\boldsymbol{x}$ and $\boldsymbol{x}$ are of good quality in the left part and right part respectively, preventing the model to be biased to always generate low quality results in the right part.

- Speech recognition requires speech data from multiple speakers in order to generalize to unseen speakers during inference. The multi-speaker speech data in speech recognition do not need to be as high-quality as that in speech synthesis, but the data amount is usually an order of magnitude larger. The speech data for speech recognition is called multi-speaker low-quality data①. Optionally, a speech recognition model can first recognize a speech sequence into a phoneme sequence, and then convert it into a character sequence with the pronunciation lexicon as used in speech synthesis.

  In addition to paired speech and text data, speech synthesis and recognition systems may also leverage unpaired speech and text data for model training.

We reduce the overall data collection cost considering that different kinds of data are of different cost:

- We only use several minutes of single-speaker high-quality paired speech text data, because this kind of data is the most costly to collect and the speech data is usually recorded in professional studios.
- We use several hours of multi-speaker low-quality paired speech text data. The low-quality paired data is relatively less costly to collect compared with the above high-quality paired data. For example, one can collect speech data from the web and ask human labelers to transcribe them.
- We use dozens of hours of multi-speaker low-quality unpaired speech data. This kind of low-quality speech without paired text is relatively easy to collect from the Web, at almost zero cost.
- We do not use high-quality speech data from the target TTS speaker, which is costly to collect.
- We do not use the pronunciation lexicon but directly take characters as the input of the TTS model and the output of the ASR model, because the lexicon is costly to obtain, especially for rare languages.

To ensure the TTS quality and ASR accuracy while reducing data cost, we develop a system for joint speech synthesis and recognition, named LRSpeech [36], which is based on three key techniques:

---

① Here low quality does not mean the transcripts are incorrect. Instead, it means that the quality of speech data is just relatively low (e.g., with background noise, incorrect pronunciations, etc.) compared with the high-quality TTS recordings.

(1) Pre-training on rich-resource languages and fine-tuning on the low-resource target language;

(2) Iterative accuracy boosting between speech synthesis and recognition through dual learning;

(3) Knowledge distillation to further improve synthesis quality and recognition accuracy.

Specifically, the training pipeline of LRSpeech contains three stages, as shown in Figure 4:

- Pre-training and fine-tuning. LRSpeech pre-trains both TTS and ASR models on rich-resource languages and then fine-tunes them on low-resource languages. Leveraging rich-resource languages in LRSpeech are based on two considerations: ① Large scale of paired data for rich-resource languages is easy to collect and might be already available in commercial speech services; ② The alignment capability between speech and text in rich-resource languages can benefit the alignment learning in low-resource languages, due to the pronunciation similarity between human languages [51].
- Dual reconstruction. Following [35], LRSpeech leverages the duality between speech synthesis and recognition and boost the accuracy of each other with unpaired speech and text data based on the principle of dual reconstruction.

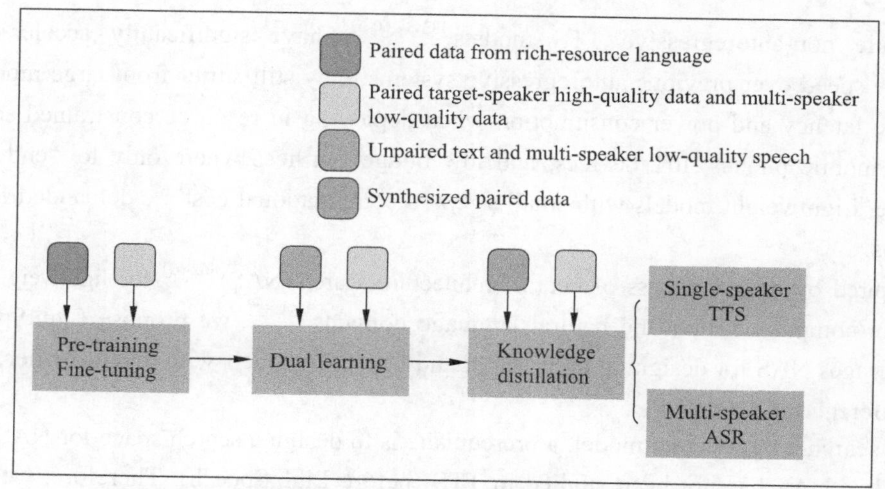

Figure 4 The three-stage pipeline of LRSpeech [36](见彩图 3)

- Post-distillation. To meet the quality requirement of commercial speech services, LRSpeech further improves the accuracy of TTS and ASR models through knowledge distillation [58-59].

Experimental results [36] show that dual learning together with pre-training (on rich-resource languages) and post-distillation can significantly improve the TTS quality and ASR accuracy. Especially, with only 5-minutes high-quality speech (plus corresponding text) from the target speaker, LRSpeech learns a TTS model that meets the quality requirement of commercial speech services. LRSpeech has been integrated in Azure TTS service to support multiple low-resource languages.①

4 Parameter Efficiency: LightSpeech and AdaSpeech Series

In this section, we introduce our work on improving parameter efficiency for neural TTS in different scenarios: LightSpeech [38] for mobile/embedded devices with limited storage and power consumption, AdaSpeech [39] for model adaptation in custom voice with only a few parameters for fine-tuning/adaptation, AdaSpeech 2 [40] for model adaptation in custom voice using only untranscribed speech data (no paired text data).

4.1 LightSpeech

While non-autoregressive TTS models [33-34,60-61] have significantly accelerated the inference speed over previous autoregressive systems, they still suffer from large model size, inference latency and power consumption when deploying in resource constrained scenarios such as mobile phones, IoT devices, and low budget services where only low-end CPU is available. Lightweight models with small size and computational cost are demanded for those scenarios.

Inspired by recent success of neural architecture search (NAS) [62-63] for lightweight model design in computer vision and natural language domains [64-65], we propose LightSpeech [38] that leverages NAS for designing lightweight and fast TTS models with much smaller size and faster inference speed on CPU.

To search a lightweight model, a prerequisite is to design a search space for NAS. To our knowledge, NAS has not been studied in TTS (before LightSpeech). Therefore, we need to

① https://techcommunity.microsoft.com/t5/azure-ai/neural-text-to-speech-previews-five-new-languages-with/ba-p/1907604.

design a search space for TTS. To design a search space, the first step of LightSpeech is to profile the model size and computational cost of each component of a state-of-the-art neural TTS model [33]. Second, based on the profiling results, we design a novel neural architecture space to search lightweight TTS models. Third, we adopt GBDT-NAS [66] for model search considering its promising performance and fitness in our task (the chain-structure of our search space is well fitted by GBDT-NAS). In the remaining of this subsection, we first introduce details of those steps and then report empirical results.

4.1.1 Model Profiling

We profile each component of a FastSpeech 2 model to identify the bottlenecks of memory (model size) and inference speed. A FastSpeech 2 model contains 5 components: the text encoder, the speech decoder, the duration predictor, the pitch predictor and theenergy predictor. The encoder and the decoder consists of 4 feed-forward Transformer blocks respectively. The duration predictor is a 2-layer 1D-convolution neural network with kernel size 3. The pitch predictor is a 5-layer 1D-convolution neural network with kernel size 5. The energy predictor has the same structure as the pitch predictor. The size (i.e., number of parameters) and the inference speed of each component are shown in Table 1.

Table 1 Profiling of the model size and the inference latency of components in FastSpeech 2 model

Name	Structure	#Params	RTF
FastSpeech 2	–	27.02M	6.1×10^{-2}
Encoder	4 Trans Block	11.56M	8.1×10^{-3}
Decoder	4 Trans Block	11.54M	4.1×10^{-2}
Duration Predictor	2 Conv Layer	0.40M	4.2×10^{-4}
Pitch Predictor	5 Conv Layer	1.64M	4.8×10^{-3}
Energy Predictor	5 Conv Layer	1.64M	4.6×10^{-3}

Note: RTF denotes the real-time factor which is the time (in seconds) required for the system to synthesize one second waveform. The RTF of each component iscalculated by measuring the inference latency of the component and divided by the total generated audio length. The latency is measured with a single thread and a single core on an Intel Xeon CPU E5-2690 v4 @ 2.60 GHz, with 256 GB memory, and a batch size of 1.

We have several observations from Table 1: ① The encoder and the decoder takes the most of the model size and inference time. Therefore we mainly aim to reduce the encoder and the decoder size and perform architecture search to discover better and efficient architectures. ② The predictors take only about 1/10 of the total size. Therefore we manually re-design those predictors with lightweight operations rather than searching the architectures.

4.1.2 Search Space Design

There are 4 feed-forward Transformer blocks in both the encoder and the decoder in [33], where each feed-forward Transformer block contains a multi-head self-attention (MHSA) layer and a feed-forward network (FFN). We use this encoder-decoder framework as our network backbone and set the number of layers in both the encoder and the decoder to 4. For the variance predictors (duration, pitch and energy) in FastSpeech 2, our preliminary experiments show that removing the energy predictor leads only marginal performance drop in terms of voice quality. Accordingly, we directly remove the energy predictor in our design.

After setting the number of layers in each component, we search for different combinations of diverse architectures in the encoder and the decoder. We carefully design the candidate operations for our task: ① LSTM is not included due to its slow inference speed. ② We decouple the original Transformer block to MHSA and FFN as separate operations. We consider MHSA with different numbers of attention heads $\{2,4,8\}$. ③ Considering that depth-wise separable convolution (SepConv) [67] is much more memory and computation efficient compared with vanilla convolution, we use SepConv to replace vanilla convolution. The parameter size of vanilla convolution is $K \times I_d \times O_d$ where K is the kernel size, I_d is the input dimension and O_d is the output dimension. The size of SepConv is $K \times I_d + I_d \times O_d$. Following [68], we adopt different kernel sizes of $\{1,5,9,13,17,21,25\}$. Finally, our candidate operations include $3+7+1=11$ different choices: MHSA with number of attention head in $\{2,4,8\}$, SepConv with kernel size in $\{1,5,9,13,17,21,25\}$ and FFN. This yields a search space of $11^{4+4}=11^8=214358881$ different candidate architectures.

To reduce the model size and latency of the variance predictors (the duration predictor and the pitch predictor), we directly replace the convolution operation in them with SepConv with the same kernel size, without searching for other operations.

4.1.3 Architecture Search

There have been many methods for searching neural architectures [62-63,66,69]. For our task, we adopt a very recent method [66] which is based on accuracy prediction. It is efficient and effective, and well fits our task (the chain-structure search space). Specifically, it uses a gradient boosting decision tree (GBDT) trained on some architecture-accuracy pairs to predict the accuracy of other candidate architectures. Then the architectures with top predicted accuracy are further evaluated by training on the training set and then evaluating on a held-out dev set. Finally, the architecture with the best evaluated accuracy is selected.

In our task, since the evaluation of the voice quality of a TTS system involves human labor, it is impractical to evaluate each candidate architecture during the search. We use the validation loss on the dev set as a proxy of the accuracy to guide the search①. Therefore, we search architectures with as small validation loss as possible.

4.1.4 Results

Experiments show that compared with the original FastSpeech 2 model [33], architecture discovered by our LightSpeech achieves 15x compression ratio (1.8M vs. 27M), 16x less MACs (0.76G vs. 12.50G) and 6.5x inference time speedup on CPU.

4.2 AdaSpeech

Custom voice, an important service in commercial speech platforms such as Microsoft Azure, Amazon AWS and Google Cloud, aims to adapt a source TTS model to a personalized voice with few adaptation data.

The main challenge in custom voice is parameter efficiency: while adapting more parameters usually result in better voice quality, it increases storage and serving cost. For example, to support one million users in a cloud speech service, if the size of each custom voice model is 100MB, the total memory storage would be about 100PB, which is a big serving cost. Therefore, when adapting the source TTS model to a new voice, we need to ensure voice quality and at the same time fine-tune as few parameters aspossible. Another challenge is that the recordings of custom voice users are usually of different acoustic conditions from the source speech data (the data to train the source TTS model). For example, the adaptation data is usually recorded with diverse speaking prosodies, styles, emotions, accents and recording environments. The mismatch in these acoustic conditions makes the source model difficult to generalize and leads to poor adaptation quality.

We design AdaSpeech [39], an adaptive TTS model for high-quality and efficient customization of new voice. AdaSpeech employs a three-stage pipeline for custom voice: pre-training, fine-tuning, and inference. During the pre-training stage, a source TTS model is trained on large-scale multi-speaker datasets, which can ensure the TTS model to cover diverse text and speaking voices that is helpful for adaptation. During the fine-tuning stage, the source TTS model is adapted on a new voiceby fine-tuning part of its parameters on limited

① In non-autoregressive models (e.g., FastSpeech and FastSpeech 2), the valid loss on the dev set is highly correlated with the final quality, while in autoregressive models it is not.

adaptation data. During the inference stage, both the unadapted part (parameters shared by all custom voices) and the adapted part (each custom voice has specific adapted parameters) of the TTS model are used to synthesize speech. We build AdaSpeech based on FastSpeech 2 [33] and further design two techniques to address the challenges in custom voice:

- Conditional layer normalization. To fine-tune as few parameters as possible while ensuring adaptation quality, we modify the layer normalization [70] in the mel-spectrogram decoder in pre-training, by using speaker embedding as conditional information to generate the scale and bias vector in layer normalization. In fine-tuning, we only adapt the parameters related to the conditional layer normalization. In this way, we largely reduce adaptation parameters and thus memory storage compared with fine-tuning the whole model, and maintain high quality of adapted voice.
- Acoustic condition modeling. In order to handle different acoustic conditions for adaptation, we model the acoustic conditions in both utterance and phoneme level in pre-training and fine-tuning. Specifically, we use two acoustic encoders to extract an utterance-level vector and a sequence of phoneme-level vectors from the target speech, which are taken as the input of the mel-spectrogram decoder to represent the global and local acoustic conditions respectively. Doing so the decoder can predict speech in different acoustic conditions based on these acoustic information. Otherwise, the model would memorize the acoustic conditions and cannot generalize well. In inference, we extract the utterance-level vector from a reference speech and use another acoustic predictor that is built upon the phoneme encoder to predict the phoneme-level vectors.

In the remaining of this sub section, we first describe the overall design of our proposed AdaSpeech, and then introduce the two key techniques to address the challenges in custom voice. At last, we list the pre-training, fine-tuning and inference pipeline of AdaSpeech for custom voice, and report empirical results.

4.2.1 Overall design

The model structure of AdaSpeech is shown in Figure 5. We adopt FastSpeech 2 [33] as the model backbone considering the FastSpeech [33-34] series are one of the most popular models in non-autoregressive TTS. The basic model backbone consists of a phoneme encoder, a mel-spectrogram decoder, and a variance adaptor which provides variance information including duration, pitch and energy into the phoneme hidden sequence following [33]. As shown

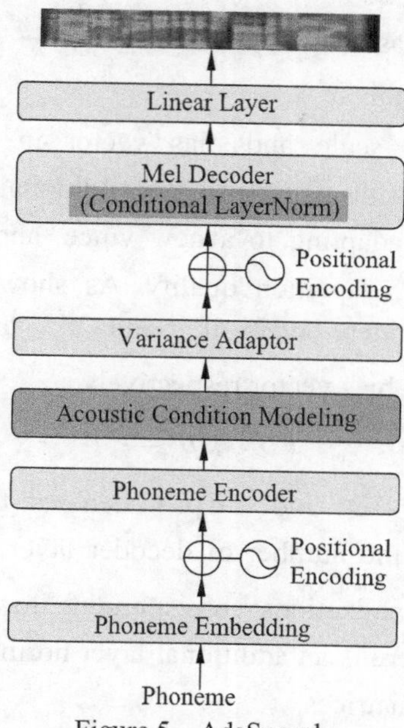

Figure 5 AdaSpeech

in Figure 5, we design two additional components to address the distinctive challenges in custom voice: ① To support diverse customers, we use acoustic condition modeling to capture the diverse acoustic conditions of target speech in different granularities; ② To support a large number of customers with affordable memory storage, we introduce conditional layer normalization in decoder for efficient adaptation with only a few parameters for fine-tuning.

4.2.2 Conditional Layer Normalization

Achieving high adaptation quality while using a few adaptation parameters is challenging. Previous works use zero-shot adaptation with speaker encoder [71-73] or only fine-tune the speaker embedding cannot achieve satisfied quality. Can we greatly increase the voice quality at the cost of slightly more but negligible parameters? To this end, we analyze the model parameters of FastSpeech 2 [33], which is basically built upon the structureof Transformer [42], with a self-attention network and a feed-forward network in each Transformer block. Both the matrix multiplications in the query, key, value and output of self-attention and two-layer feed-forward networks areparameter-intensive, which is not efficient to adapt. We find that layer normalization [70] is adopted in each self-attention and feed-forward network in decoder, which can greatly influence the hidden activation and final prediction with a light-weight learnable scale vector γ and bias vector $\boldsymbol{\beta}$: $\mathrm{LN}(\boldsymbol{x})=\gamma\frac{x-\mu}{\sigma}+\boldsymbol{\beta}$, where $\boldsymbol{\mu}$ and $\boldsymbol{\sigma}$ are the mean and variance of hidden vector $\boldsymbol{x}$.

If we can determine the scale and bias vector in layer normalization with the corresponding speaker characteristics using a small conditional network, then we can fine-tune this conditional network when adapting to a new voice, and greatly reduce the adaptation parameters while ensuring the adaptation quality. As shown in Figure 6, the conditional network consists of two simple linear layers W_c^{γ} and W_c^{β} that take speaker embedding E^{s} as input and output the scale and bias vector respectively:

$$\gamma_c^{\mathrm{s}}=E^{\mathrm{s}}W_c^{\gamma},\ \beta_c^{\mathrm{s}}=E^{\mathrm{s}}W_c^{\beta} \tag{2}$$

where s denotes the speaker ID, and $c\in[C]$ denotes there are C conditional layer normalizations in the decoder (the number of decoder layer is $(C-1)=2$ since each layer has two conditional layer normalizations corresponding to self-attention and feed-forward network in Transformer, and there is an additional layer normalization at the final output) and each uses different conditional matrices.

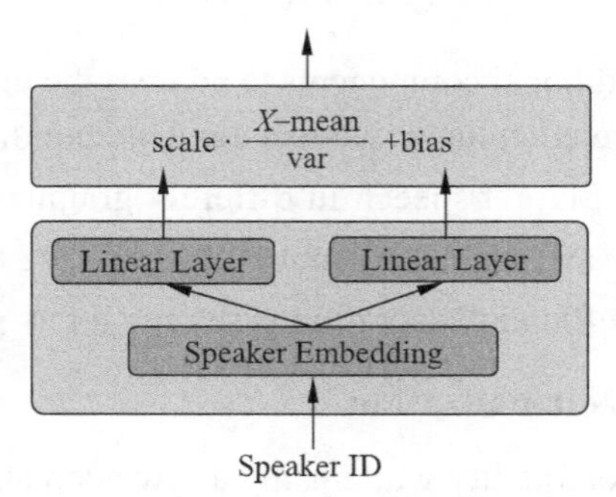

Figure 6 Conditional Layer Normalization

4.2.3 Acoustic Condition Modeling

In custom voice, the adaptation data can be spoken with diverse prosodies, styles, accents, and can be recorded under various environments, which can make the acoustic conditions far different from that in source speech data. This presents great challenges to adapt the source TTS model, since the source speech cannot cover all the acoustic conditions in custom voice. A practical way to alleviate this issue is to improve the adaptability (generalizability) of source TTS model. In text to speech, since the input text lacks enough acoustic conditions (such as speaker timbre, prosody and recording environments) to predict the target speech, the model tends to memorize and overfit on the training data [33], and has poor generalization during adaptation. A natural way to solve such problem is to provide corresponding acoustic conditions as input to make the model learn reasonable text-to-speech mapping towards better generalization instead of memorizing.

To better model the acoustic conditions with different granularities, we categorize the acoustic conditions in different levels as shown in Figure 7(a): ① speaker level, the coarse-grained acoustic conditions to capture the overall characteristics of a speaker; ② utterance level, the fine-grained acoustic conditions in each utterance of a speaker; ③ phoneme level, the more fine-grained acoustic conditions in each phoneme of an utterance, such as accents on specific phonemes, pitches, prosodies and temporal environment noises①. Since speaker ID

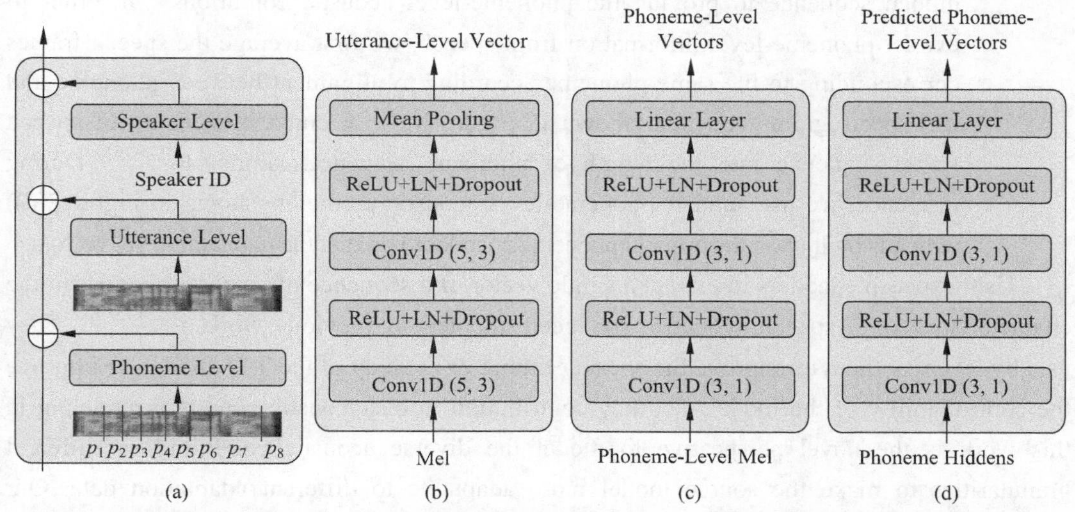

Figure 7 (a) The overall structure of acoustic condition modeling; (b) Utterance-level acoustic encoder; (c) Phoneme-level acoustic encoder, where phoneme-level mel means the mel-frames aligned to the same phoneme are averaged; (d) Phoneme-level acoustic predictor, where phoneme hidden representations are the hidden sequence from the phoneme encoder in Figure 5

"Conv1D (m , n)" means the kernel size and stride size in 1D convolution is m and n respectively. "LN" means layer normalization. As shown in Figure 2(a), the phoneme-level vectors are directly added element-wisely into the hidden sequence, and the utterance-level and speaker level vector/embedding are first expanded to the same length and then added element-wisely into the hidden sequence

① Generally, more fine-grained frame-level acoustic conditions [77] exist, but have marginal benefits considering their prediction difficulty. Similarly, more coarse-grained language level conditions also exist, but we do not consider multilingual setting in this work and leave it for future work.

(embedding) is widely used to capture speaker-level acoustic conditions in multi-speaker scenario [74], speaker embedding is used by default. We describe the utterance-level and phoneme-level acoustic condition modeling as follows.

- Utterance Level. We use an acoustic encoder to extract a vector from a reference speech, similar to [71-73], and then expand and add it to the phoneme hidden sequence to provide the utterance-level acoustic conditions. As shown in Figure 7(b), the acoustic encoder consists of several convolutional layers and a mean pooling layer to get a single vector. The reference speech is the target speech during training, while a randomly chosen speech of this speaker during inference.
- Phoneme Level. We use another acoustic encoder (shown in Figure 7(c)) to extract a sequence of phoneme-level vectors from the target speech and add it to the phoneme hidden sequence to provide the phoneme-level acoustic conditions①. In order to extract phoneme-level information from speech, we first average the speech frames corresponding to the same phoneme according to alignment between phoneme and mel-spectrogram sequence (shown in Figure 7(a)), to convert to length of speech frame sequence into the length of phoneme sequence, similar to [75-76]. During inference, we use another phoneme-level acoustic predictor (shown in Figure 7(d)) which is built upon the original phoneme encoder to predict the phoneme-level vectors.

Using speech encoders to extract a single vector or a sequence of vectors to represent the characteristics of a speech sequence has been adopted in previous works [72-74,75-76]. They usually leverage them to improve the speaker timbre or prosody of the TTS model, or improve the controllability of the model. The key contribution in our acoustic condition modeling in this work is the novel perspective to model the diverse acoustic conditions in different granularities to make the source model more adaptable to different adaptation data. Our experiments show that utterance-level and phoneme-level acoustic modeling indeed helps the learning of acoustic conditions and is critical to ensure the adaptation quality.

4.2.4 Pipeline of AdaSpeech

We list the pre-training, fine-tuning and inference pipeline of AdaSpeech in Algorithm 1.

---

① Note that although the extracted vectors can contain all phoneme-level acoustic conditions ideally, we still use pitch and energy in the variance adaptor (shown in Figure 5) as additional input following [33], in order to ease the burden of acoustic condition learning and focus on learning other acoustic conditions. We also tried to remove pitch and energy but found it causes worse adaptation quality.

Algorithm 1 Pre-training, fine-tuning and inference of AdaSpeech

1. **Pre-training**: Train the AdaSpeech model θ with source training data D.
2. **Fine-tuning**: Fine-tune W_c^{γ} and W_c^{β} in each conditional layer normalization $c \in [C]$ and speaker embedding E^s with the adaptation data D^s for each custom speaker/voice s.
3. **Inference**:
4. Calculate γ_c^s, β_c^s in each conditional layer normalization $c \in [C]$, and get the parameters $\theta^s = \{\{\gamma_c^s, \beta_c^s\}_{c=1}^C, E^s\}$ for speaker s.
5. Deploy the shared model parameters $\tilde{\theta}$ (not fine-tuned in θ during adaptation) and speaker specific parameters θ^s for s
6. Use $\tilde{\theta}$ and θ^s to synthesize custom voice for speaker s.

During fine-tuning, we only fine-tune the two matrices W_c^{γ} and W_c^{β} in each conditional layer normalization in decoder and the speaker embedding E^s, fixing other model parameters including the utterance-level and phoneme-level acoustic encoders and phoneme-level acoustic predictor. During inference, we do not directly use the two matrices W_c^{γ} and W_c^{β} in each conditional layer normalization since they still have large parameters. Instead, we use the two matrices to calculate each scale and bias vector γ_c^s and β_c^s from speaker embedding E_s according to Eqn. (2) considering E_s is fixed in inference. Dong so we can save a lot of memory storage①.

4.2.5 Results

Experiments with different adaptation settings show that AdaSpeech achieves better adaptation quality in terms of MOS (mean opinion score) and SMOS (similarity MOS) than baseline methods, with only about 5K user-specific parameters for each speaker, demonstrating its effectiveness for custom voice.

4.3 AdaSpeech 2

Most previous works [39,71,73,78] including AdaSpeech [39] on TTS adaptation require paired data

① Assume the dimension of speaker embedding and hidden vector are both h, the number of conditional layer normalization is C. Therefore, the number of adaptation parameters are $2h^2C + h$, where the first 2 represents the two matrices for scale and bias vectors, and the second term h represents the speaker embedding. If $h = 256$ and $C = 9$, the total number of parameters are about 1.2M, which is much smaller compared the whole model (31M). During deployment for each custom voice, the total additional model parameters for a new voice that need to be stored in memory becomes $2hC + h$, which is extremely small (4.9K in the above example).

(speech and its transcripts/text). However, paired data is harder to obtain than untranscribed speech data, which can be broadly accessible through conversations, talks, public speech, etc. Therefore, leveraging untranscribed speech data for TTS adaptation can greatly extend its application scenarios.

A straightforward method to use untranscribed data is to first use an ASR system to transcribe the speech into text and then conduct adaptation as in previous works. However, this additional ASR system may not be available in some scenarios, and its recognition accuracy is not high enough which will generate incorrect transcripts and hurt adaptation.

We propose AdaSpeech 2 [40] that leverages untranscribed speech data for adaptation. We use the basic structure of AdaSpeech [39] as our source TTS model backbone, which contains a phoneme encoder and a mel-spectrogram decoder. To enable untranscribed speech adaptation, we additionally introduce a mel-spectrogram encoder into the well-trained source TTS model for speech reconstruction together with the mel-spectrogram decoder. At the same time, we constrain the output sequence of the mel encoder to be close to that of the phoneme encoder with an L2 loss. After training this additional mel encoder, we use untranscribed speech data for speech reconstruction and only fine-tune a part of the model parameters in the mel decoder following [39], while the mel encoder and the phoneme encoder remain unchanged. In inference, the fine-tuned mel-spectrogram decoder together with the unchanged phoneme encoder forms a TTS model that can synthesize custom voicefor target speaker.

AdaSpeech 2 has two main advantages: ① Pluggable. As the essence of our method is to add an additional encoder to a trained TTS pipeline, it can be easily applied to existing TTS models without re-training, improving reusability and extendibility. ② Effective. It achieves on-par voice quality with the transcribed TTS adaptation, with the same amount of untranscribed data.

As shown in Figure 8, AdaSpeech 2 has two main modules: a TTS model pipeline with a phoneme encoder and a mel-spectrogram decoder, which is based on the backbone of AdaSpeech [39], and an additional mel-spectrogram encoder, which is introduced to leverage untranscribed speech data for adaptation.

As shown in Figure 9, the adaptation pipeline consists of the following steps: ① Source model training, where we need to train a multi-speaker TTS model and use it for adaptation. ② Mel-spectrogram encoder aligning, where we need to train a mel-spectrogram encoder and use an alignment loss to make the output space of the mel-spectrogram encoder to be close to that of

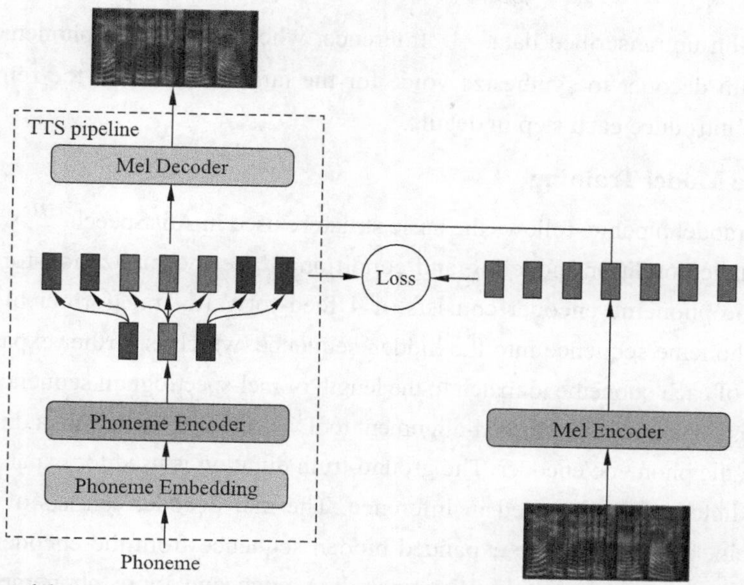

Figure 8 The model structure of AdaSpeech 2, where the TTS pipeline in the figure follows AdaSpeech [39]
Note that we also use the acoustic condition modeling and conditional layer normalization as in AdaSpeech, but do not show here mainly for simplicity

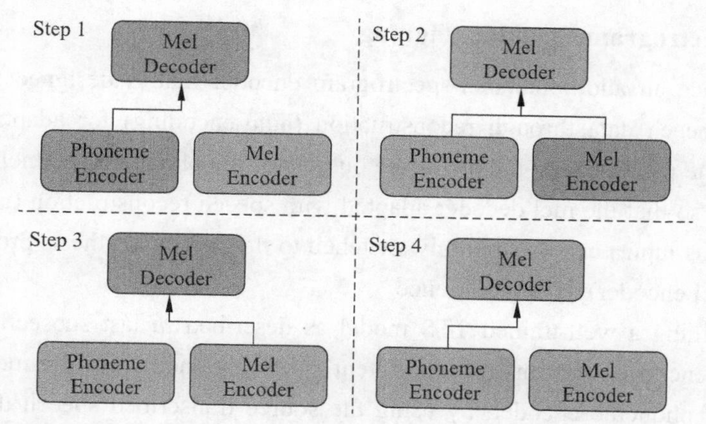

Figure 9 The four-step adaptation pipeline in AdaSpeech 2, where the first three steps represent the training process (source model training, mel-spectrogram encoder aligning and untranscribed speech adaptation) and the last step represents the inference process（见彩图 4）
Orange blocks represent the modules that are optimized while grey blocks represent the modules unchanged in the current step

the phoneme encoder. ③ Untrancribed speech adaptation, where the mel-spectrogram decoder is adapted to the target speaker with the help of the mel-spectrogram encoder through auto-encoding on untranscribed data. ④ Inference, where we use the phoneme encoder and mel-spectrogram decoder to synthesize voice for the target speaker. In the remaining of this subsection, we introduce each step in details.

4.3.1 Source Model Training

Our TTS model pipeline follows the basic structure used in AdaSpeech [39], with specifically designed acoustic condition modeling and conditional layer normalization for efficient TTS adaptation. The phoneme encoder consists of 4 feed-forward Transformer blocks [34,42] and converts the phoneme sequence into the hidden sequence, which is further expanded according to the duration of each phoneme to match to the length of mel-spectrogram sequence. The ground-truth duration is obtained from a forced-alignment tool [47], and is used to train a duration predictor that built upon the phoneme encoder. The ground-truth duration is used for expansion in training while the predictedvalues are used in inference. The mel decoder consists of 4 feed-forward Transformer blocks and takes the expanded hidden sequence from the encoder as input. The decoder adds more variance information including pitch and more phoneme-level acoustic condition information as used in [39]. The phoneme-level acoustic vectors are also expanded to the length of mel-spectrogram sequence and added into the decoder input with a dense layer.

4.3.2 Mel-spectrogram Encoder Aligning

We introduce an additional mel-spectrogram encoder that is designed to make use of untranscribed speech data through reconstruction (auto-encoding) for adaptation. From the perspective of the mel decoder, it expects the outputs of the phoneme and mel encoders are in the same space, so that the mel decoder adapted with speech reconstruction (taking the output of mel encoder as input) can be smoothly switched to the speech synthesis process (taking the output of the mel encoder) during inference.

After obtaining a well-trained TTS model as described in last subsection, we add this additional mel encoder and constrain the latent space obtained from the mel encoder to be close to that of phoneme encoder, by using the source transcribed speech data. The hidden sequence generated by the mel encoder is constrained with an L2 loss to match to the hidden sequence from the phoneme encoder, where two hidden sequences are of the same length since the phoneme hidden sequence has been expanded according to the phoneme duration. The mel encoder also consists of 4 feed-forward Transformer blocks, considering the symmetry of the system.

The alignment process above distinguishes our work from previous works [79] that also adopt hidden sequence alignment to align the phoneme and mel encoder. We fix the parameters of the source TTS model (phoneme encoder and mel decoder)and only update the parameters of the mel encoder, which can form a plug-and-play way to leverage any well-trained source TTS model. On the contrary, previous works usually requires the training of source TTS model and adaptation process at the same time, which is not pluggable and may limit the broad usage in TTS adaptation.

4.3.3 Untranscribed Speech Adaptation

The untranscribed data from the target speaker is used to fine-tune the model through the way of speech reconstruction with the mel encoder and decoder. To fine-tune as small amount of parameters as possible while ensure the adaptation quality, we adapt the parameters related to the conditional layer normalization following AdaSpeech [39].

4.3.4 Inference

After the aligning and adaptation process mentioned above, the TTS pipeline can imitate the custom voice for speech synthesis. We use the original unadapted phoneme encoder and the partial adapted mel decoder to form the personalized TTS system to generate voice for each specific speaker.

4.3.5 Results

To evaluate AdaSpeech 2 for custom voice, we train the source TTS model on LibriTTS dataset and adapt the model on VCTK and LJSpeech datasets with different adaptation settings. The average MOS of our results is 3.38 on VCTK and 3.42 on LJSpeech. The avarage SMOS of our results is 3.87 while that of method using transcribed data is 3.96. Besides, we also use our internal spontaneous ASR speech data to fine-tune the model. The results show that AdaSpeech 2 achieves better adaptation quality than baseline methods and close to the transcribed upper bound (AdaSpeech).

5 Summary

In this article, we summarized our recent works on efficient neural TTS, including FastSpeech [34] and FastSpeech 2/2s [33] to improve inference efficiency for neural TTS, DualSpeech [35] and LRSpeech [36] to improve data efficiency, LightSpeech [38] and AdaSpeech

1/2 [39-40] to improve parameter efficiency. Many of those techniques have been shipped to Microsoft Azure TTS services to improve final user experiences.

Improving computation efficiency (i.e., speeding up training and inference), data efficiency (i.e., using less labeled data or better using unlabeled data), parameter efficiency (i.e., designing lightweight models with fewer parameters and less power consumption) is a long lasting focus of machine learning research. While our research makes good progress along this direction, there is definitely huge potential to further improve the efficiency of neural TTS. We hope our works can help researchers to better understand the problem and inspire them to design better TTS algorithms/models/systems. We also believe our models/systems are helpful for industry practitioners to sovle the real-world TTS tasks in their hands.

References

[1] PAUL T. Text-to-speech synthesis[M]. Cambridge university press, 2009.

[2] RONALD B A, GEORGE R R, ALEXANDRE S. Understanding human communication[M]. Holt, Rinehart and Winston Chicago: Oxford University Press, 1991.

[3] DAN J. Speech & language processing[M]. Pearson Education India, 2000.

[4] CHRISTOPHER M, HINRICH S. Foundations of statistical natural language processing[M]. MIT press, 1999.

[5] STUART R, PETER N. Artificial intelligence: a modern approach[M]. 2002.

[6] JONATHAN A, SHARON H, ROLF C, et al. Mitalk-79: The 1979 MIT text-to-speech system[J]. The Journal of the Acoustical Society of America, 1979, 65(S1):S130-S130.

[7] DENNIS H KLATT. Software for a cascade/parallel formant synthesizer[J]. the Journal of the Acoustical Society of America, 67(3):971-995, 1980.

[8] DENNIS H KLATT. Review of text-to-speech conversion for english[J]. The Journal of the Acoustical Society of America, 82(3):737-793, 1987.

[9] CHRISTINE H S, ROBERT I D. Prospects for articulatory synthesis: A position paper[C]//Proceedings of 4th ISCA Tutorial and Research Workshop (ITRW) on Speech Synthesis, 2001.

[10] ALAN B, PAUL T, RICHARD C, et al. The festival speech synthesis system[M]. 1998.

[11] ANDREW J H, ALAN W B. Unit selection in a concatenative speech synthesis system using a large speech database[C]//Proceedings of 1996 IEEE International Conference on Acoustics, Speech, and Signal Processing Conference Proceedings. IEEE, 1996, 1: 373-376.

[12] ERIC M, FRANCIS C. Pitch-synchronous waveform processing techniques for text-to-speech synthesis using diphones[J]. Speech communication, 1990, 9(5-6): 453-467.

[13] YOSHINORI S, NOBUYOSHI K, NAOTO I, et al. Atr μ-talk speech synthesis system[C]//Proceedings of Second International Conference on Spoken Language Processing, 1992.

[14] KEIICHI T, YOSHIHIKO N, TOMOKI T, et al. Speech synthesis based on hidden markov models[J]. Proceedings of the IEEE, 2013, 101(5):1234-1252.

[15] KEIICHI T, TAKAYOSHI Y, TAKASHI M, et al. Speech parameter generation algorithms for hmm-based speech synthesis[C]//Proceedings of 2000 IEEE International Conference on Acoustics, Speech, and Signal Processing. IEEE, 2000, 3: 1315-1318.

[16] TAKAYOSHI Y, KEIICHI T, TAKASHI M, et al. Simultaneous modeling of spectrum, pitch and duration in hmm-based speech synthesis[C]//Proceedings of 6th European Conference on Speech Communication and Technology, 1999.

[17] ZEN H G, KEIICHI T, ALAN W B. Statistical parametric speech synthesis[J]. Speech Communication, 2009, 51(11):1039-1064.

[18] FAN Y C, QIAN Y, XIE F L, et al. TTS synthesis with bidirectional LSTM based recurrent neural networks[C]//Proceedings of 15th annual conference of the international speech communication association, 2014.

[19] LI H, KANG Y G, WANG Z Y. Emphasis: An emotional phoneme-based acoustic model for speech synthesis system[Z]. arXiv preprint arXiv:1806.09276, 2018.

[20] QIAN Y, FAN Y C, HU W P, et al. On the training aspects of deep neural network (dnn) for parametric TTS synthesis[C]//Proceedings of 2014 IEEE International Conference on Acoustics, Speech and Signal Processing (ICASSP). IEEE, 2014: 3829-3833.

[21] AÄRON VAN DEN O, SANDER D, HEIGA Z, et al. Wavenet: A generative model for raw audio[C]//Proceedings of 9th ISCA Speech Synthesis Workshop. 125-125.

[22] WANG W F, XU S, XU B. First step towards end-to-end parametric TTS synthesis: Generating spectral parameters with neural attention[J]. Interspeech, 2016: 2243-2247.

[23] WANG Y X, SKERRY-RYAN R J, DAISY S, et al. Tacotron: Towards end-to-end speech synthesis[Z]. arXiv preprint arXiv:1703.10135, 2017.

[24] ZEN H G, HA S S. Unidirectional long short-term memory recurrent neural network with recurrent output layer for low-latency speech synthesis[C]//Proceedings of 2015 IEEE International Conference on Acoustics, Speech and Signal Processing (ICASSP). IEEE, 2015: 4470-4474.

[25] ZEN H G, ANDREW S, MIKE S. Statistical parametric speech synthesis using deep neural networks[C]//Proceedings of 2013 IEEE international conference on acoustics, speech and signal processing. IEEE, 2013: 7962-7966.

[26] JONATHAN S, RUOMING P, RON J W, et al. Natural TTS synthesis by conditioning wavenet on mel spectrogram predictions[C]//Proceedings of 2018 IEEE International Conference on Acoustics, Speech and Signal Processing (ICASSP). IEEE, 2018: 4779-4783.

[27] PING W, PENG K N, ANDREW G, et al. Deep voice 3: 2000-speaker neural text-to-speech[C]//Proceedings of International Conference on Learning Representations, 2018.

[28] PING W, PENG K N, CHEn J T. Clarinet: Parallel wave generation in end-to-end text-to-speech[C]//Proceedings of International Conference on Learning Representations, 2019.

[29] DANIEL G, JAE L. Signal estimation from modified short-time fourier transform[J]. IEEE Transactions on Acoustics, Speech, and Signal Processing, 1984, 32(2): 236-243.

[30] AARON VAN DEN O, LI Y Z, IGOR B, et al. Parallel wavenet: Fast high-fidelity speech synthesis[Z]. arXiv preprint arXiv:1711.10433, 2017.

[31] RyAN P, RAFAEL V, BRYAN C. Waveglow: A flow-based generative network for speech synthesis[C]//Proceedings of ICASSP 2019-2019 IEEE International Conference on Acoustics, Speech and Signal Processing (ICASSP). IEEE, 2019: 3617-3621..

[32] WU Z Z, OLIVER W, SIMON K. Merlin: An open source neural network speech synthesis system[J]. SSW, 2016: 202-207, 2016.

[33] REN Y, HU C X, TAN X, et al. Fastspeech 2: Fast and high-quality end-to-end text to speech[C]//Proceedings of International Conference on Learning Representations, 2021.

[34] REN Y, RUAN Y J, TAN X, et al. Fastspeech: Fast, robust and controllable text to speech[J]. NeurIPS, 2019.

[35] REN Y, TAN X, QIN T, et al. Almost unsupervised text to speech and automatic speech recognition[C]//Proceedings of International Conference on Machine Learning, PMLR, 2019: 5410-5419.

[36] XU J, TAN X, REN Y, et al. Lrspeech: Extremely low-resource speech synthesis and recognition[C]//Proceedings of the 26th ACM SIGKDD International Conference on Knowledge Discovery & Data Mining. 2020: 2802-2812.

[37] QIN T. Dual learning[M]. Springer, 2020.

[38] LUO R Q, TAN X, WANG R, et al. Lightspeech: Lightweight and fast text to speech with neural architecture search[C]//Proceedings of 2021 IEEE International Conference on Acoustics, Speech and Signal Processing (ICASSP). IEEE, 2021.

[39] CHEN M J, TAN X, LI B H, et al. Adaspeech: Adaptive text to speech for custom voice[C]//Proceedings of International Conference on Learning Representations, 2021.

[40] YAN Y Z, TAN X, LI B H, et al. Adaspeech 2: Adaptive text to speech with untranscribed data[C]//Proceedings of 2021 IEEE International Conference on Acoustics, Speech and Signal Processing (ICASSP). IEEE, 2021.

[41] SAMY B, ORIOL V, NAVDEEP J, et al. Scheduled sampling for sequence prediction with recurrent neural networks[J]. Advances in Neural Information Processing Systems, 2015: 1171-1179.

[42] ASHISH V, NOAM S, NIKI P, et al. Attention is all you need[J]. Advances in Neural Information Processing Systems, 2017: 998-6008.

[43] JONAS G, MICHAEL A, DAVID G, et al. Convolutional sequence to sequence learning. [C]// Proceedings of the 34th International Conference on Machine Learning. 2017, 70: 1243-1252.

[44] ZEYU J, ADAM F, GAUTHAM J M, et al. Fftnet: A real-time speaker-dependent neural vocoder[C]// Proceedings of IEEE International Conference on Acoustics, Speech and Signal Processing (ICASSP). IEEE, 2018: 2251-2255.

[45] LI N H, LIU S J, LIU Y W, et al. Close to human quality TTS with transformer[Z]. arXiv preprint arXiv:1809.08895, 2018.

[46] MICHAEL G, MAXIMILIAN T, DORIT M, et al. An asymetric cycle-consistency loss for dealing with many-to-one mappings in image translation: A study on thigh MR scans[Z]. arXiv preprint arXiv: 2004.11001, 2020.

[47] MICHAEL M, MICHAELA S, SARAHMIHUC, M, et al. Montreal forced aligner: Trainable text-speech alignment using KALDI[J]. Interspeech, 2017:498-502.

[48] SERCAN O A, MIKE C, ADAM C, et al. Deep voice: Real-time neural text-to-speech[Z]. arXiv preprint arXiv:1702.07825, 2017.

[49] ANDREW G, SERCAN A, GREGORY D, et al. Deep voice 2: Multi-speaker neural text-to-speech[J]. Advances in neural information processing systems, 2017: 2962-2970.

[50] KEIKICHI H, TAO J H. Speech Prosody in Speech Synthesis: Modeling and generation of prosody for high quality and flexible speech synthesis[M]. Springer, 2015.

[51] ANTTI S S, DANIEL, TUOMO R, et al. Wavelets for intonation modeling in hmm speech synthesis[C]// Proceedings of 8th ISCA Workshop on Speech Synthesis. ISCA, 2013.

[52] JESSE E, LAMTHARN H, CHENJIE G, et al. Ddsp: Differentiable digital signal processing[Z]. arXiv preprint arXiv:2001.04643, 2020.

[53] RYUICHI Y, EUNWOO S, JAE-MIN K. Parallel wavegan: A fast waveform generation model based on generative adversarial networks with multi-resolution spectrogram[C]//Proceedings of ICASSP 2020-2020 IEEE International Conference on Acoustics, Speech and Signal Processing (ICASSP). IEEE, 2020: 6199-6203.

[54] AARON VAN DEN O, NAL K, LASSE E, et al. Conditional image generation with pixelcnn decoders[J]. Advances in neural information processing systems, 2016: 4790-4798.

[55] HE D, XIA Y C, QIN T, et al. Dual learning for machine translation[J]. Advances in Neural Information Processing Systems, 2016: 820-828.

[56] SUN H, TAN X, GAN J W, et al. Token- level ensemble distillation for grapheme-to-phoneme conversion[J]. INTERSPEECH, 2019.

[57] JAN W. The evolutionary history of the human speech organs[J]. Studies in language origins, 1989, 1: 173-197.

[58] YOON K, ALEXANDER M R. Sequence-level knowledge distillation[Z]. arXiv preprint arXiv:1606.07947, 2016.

[59] TAN X, REN Y, HE D, et al. Multilingual neural machine translation with knowledge distillation[C]//Proceedings of International Conference on Learning Representations, 2019.

[60] DAN L, WON J, O G, et al. JDI-T: Jointly trained duration informed transformer for text-to-speech without explicit alignment[J]. Proceedings of Interspeech, 2020: 4004-4008.

[61] MIAO C F, LIANG S, CHEN M C, et al. Flow-TTS: A non-autoregressive network for text to speech based on flow[J]. ICASSP, 2020: 7209-7213.

[62] LUO R Q, TIAN F, QIN T, et al. Neural architecture optimization[J]. Advances in neural information processing systems, 2018.

[63] BARRET Z, VIJAY V, JONATHON S, et alLearning transferable architectures for scalable image recognition[J]. CVPR, 2018.

[64] CHEN D Y, LI Y L, QIU M H, et al. Adabert: Task-adaptive bert compression with differentiable neural architecture search[Z]. arXiv preprint arXiv:2001.04246, 2020.

[65] TAN M X, QUOC L. Efficientnet: Rethinking model scaling for convolutional neural networks[J]. ICML, 2019: 6105-6114.

[66] LUO R Q, TAN X, WANG R, et al. Neural architecture search with GBDT[Z]. arXiv preprint arXiv: 2007.04785, 2020.

[67] LUKASZ K, AIDAN N G, FRANCOIS C. Depthwise separable convolutions for neural machine translation[J]. ICLR, 2018.

[68] LUO R Q, TAN X, WANG R, et al. Semi-supervised neural architecture search[J]. Advances in Neural Information Processing Systems, 2020, 33.

[69] HIEU P, MELODY G, BARRET Z, et al. Efficient neural architecture search via parameter sharing[J]. ICML, 2018: 4092-4101.

[70] JIMMY L B, JAMIE R K, GEOFFREY E H. Layer normalization[Z]. arXiv preprint arXiv:1607.06450, 2016.

[71] SERCAN Ö A, CHEN J T, PENG K N, et al. Neural voice cloning with a few samples[C]//Proceedings of the 32nd International Conference on Neural Information Processing Systems. 2018: 10040-10050.

[72] ERICA C, LAI C I, YUSUKE Y, et al. Zero-shot multi-speaker text-to-speech with state-of-the-art neural speaker embeddings[C]//Proceedings of ICASSP 2020-2020 IEEE International Conference on Acoustics, Speech and Signal Processing (ICASSP). IEEE, 2020: 6184-6188.

[73] JiA Y, ZHANG Y, RON J W, et al. Transfer learning from speaker verification to multispeaker text-to-speech synthesis[C]//Proceedings of the 32nd International Conference on Neural Information Processing Systems. 2018: 4485-4495.

[74] CHEN M J, TAN X, REN Y, et al. Multispeech: Multi-speaker text to speech with transformer[J]. INTERSPEECH, pages 4024-4028, 2020.

[75] SUN G Z, ZHANG Y, RON J W, et al. Generating diverse and natural text-to-speech samples using a quantized fine-grained vae and autoregressive prosody prior[C]//Proceedings of ICASSP 2020-2020

IEEE International Conference on Acoustics, Speech and Signal Processing (ICASSP). IEEE, 2020: 6699-6703.

[76] ZENG Z, WANG J Z, CHENG N, et al. Prosody learning mechanism for speech synthesis system without text length limit[Z]. arXiv preprint arXiv:2008.05656, 2020.

[77] ZHANG C, REN Y, TAN X, et al. Denoispeech: Denoising text to speech with frame-level noise modeling[C]//Proceedings of 2021 IEEE International Conference on Acoustics, Speech and Signal Processing (ICASSP). IEEE, 2021.

[78] CHEN Y T, YANNIS A, BRENDAN S, et al. Sample efficient adaptive text-to-speech. [C]//Proceedings of International Conference on Learning Representations, 2018.

[79] HIEU-THI L, JUNICHI Y. Nautilus: a versatile voice cloning system[J]. IEEE/ACM Transactions on Audio, Speech, and Language Processing, 2020, 28:2967-2981.

面向开放世界的分类器学习

刘成林

（中国科学院自动化研究所模式识别国家重点实验室，北京 100190）

1 引言

模式识别是人工智能学科的主要分支之一，研究如何使机器（包括计算机）模拟人的感知功能，从环境感知数据中检测、识别和理解目标、行为、事件等模式。模式是感知数据（如图像、视频、语音、文本）中具有一定特点的目标、行为或事件，具有相似特点的模式组成类别(class, category)。模式识别的研究内容包括模式（目标）检测、分割、特征提取、分类、描述等，这是从一个模式识别系统流程的角度来说的。分类是模式识别的核心任务，为此提出了大量的模型和方法。相关的研究问题还包括特征提取与选择、概率密度估计、聚类分析（无监督分类）等。分类器的设计与优化、特征表示等同时也是机器学习领域的主要研究内容。因此，20 世纪 60 年代以来，模式识别与机器学习经常不加区分地相提并论[1-2]。虽然两个领域研究的大部分内容重叠，也有一些侧重点的区别，如模式识别研究强调面向工程应用（机器感知、感知信息处理），而学习理论（收敛性、泛化性理等）的研究是机器学习领域特有的。模式识别因面向感知应用，有大量工作针对不同感知数据的具体处理环节，如语音、图像和视频处理、分割等。针对视觉信息（图像视频）处理的研究工作又形成了计算机视觉领域，当然其中也普遍使用模式分类和机器学习方法。

20 世纪 50 年代以来，模式识别领域提出了大量有效的方法。从分类器模型的角度，模式识别方法包括统计模式识别（统计决策）[3-5]、句法（结构）模式识别[6]、人工神经网络[7]、支持向量机[8]、多分类器系统（集成学习）[9-10]等。从分类器学习的角度，主要有监督学习、无监督学习、半监督学习[11]、迁移学习[12]、多标记学习[13]、多任务学习[14]、在线学习、增量学习[15]等。

近年来，深度学习（深度神经网络方法）[16]逐渐成为模式识别的主流和统治性方法，在多数模式识别应用任务中大幅超越传统模式识别方法（基于人工特征提取的分类方法）的性能。深度学习的方法最早发表在 2006 年[17]，针对深度神经网络难以收敛的问题，提出了逐层训练的方法，在语音识别中明显提升了识别精度。后来陆续提出了一系列改进训练收敛性和泛化性能的深度神经网络模型和训练算法，包括不同的训练方法或正则化方法（如 Dropout, Batch Normalization 等）、不同的卷积神经网络结构（如 AlexNet、GooglNet、VGGNet、ResNet、DenseNet、全卷积网络等）、循环神经网络（如 LSTM、CRNN）、self-attention 网络、图卷积网络等。2012 年深度卷积神经网络在大规模图像分类竞赛 ImageNet 中取得巨大成功[18]，从此推动深度学习的研究和应用进入高潮。深度学习的优越性能从视觉领域延伸到自然语言处理领域，开始在机器翻译、阅读理解、自动问答等语言理解任务中大幅超越基于统计语言模型的方法。

目前成熟的（成功应用的）深度神经网络和过去传统的统计模式识别方法一样，都按照这样一种范式设计：①对预先定义的已知类别收集大量样本，用一部分样本作为训练样本，剩下的作为测试样本；②在训练样本集上估计（学习）分类器（模型结构预先人为指定或通过交叉验证进行选择）的参数；③在测试样本集上评估分类器的性能。这种设计范式一般都隐含了三个基本假设：①封闭世界假设：类别集是预先定义且固定不变的，假设识别对象总是来自预定类别集中某一类。②独立同分布（independently and identically distributed, i.i.d.）假设：样本之间相互独立且训练样本和测试样本服从相同分布。③大数据假设：有足够多的样本用于训练分类器（估计参数），分类器逼近贝叶斯错误率（最小错误率）的前提是有无穷多样本估计得到准确的条件概率密度，基于经验风险最小化设计的分类器期望误差随样本数增多而逐渐降低。这三个假设在一些有限制的应用场景（环境可控、类别固定）可近似满足，但到了开放场景（环境不可控、分布变化、类别变化等）经常是不满足的。比如在文字识别的场合，汉字的类别非常多，很难一次收集所有类别的样本训练分类器，文档中相邻字符之间条件相关（不独立），测试样本可能是新的书写风格或字体（分布变化），有些类别（如生僻字）难以获得足够多的训练样本（图 1）。文献[19]对这三个假设不成立情况下的模式识别问题进行了总结和研究进展综述（图 2）。

本文主要讨论类别数不固定、只有部分类别有训练样本的场合（开放世界）的分类器学习问题，这种情况又被称为开放集识别或开放集分类[20-21]。第 2 节对开放世界模式分类和学习问题做基本介绍，第 3 节介绍面向开放集的分类决策规则，第 4 节介绍面向开放集的分类器设计与学习方法，第 5 节介绍一种面向开放集的深度学习模型——卷积原型网络，第 6 节是讨论和小节。

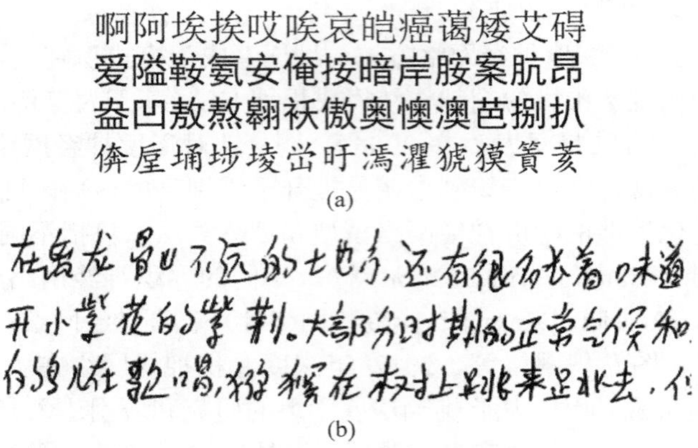

啊阿埃挨哎唉哀皑癌蔼矮艾碍
爱隘鞍氨安俺按暗岸胺案肮昂
盎凹敖熬翱袄傲奥懊澳芭捌扒
侪厔埇坜垵岜旰溤濯猇獏箦荄

(a)

(b)

图 1　汉字类别数大、字体和书写风格变化多
（a）类别数大，有些类别样本少；（b）实际文档中字符之间条件相关

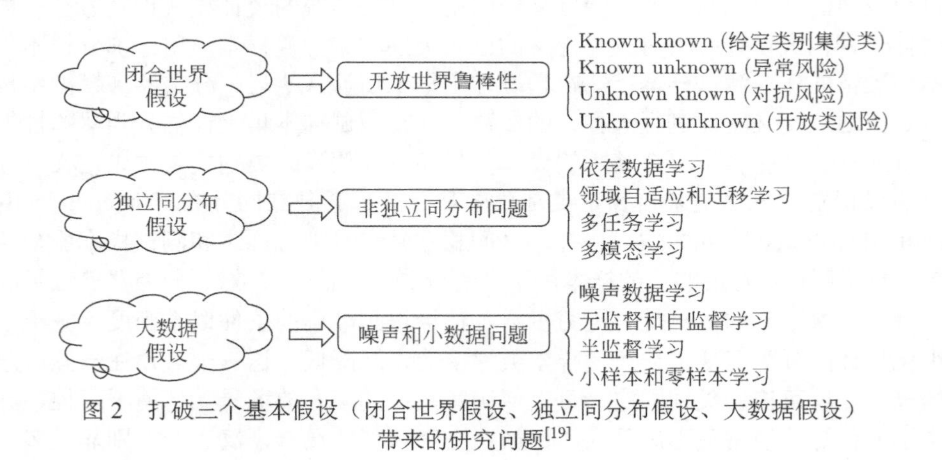

图 2　打破三个基本假设（闭合世界假设、独立同分布假设、大数据假设）带来的研究问题[19]

2　开放世界的模式分类和学习问题

模式识别问题中一般假设输入模式（识别对象）属于预先定义（已知）的 C 个类别 $\omega_i, i=1,2,\cdots,C$ 之一，模式（一个具体的模式又被称为样本或样例）的特征用特征向量 $\boldsymbol{x}=[x_1,x_2,\cdots,x_d]\in R^d$ 表示。如果知道每个类别的先验概率 $P(\omega_i)$，并且通过从数据（训练样本）估计得到每个类别的条件概率密度 $p(\boldsymbol{x}\mid\omega_i)$，对于一个输入模式 $\boldsymbol{x}$，可根据贝叶斯公式计算其属于每个类别的后验概率：

$$P(\omega_i|\boldsymbol{x})=\frac{p(\boldsymbol{x}\,|\,\omega_i)P(\omega_i)}{p(\boldsymbol{x})}=\frac{p(\boldsymbol{x}\,|\,\omega_i)P(\omega_i)}{\sum_{j=1}^{C}p(\boldsymbol{x}\,|\,\omega_j)P(\omega_j)} \tag{1}$$

这个后验概率满足 $\sum_{i=1}^{C}P(\omega_i\,|\,\boldsymbol{x})=1$，即符合闭合世界假设。在后验概率的基础之上，可以实现最小错误率决策（最大后验概率决策）。

上述闭合世界分类器在应用中显然有局限：当输入模式不属于任何一个已知类别时（这种情况被称为异常模式(novelty, outlier)），分类器或者不能识别，或者根据最大后验概率强制判别其属于一个已知类别，这就产生了识别错误。虽然分类器也可以根据最大后验概率（类别最大判别函数值）是否小于一个阈值对输入模式拒识[22]，由于多数基于判别学习（discriminative learning）的分类器（尤其是深度神经网络）倾向于对训练数据过拟合，对模式的输出概率（或判别函数）过自信（over-confidence），即使对异常模式也会得到很高的概率置信度（如大于 0.99），因而将明显非正常的模式误识为已知类别[23]。相比而言，这种情况下人的处理方式是：在见到没有见过（没有学习过）的模式时，会自然地判为“不认识”（拒识）。因此，当今主流的模式分类方法在开放类别集的情况下识别性能与人相比还有比较大的差距，尤其是对异常模式的处理。

开放集模式识别就是这样的问题：分类器用已知类别的样本训练后，对属于已知类别的输入模式希望给出分类结果，而对异常模式希望做出拒识的决策。对开放集识别的性能评价要同时考虑已知类模式的分类精度和异常模式拒识的正确率。由于拒识异常模式的同时也会错误地拒识一部分正常（已知类别）模式，因此评价异常拒识性能要同时给出异常模式错误接受率（false positive rate, FPR）和正常模式正确接受率（true positive rate, TPR）。或者更全面地，给出拒识阈值可变情况下的 FPR-TPR 变化曲线（receiver operating characteristic curve, ROC curve）。

基于最大后验概率的拒识规则不能有效拒识异常模式，这是因为在闭合世界假设下，已知类别后验概率之和被强制等于 1，一般都会有某一个类别的后验概率比较大（比如大于 0.5），即使输入模式并不是来自这个类别，也会被判为这个类别，且根据最大后验概率拒识规则也不能拒识。相比而言，基于类条件概率密度或距离的拒识规则[24]更适合异常拒识，但其拒识性能跟分类器结构（函数形式）和参数学习方法关系很大，也不保证能充分拒识异常模式。

闭合世界假设下基于后验概率的拒识规则是为拒识歧义模式（不确定属于已知类别的哪一类）而设计的，因此又被称为歧义拒识；基于条件概率密度或距离的拒识规则是专为拒识异常模式而设计的，因此又被称为异常拒识。开放集模式识别的场合存在两种拒识（歧义拒识、异常拒识）（图 3）。

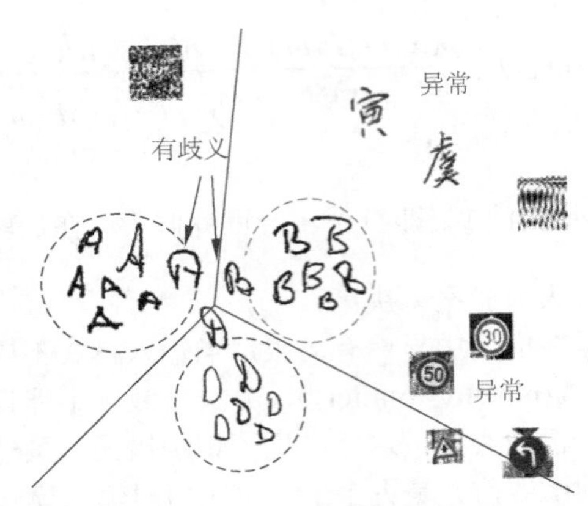

图 3　开放集识别问题示例：3 个已知类别，有歧义(ambiguity)拒识和异常(outlier)拒识（见彩图 5）
红线表示三个类别之间的分类边界（决策面），虚线相当于生成模型的决策边界（把一类模式与所有其他模式区分开）

上述分析显示，开放集识别（同时进行已知类模式分类和异常拒识）的性能受闭合世界假设和拒识规则影响。除此之外，开放集识别性能更加受分类器模型结构和学习算法影响。

分类器模型：模型结构决定了分类器判别函数的形式，如贝叶斯分类器的判别函数为 $P(\omega_i)p(\boldsymbol{x}\mid\omega_i)$ 或 $\log P(\omega_i)p(\boldsymbol{x}\mid\omega_i)$（对数似然度），在高斯密度假设下简化为二次判别函数或线性判别函数；单层神经网络输出线性判别函数，但实际性能与高斯密度假设下的线性判别函数性能有区别；多层神经网络输出非线性判别函数等。基于条件概率密度函数或对数似然度的分类决策可保证异常拒识性能，且可以很方便地扩充类别，因为这种分类器的模型参数是每类分别用一类样本进行估计的（生成模型一般都具有这种特点，概率密度函数是生成模型的一种）。多层神经网络、支持向量机等模型是判别模型，一般是在闭合世界假设下，通过判别学习（分类损失或类似分类损失的经验风险最小化）估计参数，因此不适合开放集识别。由图 3 可以看出，判别模型的决策面把整个特征空间划分成已知类别的区域，相当于把未知类别和异常模式的空间区域都分到了已知类别，因此不适合拒识异常模式。图 3 也显示，生成模型的决策边界（每类概率密度或判别函数减去一个阈值）把一个类别的模式与所有其他模式（包括异常）区分开，因此适合异常拒识。

学习算法：判别模型隐含了闭合世界假设，因此不适合开放集识别。学习过程中的正则化（如权值 L2 正则化、函数平滑度约束）可以改善分类性能和开放集识别性能，但不能从根本上克服闭合世界假设带来的问题。生成模型在学习中用一类样本估计一类模型的参数，没有考虑不同类别的区分性，因此分类性能一般不如判别模型。过去已经有很多研究工作将判别学习方法用于生成模型的学习，即通过分类损失最小化来估计生成模型的参数，如判别学习二次分类器[25]。这种方法又被称为混合判别-生成学习，既可保持生成模型的函数特点，又能提高分类精度，因此适合开放集识别。

第 3 节和第 4 节分别介绍开放集识别的分类决策规则和模型设计与学习算法。

3 面向开放集的分类决策规则

对于 C 个已知类别的分类问题，用 $f(\boldsymbol{x},\omega_i)$ 或 $f_i(\boldsymbol{x})$ 表示每个类别的判别函数。对于贝叶斯分类器，有 $f(\boldsymbol{x},\omega_i)=\log P(\omega_i)p(\boldsymbol{x}\mid\omega_i)$（这种形式也被称为对数似然度）。$f(\boldsymbol{x},\omega_i)$ 可看作一个广义的相似度（输入模式 x 属于类别 ω_i 的似然度）。对于基于距离的分类器，可以认为 $f(\boldsymbol{x},\omega_i)=-d(\boldsymbol{x},\omega_i)$，其中 $d(\boldsymbol{x},\omega_i)$ 表示输入模式 $\boldsymbol{x}$ 与类别 ω_i 的模型/模板之间的匹配距离。

对于闭合世界分类问题，由 $f(\boldsymbol{x},\omega_i)=\log P(\omega_i)p(\boldsymbol{x}\mid\omega_i)$ 可以计算后验概率为

$$P(\omega_i|\boldsymbol{x})=\frac{p(\boldsymbol{x}\mid\omega_i)P(\omega_i)}{\sum_{j=1}^{C}p(\boldsymbol{x}\mid\omega_j)P(\omega_j)}=\frac{\mathrm{e}^{f(\boldsymbol{x},\omega_i)}}{\sum_{j=1}^{C}\mathrm{e}^{f(\boldsymbol{x},\omega_j)}} \tag{2}$$

这就是常说的 soft-max 形式，跟公式(1)的贝叶斯公式相似，隐含了闭合世界假设，这是因为混合概率密度 $p(\boldsymbol{x})$ 是 C 个已知类别概率密度的和，没有考虑未知类别和异常模式的分布。

开放集识别中假设未知类别和异常模式是没有训练样本的，因为实际应用中这样的样本难以获取。这样，对未知类别和异常模式的处理通常有两种方式：

（1）将未知类别和异常模式合并为一个“异常类”加到分类器中，得到 $\sum_{i=1}^{C+1}P(\omega_i\mid\boldsymbol{x})=1$，其中 $P(\omega_{C+1}\mid\boldsymbol{x})$ 表示输入模式属于异常类的概率。这种表示中，仍需要异常类的训练样本估计 $f(\boldsymbol{x},\omega_{C+1})$ 的参数。有些方法通过合成样本的方式来实现[26-27]。

（2）假设未知类别和异常模式在特征空间中一定区间内任意随机出现，即 $P(\boldsymbol{x}\mid\omega_{C+1})=p_0$ 为均匀分布。这样，相当于异常类 ω_{C+1} 的判别函数为常数 $f_0=\log P(\omega_{C+1})p_0$，此时 C+1 个类别的后验概率为

$$P(\omega_i|\boldsymbol{x})=\begin{cases}\dfrac{\mathrm{e}^{f(\boldsymbol{x},\omega_i)}}{\sum_{j=1}^{C}\mathrm{e}^{f(\boldsymbol{x},\omega_j)}+\mathrm{e}^{f_0}}, & i=1,2,\cdots,C\\[2ex] \dfrac{\mathrm{e}^{f_0}}{\sum_{j=1}^{C}\mathrm{e}^{f(\boldsymbol{x},\omega_j)}+\mathrm{e}^{f_0}}, & i=C+1\end{cases} \tag{3}$$

基于 C+1 个类别的后验概率，开放集识别的分类决策规则为

$$\boldsymbol{x} \in \begin{cases} \omega_k, & \text{如果} P(\omega_k \mid \boldsymbol{x}) = \max\limits_{i=1,2,\cdots,C} P(\omega_i \mid \boldsymbol{x}) \text{且} P(\omega_k \mid \boldsymbol{x}) > P(\omega_{C+1} \mid \boldsymbol{x}) \\ \omega_{C+1}, & \text{否则} \end{cases} \tag{4}$$

根据后验概率与判别函数的关系式（2），决策规则等价为

$$\boldsymbol{x} \in \begin{cases} \omega_k, & \text{如果} f(\boldsymbol{x},\omega_k) = \max\limits_{i=1,2,\cdots,C} f(\boldsymbol{x},\omega_i) \text{且} f(\boldsymbol{x},\omega_k) > f_0 \\ \omega_{C+1}, & \text{否则} \end{cases} \tag{5}$$

根据此，异常拒识规则为

$$\text{当} \max_{i=1,2,\cdots,C} f(\boldsymbol{x},\omega_i) < f_0 \text{，} x \text{为异常} \tag{6}$$

其中，f_0 为一个人工可调的阈值。根据 $f(\boldsymbol{x},\omega_i)$ 与对数似然度和条件概率密度之间的关系，式（6）也被称为基于条件概率密度的拒识规则。对于基于距离的分类器，由于 $f(\boldsymbol{x},\omega_i) = -d(\boldsymbol{x},\omega_i)$，异常拒识规则为

$$\text{当} \min_{i=1,2,\cdots,C} d(\boldsymbol{x},\omega_i) > t_1 \text{，} \boldsymbol{x} \text{为异常} \tag{7}$$

这在文献[27]又被称为基于距离的拒识规则。

开放集识别中也经常用闭合集最大后验概率规则来拒识异常模式。这个规则的后验概率按式(1)或式(2)计算：

$$\text{当} \max_{i=1,2,\cdots,C} P(\omega_i \mid \boldsymbol{x}) < \lambda_0 \text{，} \boldsymbol{x} \text{为拒识} \tag{8}$$

这个经典的拒识规则是为拒识歧义模式设计的[22]。$\max\limits_{i=1,2,\cdots,C} P(\omega_i \mid \boldsymbol{x}) < \lambda_0$ 意味着 $\boldsymbol{x}$ 在特征空间中处于两个或多个类别的分类决策面附近（也就是两个或多个类别的判别函数值相近），因此不确信属于哪一类。这个规则之所以可以拒识异常模式，是因为异常模式经常处于已知类别分类决策面附近[28]。

拒识规则 $\max\limits_{i=1,2,\cdots,C} P(\omega_i \mid \boldsymbol{x}) < \lambda_0$ 可以简化为两个类别判别函数之差的形式：假设分类器输出的判别函数从大到小排序为 $f(\boldsymbol{x},\omega_{r_1}) > f(\boldsymbol{x},\omega_{r_2}) > \cdots > f(\boldsymbol{x},\omega_{r_C})$，且前两个候选类别（top-2 ranks）的判别函数值明显大于其他类别，则后验概率可近似为

$$P(\omega_i \mid \boldsymbol{x}) = \begin{cases} \dfrac{\mathrm{e}^{f(\boldsymbol{x},\omega_i)}}{\mathrm{e}^{f(\boldsymbol{x},\omega_{r_1})} + \mathrm{e}^{f(\boldsymbol{x},\omega_{r_2})}}, & i = r_1, r_2 \\ 0, & \text{否则} \end{cases} \tag{9}$$

最大后验概率则变成

$$P(\omega_{r_1}|\boldsymbol{x})=\frac{\mathrm{e}^{f(x,\omega_{r_1})}}{\mathrm{e}^{f(x,\omega_{r_1})}+\mathrm{e}^{f(x,\omega_{r_2})}}=\frac{1}{1+\mathrm{e}^{-[f(x,\omega_{r_1})-f(x,\omega_{r_2})]}} \tag{10}$$

这就是常见的 sigmoid 形式。相应地，基于后验概率的歧义拒识规则变成

$$\text{当 } f(\boldsymbol{x},\omega_{r_1})-f(\boldsymbol{x},\omega_{r_2})<t_2\text{，}\boldsymbol{x}\text{ 为拒识} \tag{11}$$

上述拒识规则（式(6)和式(11)）分别为异常拒识和歧义拒识而设计，但实际上每个拒识规则对两种拒识类型（异常，歧义）都有效果。我们早年的工作对这两个拒识规则在多种分类器结构上做了实验比较[29]。实验中针对手写数字识别设计了五种分类器：多层神经网络（MLP）、径向基函数（RBF）神经网络、多项式分类器（PC）[30]、学习矢量量化（LVQ）[31-32]、修正二次判别函数（MQDF）[33]。MLP，RBF 和 PC 三种神经网络分类器都是判别模型，输出 sigmoid 概率值（判别函数仍取 sigmoid 函数之前的线性加权和），通过随机梯度最小化（SGD）对平方误差损失最小化进行学习。LVQ 是最近原型分类器（每类一个或多个原型，用最近原型距离规则分类），原型用判别学习准则(minimum classification error, MCE)[31, 22]学习。原型可以看作简单的生成模型，LVQ 可以看作混合判别-生成模型。MQDF 是基于高斯概率密度假设下贝叶斯分类器（即二次判别函数 QDF）的改进：将每类协方差矩阵的较小本征值用常数代替，判别函数中可省略一些本征向量的计算，同时可提升分类性能[33]。显然，MQDF 是基于概率密度估计的生成模型。文献[29]的实验结果表明，判别模型 MLP，RBF 和 PC 比 LVQ 和 MQDF 得到的分类精度更高。实验中用不同阈值进行歧义拒识，以拒识率-错误率曲线来表示歧义拒识性能，实验结果表明基于后验概率的近似规则（式（11））更适合歧义拒识。

在文献[29]的异常拒识实验中，以合成的非字符样本（图 4）作为异常模式的测试样本。神经网络（MLP，RBF，PC）也采用非字符样本进行训练（将学习准则中的目标输出全部设为 0），得到的分类器表示为 EMLP，ERBF，EPC。异常拒识的性能用数字样本错误拒识的 FNR（false negative rate）和异常样本错误接受的 FPR（false positive rate）的变化曲线评价。分别用两个拒识规则 Rule 1（式（6））和 Rule 2（式（11））及不同阈值实验，结果如图 5 所示。可以看出，基于条件概率密度的拒识规则（Rule 1）明显更适合异常拒识（给出较低的 FNR 和 FPR）。比较不同的分类器可以看出，判别模型（MLP，RBF，PC）在没有用异常模式样本训练时，其异常拒识能力非常差（也就是开放集识别性能很差）。当用异常样本训练后（EMLP，ERBF，EPC），异常拒识能力得到大幅提升。对比之下，生成模型 MQDF 没有用异常样本训练，其异常拒识性能天然地非常出色。混合判别-生成模型 LVQ 也没有用异常样本训练，其异常拒识性能明显优于神经网络分类器 MLP，RBF

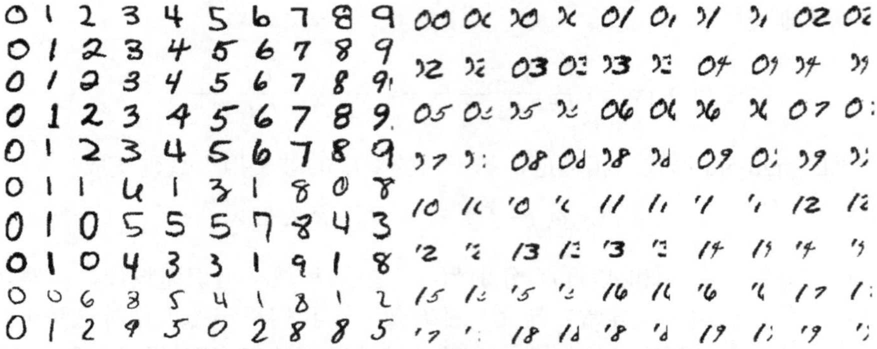

图 4　手写数字样本（左）和合成的非字符样本（异常样本）（右）

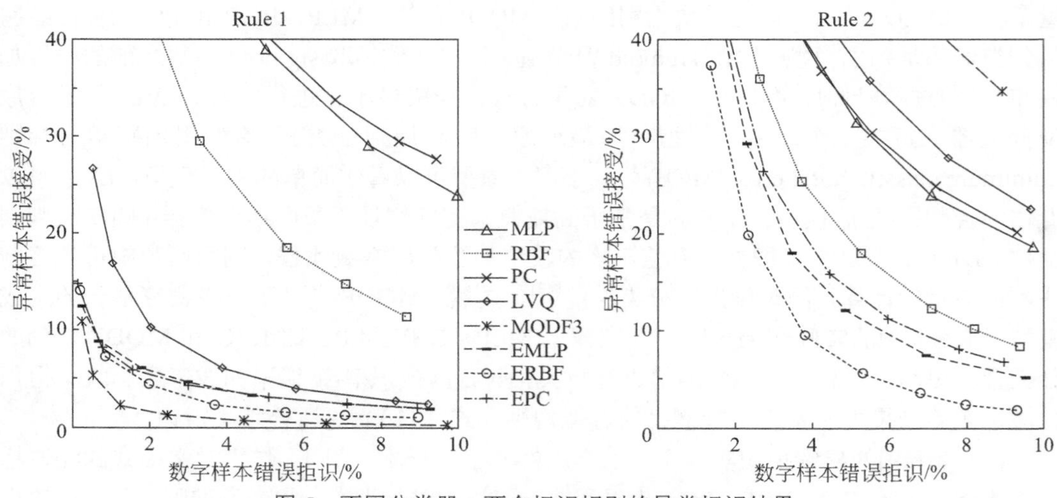

图 5　不同分类器、两个拒识规则的异常拒识结果

和 PC。这验证了生成模型非常适合异常拒识，也就是开放集识别。

对于具有类似 $f(\boldsymbol{x},\omega_i)=\log P(\omega_i)p(\boldsymbol{x}\mid\omega_i)$ 特点的生成模型，拒识规则（式（6））等价为

$$p(\boldsymbol{x}\mid\omega_i)<\frac{\mathrm{e}^{f_0}}{P(\omega_i)},\quad i=1,2,\cdots,C \tag{12}$$

因此，每个类别自身的决策区域是 $p(\boldsymbol{x}\mid\omega_i)\geqslant\dfrac{\mathrm{e}^{f_0}}{P(\omega_i)}$。如果 $p(\boldsymbol{x}\mid\omega_i)$ 是中心集中的单模态

分布（如高斯概率密度），则决策区域是闭合区域（如图 3 中虚线表示的椭圆区域），其他类别和异常模式都在这个区域之外，因此这样的生成模型适合异常拒识和开放集识别。

4 面向开放集的分类器设计与学习

4.1 分类器设计

上面说到生成模型比判别模型更适合开放集识别。生成模型一般是对每一类样本数据建立一个表示模型，表示这个类别的内在特点（特征、分布、结构等），因此是 one-class 模型。这个模型可以用来生成数据（产生符合类别特点的样本），通过计算输入模式与每个类别模型之间的相似度或距离，可以进行分类和异常拒识。

生成模型和判别模型的不同特性在图 6 中示意说明。生成模型（如概率密度模型）表示每个类别的分布，根据相似度或距离可把不同类别的样本很好地分开。同时，异常模式由于距离已知类别分布区域比较远，很容易根据相似度或距离予以拒识。另一方面，判别模型直接对闭合世界后验概率进行估计（或输出隐含闭合世界假设的判别函数），虽然可以很好地区分不同已知类别，但异常模式距离已知类别区域越远，有一类后验概率越接近 1，因此根据后验概率不能拒识。

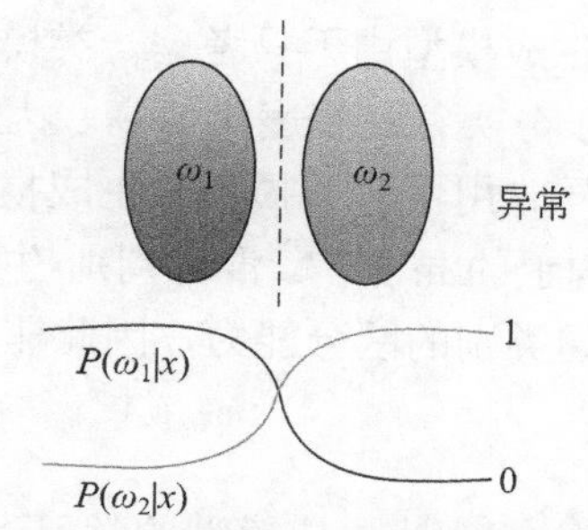

图 6　生成模型和判别模型示意图

生成模型有很多不同的具体形式，现在来简单分析一下主要的生成模型类型：

（1）模板匹配。模板匹配是一种类人思维的模式识别方式：人脑中记住了每个已知类别的主要特征或模板，将识别对象与每个类别的特征或模板匹配，得到相似度或距离，基于最小距离/最大相似度分类，如果最小距离大于一个主观阈值，则拒识为异常（不认识）。模板匹配或基于距离的分类器性能取决于模板和距离度量如何设计。简单的刚性模板对模式形变（类内变化）适应性较差，提取不变性特征有助于提升分类性能。结构模式匹配具有很强的形变适应性和分类性能，但结构匹配（如图匹配）、结构模板学习都是较难的技术问题。

（2）基于概率密度的生成模型。概率密度估计是统计模式识别中的经典问题，贝叶斯分类即依赖于类条件概率密度估计。准确的概率密度估计（包括参数估计、非参数估计、半参数估计）都需要无穷多的训练样本和足够复杂的概率密度函数。在样本有限的情况下，往往假设限定形式的概率密度函数（如高斯函数），也能得到比较好的分类性能。

（3）深度生成模型。人工神经网络中有些模型是生成模型，如矢量量化（VQ）、RBM、auto-encoder、GAN (generative adversarial network)。VQ 主要用于数据压缩，用作原型分类器时，由于没有经过判别学习，其分类精度不高。RBM 过于简单，接近线性模型，数据表示和分类性能有限。Auto-encoder 对每个类别学习一个低维流形，样本重构误差可作为分类的距离度量。Auto-encoder 有很多扩展，如多层 auto-encoder、denoising auto-encoder、卷积 auto-encoder 等。GAN 主要为数据生成设计，还没看到在分类中的应用。总的来说，深度生成模型不是专为模式分类设计，但其生成或重构的思想可结合到分类器设计中，提升分类器的类别表示和异常拒识能力，如在卷积神经网络（CNN）学习中结合重构损失用于开放集识别[34]。现在深度神经网络（DNN）得到广泛应用，但很多传统的生成模型是非深度的（建立在人工提取特征基础之上），可以利用深度神经网络提取特征，然后在深度特征基础上建立生成模型。深度特征+生成模型的组合方式可以很灵活，学习方式可以是端到端(end-to-end)，也可以非端到端。

（4）混合判别-生成模型。生成模型由于只考虑一类模式数据的内在特点，没有考虑不同类别之间的区别，直接用于分类往往精度不够高，因此通过判别学习调整参数以提高分类性能。这个学习过程中同时用了分类损失和生成损失（如重构损失）。混合判别-生成模型也有其他形式，如结构上的混合[35]。混合判别-生成模型既保留了生成模型的结构，又通过判别学习增强对不同类别的区分能力，因此非常适合开放集识别。

4.2 学习算法

面向开放集识别，下面对混合判别-生成模型的学习进行介绍。

用 $f(\boldsymbol{x},\theta)$ 表示分类器函数，其中 θ 表示分类器的全部参数集合。给定一个标记样本集合 $\{(\boldsymbol{x}_n,y_n)\mid n=1,2,\cdots,N\}$，分类器学习（参数估计）目标是优化一个准则函数（经验风险）：

$$\min_{\theta} R_{\text{emp}}=\frac{1}{N}\sum_{n=1}^{N}L(y_n,f(\boldsymbol{x}_n,\theta)) \tag{13}$$

其中，$L(y_n,f(\boldsymbol{x}_n,\theta)$ 表示样本 $\boldsymbol{x}_n$ 的分类损失。对于判别学习，损失函数一般是 $\boldsymbol{x}_n$ 的分类错误率的平滑、界或其他变换形式。对于生成模型，学习目标一般是对一类数据样本最大化似然度或对数似然度：

$$\max_{\theta_i} LL_i(\theta_i) = \log p(X_i \mid \theta_i) = \sum_{n=1}^{N_i} \log p(\boldsymbol{x}_n \mid \theta_i) \tag{14}$$

这个参数解被称为最大似然(maximum likelihood, ML)估计。对于基于距离的分类器，参考高斯概率密度分布下对数似然度与距离之间的关系，假设$\log p(\boldsymbol{x}_n \mid \theta_i) \propto -d(\boldsymbol{x}_n, \omega_i)$，最大似然估计近似地等价于

$$\max_{\theta_i} LL_i(\theta_i) = \log p(X_i \mid \theta_i) = \sum_{n=1}^{N_i} \log p(\boldsymbol{x}_n \mid \theta_i) \propto -\sum_{n=1}^{N_i} d(\boldsymbol{x}_n, \omega_i) \tag{15}$$

这个效果相当于使每类样本相互之间距离和最小化，也就是每类样本分布尽量紧凑。

在生成模型的判别学习中，为了使模型参数不偏离最大似然估计太多，用公式(15)的最大似然准则作为正则项：

$$\min_{\theta} R_{\text{emp}} = \frac{1}{N} \sum_{n=1}^{N} [L(y_n, f(\boldsymbol{x}_n, \theta)) + \lambda d(\boldsymbol{x}_n, \omega_{y_n})] \tag{16}$$

其中，λ 为正则化权重。这种最大似然正则化的思想[25, 36]早年曾用于语音识别（被称为 I-smoothing[37]）。

下面以原型分类器的判别学习为例，讨论 ML 正则化的学习效果。原型学习用判别学习算法（如 MCE[32]和 LOGM[38]）学习时，正确类别（训练样本的标记类别）的原型趋于向样本靠近，其他类别的原型趋向远离样本，因而基于最近原型分类规则，训练样本上的分类错误率会逐渐降低。但如果没有正则化对原型的移动进行限制，所有原型都会离样本分布中心区域越来越远（这样会造成异常样本被错误接受，而且原型移动太多也影响已知类别分类的泛化性能）。用式(15)和式(16)的 ML 正则，会使每个类别的原型在判别学习过程中尽量靠近本类样本分布中心（或聚类中心），如图 7 所示，因而有助于提升泛化性能和异常拒识能力。

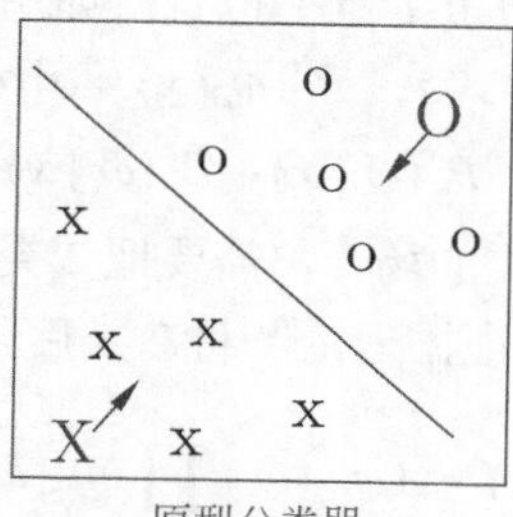

图 7　原型分类器判别学习中加入 ML 正则可保证原型（红色的 o/x）尽量接近样本（黑色的 o/x）分布中心或聚类中心（见彩图 6）

4.3 One-vs-all 学习

除了上述一般的多类分类器判别学习方法，one-vs-all（OVA，表示把每一类与所有其他类的集合相区分）方法也经常被用于学习多类分类器，比如支持向量机和 Boosting 一般就是通过多个 one-vs-all 分类器结合实现多类分类。One-vs-all 分类器是一个二值(binary)分类器，把输入模式判为正类（标记为+1）或负类（标记为–1）。OVA 分类器与一类(one-class)分类器的不同之处在于：一类分类器只用正类样本训练，而 OVA 分类器同时需要两类的样本。当然，一类分类器也可用于 OVA 分类，但相比二类判别学习的 OVA 分类器，其分类性能要弱一些。

OVA 分类器由于把一类样本与多个其他类别样本分开的特点，也适合用于异常拒识或开放集分类。这是因为，一类以外的多个其他类合并成一个负类，这个负类的决策区域可能包含异常类的分布区域（图 8），因而 OVA 分类器也能将异常模式判为负类。基于 OVA 的多类分类器中，每个已知类别有一个 OVA 二值分类器，如果所有 OVA 分类器都将异常模式判为负类，则该异常模式被成功拒识。

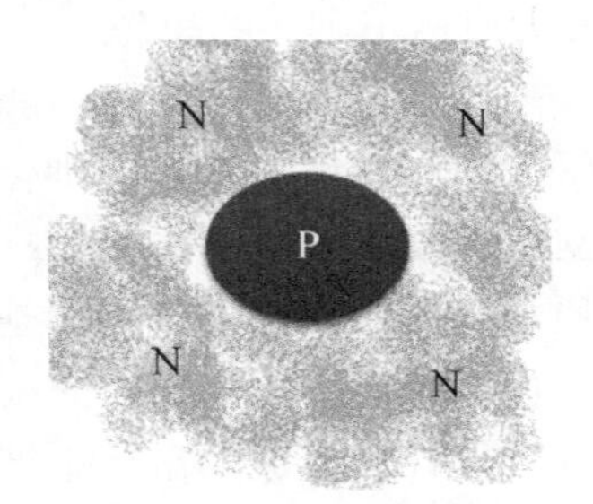

图 8　OVA 分类器的负类决策区域
P: positive，表示正类；N: negative，表示负类

对于 OVA 分类器的开放集识别能力，可以从概率论的角度进行分析。每个 OVA 二值分类器的输出判别函数可以通过 0-1 概率拟合（如用文献[39]和文献[40]的方法）转化为 2 类后验概率（sigmoid 函数形式）：$P^b(\omega_i \mid \boldsymbol{x}) = \sigma[\alpha f(\boldsymbol{x},\omega_i)+\beta]$。2 类后验概率具有“正类概率+负类概率=1”的特点：$P^b(\omega_i \mid \boldsymbol{x}) + P^b(\overline{\omega}_i \mid \boldsymbol{x}) = 1$，其中 $\overline{\omega}_i$ 表示 ω_i 的负类。对于组合多个 OVA 分类器用于多类分类的场合，需要把 2 类后验概率转化为多类后验概率。文献[40]给出了基于 Dempster-Shafer 证据理论[41]将 2 类后验概率转化为多类后验概率的方法：

$$P^m(\omega_i \mid \boldsymbol{x}) = z_i^m = m(\omega_i) = A \cdot m_i(\omega_i) \prod_{j=1, j\neq i}^{C} m_j(\overline{\omega}_j) = A \cdot z_i^b \prod_{j=1, j\neq i}^{C} (1 - z_j^b) \tag{17}$$

其中，$z_j^b = P^b(\omega_i \mid \boldsymbol{x})$，$A$ 是归一化因子：

$$A^{-1}=\sum_{i=1}^{C} z_i^b \prod_{j=1,j\neq i}^{C}(1-z_j^b)]+\prod_{j=1}^{C}(1-z_j^b) \tag{18}$$

由于 A^{-1} 中考虑了所有已知类别都为负的情况 $\prod_{j=1}^{C}(1-z_j^b)$，可以证明：

$$\sum_{i=1}^{C} P^m(\omega_i \mid \boldsymbol{x}) \leqslant 1 \tag{19}$$

这正是开放集识别希望得到的后验概率输出：当 x 为异常模式，会得到 $\sum_{i=1}^{C} P^m(\omega_i \mid \boldsymbol{x})<1$ 的结果，而 $1-\sum_{i=1}^{C} P^m(\omega_i \mid \boldsymbol{x})$ 就是异常类的概率。

既然 OVA 分类器适合开放集识别，那么我们希望将原本用于神经网络、支持向量机等判别模型的 OVA 学习算法推广到生成模型。要做到这一点，首先需要将生成模型的判别函数转化为类似于支持向量机的二值判别函数（特点：根据 $f(\boldsymbol{x})$是否大于 0 判别是属于正类还是负类）。假设生成模型给每类输出的判别函数为距离（或类似距离的函数），则给每类距离加一个阈值使之变为二值判别函数：$f_i(\boldsymbol{x})=-(d(\boldsymbol{x},\omega_i)-\tau_i)$。这个二值判别函数可转化为 sigmoid 形式的 2 类后验概率：

$$p_i(\boldsymbol{x})=\sigma[\xi f_i(\boldsymbol{x})]=\frac{1}{1+\mathrm{e}^{-\xi f_i(\boldsymbol{x})}} \tag{20}$$

基于 2 类损失的 OVA 学习目标函数具有这样的特点：多类损失可以分解为多个 2 类损失之和。因而训练集上的经验风险为

$$R_{\text{emp}}=\frac{1}{N}\sum_{n=1}^{N}\sum_{i=1}^{C} L[\delta(y_n,i),f_i(\boldsymbol{x}_n)]=\frac{1}{N}\sum_{i=1}^{C}\sum_{n=1}^{N} L[\delta(y_n,i),f_i(\boldsymbol{x}_n)] \tag{21}$$

其中，$L[\delta(y_n,i),f_i(\boldsymbol{x}_n)]$ 表示 OVA 二值判别函数 $f_i(\boldsymbol{x}_n)$ 对标记类别 y_n 的 2 类损失。加上 ML 正则，经验风险为

$$R_{\text{emp}}=\frac{1}{N}\sum_{n=1}^{N}\left\{\sum_{i=1}^{C} L[\delta(y_n,i),f_i(\boldsymbol{x}_n)]+\lambda d(\boldsymbol{x}_n,\omega_{y_n})\right\} \tag{22}$$

经验风险通过随机梯度下降法进行最小化从而实现分类器参数的学习。

这种生成模型的 OVA 学习算法用于原型分类器，验证了其对于 2 类分类的有效性[42]。对原型分类器的 OVA 学习考虑每个类别有多个原型 $\boldsymbol{m}_{ij}$ 的情况，每个原型设一个阈值（可与原型一起学习），二值判别函数为

$$f_i(\boldsymbol{x}) = \max_j f_{ij}(\boldsymbol{x}) = -\min_j(\| \boldsymbol{x} - \boldsymbol{m}_{ij} \|^2 - \tau_{ij}) \tag{23}$$

根据公式（22）的经验风险最小化学习原型参数，其中分类损失函数用二值交叉熵：

$$\text{CE} = -\sum_{n=1}^{N} \left\{ \sum_{i=1}^{C} [t_i^n \log p_i + (1 - t_i^n) \log(1 - p_i)] \right\} \tag{24}$$

其中，$t_i^n = \delta(y_n, i)$。加上 ML 正则，经验风险为

$$\text{CE}_1 = -\sum_{n=1}^{N} \left\{ \sum_{i=1}^{C} [t_i^n \log p_i + (1 - t_i^n) \log(1 - p_i)] - \lambda \| \boldsymbol{x}^n - \boldsymbol{m}_{y_n l} \|^2 \right\} \tag{25}$$

其中，$\boldsymbol{m}_{yl}$ 表示 y 类原型中离 $\boldsymbol{x}$ 最近的原型。通过随机梯度下降法进行经验风险最小化学习分类器参数。

文献[42]的实验中对 OVA 原型学习及代表性的判别学习算法 MCE [32]和 LOGM [38]等进行了比较，发现它们用于多类分类时分类精度相当，但用于类别检索（判断一个样本是否属于某一类）时，OVA 原型分类器的性能明显优于 MCE 和 LOGM。

为了评价 OVA 原型学习算法的开放集识别性能，对 OVA 原型分类器的异常拒识性能重新进行了评价。在 MNIST 手写数字数据集上，用训练样本在 OVA 和 LOGM 两种损失函数下分别训练原型分类器，两个分类器在测试集上的分类精度是相当的。为比较异常拒识性能，合成非字符图像作为异常的测试样本（图 9）。测试中，对最近原型距离用不同的阈值拒识异常样本，统计错误接受率（FPR）和数字样本的正确率（TPR），得到 ROC 曲线如图 10 所示。可以看出，拒识规则 Rule 1（式（7））比 Rule 2（式（11））更适合异常拒识，而 OVA 原型分类器比 LOGM 学习（其中同样加了 ML 正则）的原型分类器得到更好的异常拒识性能。

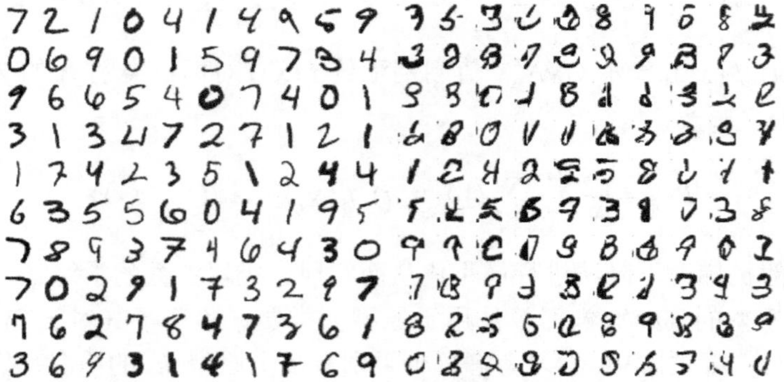

图 9　MNIST 手写数字样本（左）和合成的异常模式样本（右）

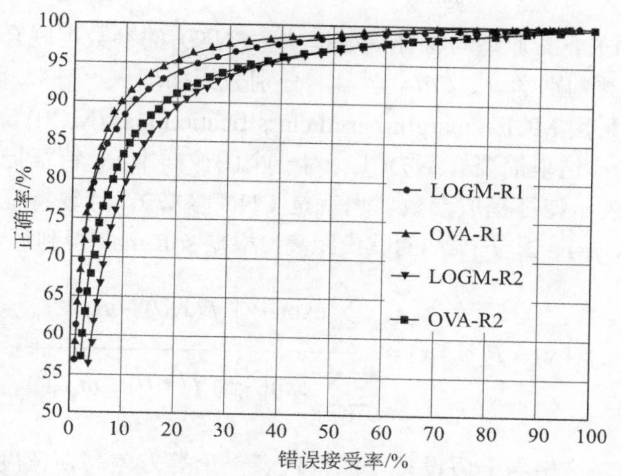

图 10 OVA 原型分类器和多类判别学习 LOGM 原型分类器的异常拒识 ROC

可以预期，OVA 学习方法可用于更多的生成模型以用于开放集识别。下面介绍一个我们最近提出来的深度生成模型：卷积原型网络。

5 面向开放集的卷积原型网络

上述原型分类器虽然具有生成模型的特点，适合开放集分类，但是其输入特征仍是人工设计的特征，不像深度神经网络自动学习特征表示，因此其分类精度跟典型的深度神经网络（如卷积神经网络 CNN）相比有明显差距。然而常规判别学习的 CNN 对噪声和异常模式敏感，不适合开放集分类。因此考虑结合 CNN 的特征学习能力和原型分类器，提出了卷积原型网络（convolutional prototype network, CPN）[28, 43]。CPN 由两部分组成（图 11）：由多个卷积层（中间有降采样层）组成的特征提取模块（可以有全连接层降维）和一个原型分类器（每个类别有一个或多个原型）。这种情况下，原型是

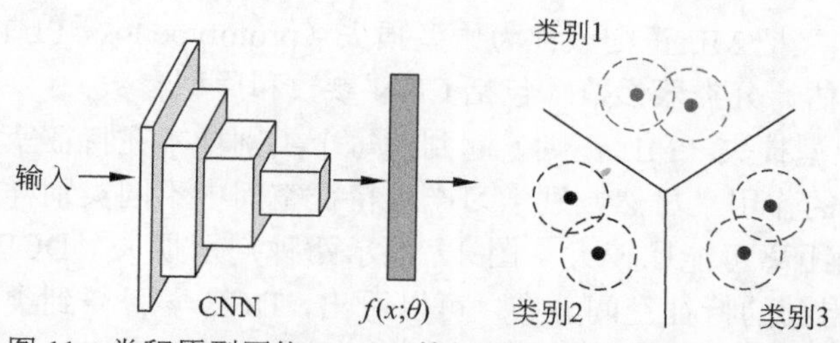

图 11 卷积原型网络：CNN 特征提取、特征空间中的原型

CNN 特征空间中的矢量，原型与特征提取模块（CNN）的参数可联合学习。我们设计了两种学习算法：多类判别学习、OVA 学习，分别介绍如下。

多类判别学习采用 MCE、margin-based classification loss (MCL)、基于距离的交叉熵（DCE）等损失函数。这里简要介绍 DCE。对一个图像样本 $\boldsymbol{x}$，特征提取模块输出特征表示 $f(\boldsymbol{x},\theta)$，这里 θ 表示特征提取参数（也就是 CNN 参数）。计算特征点到所有类别原型 $\boldsymbol{m}_{ij}$ (i=1，2，…，C，j=1，2，…，P)的欧氏距离，根据 soft-max 得到每个类别的后验概率：

$$p_i(\boldsymbol{x}) = P(\omega_i \mid \boldsymbol{x}) = \frac{\sum_{j=1}^{P} \exp(-\gamma \| f(x,\theta) - \boldsymbol{m}_{ij} \|^2)}{\sum_{k=1}^{C}\sum_{j=1}^{P} \exp(-\gamma \| f(x,\theta) - \boldsymbol{m}_{kj} \|^2)} \tag{26}$$

其中，γ 是超参数。这相当于假设每个原型代表一个等方差高斯密度函数，而每个类别服从混合高斯分布。对于标记类别为 y 的样本，DCE 损失为

$$L(\boldsymbol{x}, y) = -\log p_y(\boldsymbol{x}) \tag{27}$$

这是多类交叉熵（跟二值交叉熵不同），在机器学习中也被称为条件对数似然度[44]。

对于 OVA 学习，每个原型的距离加上一个可学习阈值，得到二值判别函数：

$$g_i(\boldsymbol{x}) = \max_j g_{ij}(\boldsymbol{x}) = -\min_j(\| f(\boldsymbol{x},\theta) - \boldsymbol{m}_{ij} \|^2 - \tau_{ij}) \tag{28}$$

转化为 2 类后验概率：$p_i(\boldsymbol{x}) = \sigma[\xi g_i(\boldsymbol{x})]$。样本$(\boldsymbol{x},y)$上的 OVA 损失函数（二值交叉熵）为

$$L(\boldsymbol{x}, y) = -\sum_{i=1}^{C}[t_i \log p_i + (1-t_i)\log(1-p_i)] \tag{29}$$

其中，$t_i = \delta(y,i)$。训练样本集上的经验风险（加上 ML 正则）为

$$R_{\text{emp}} = \frac{1}{N}\sum_{n=1}^{N}\left\{\sum_{i=1}^{C} L(\boldsymbol{x}_n, y_n) + \lambda \| f(\boldsymbol{x}_n,\theta) - \boldsymbol{m}_{y_n l} \|^2\right\} \tag{30}$$

文献[28]和文献[43]又把 ML 正则项称为原型损失（prototype loss, PL）。通过随机梯度下降最小化经验风险估计分类器参数（包括 CNN 参数和原型参数）。

CPN 学习中原型损失（ML 正则）起到使每个类别样本在特征空间中分布紧凑（相互之间距离靠近）的作用。虽然判别学习产生特征空间中不同类别样本可分的效果，但一类样本相互之间距离可能比较大。图 12 显示两种判别损失（DCE 和 OVA）和原型损失下 CPN 学习得到的特征空间分布。可以看出，DCE 学习得到类别可分特征，但类内方差大，加原型损失（ML 正则）后，每类样本分布变得很紧凑。在没有 ML 正则时，

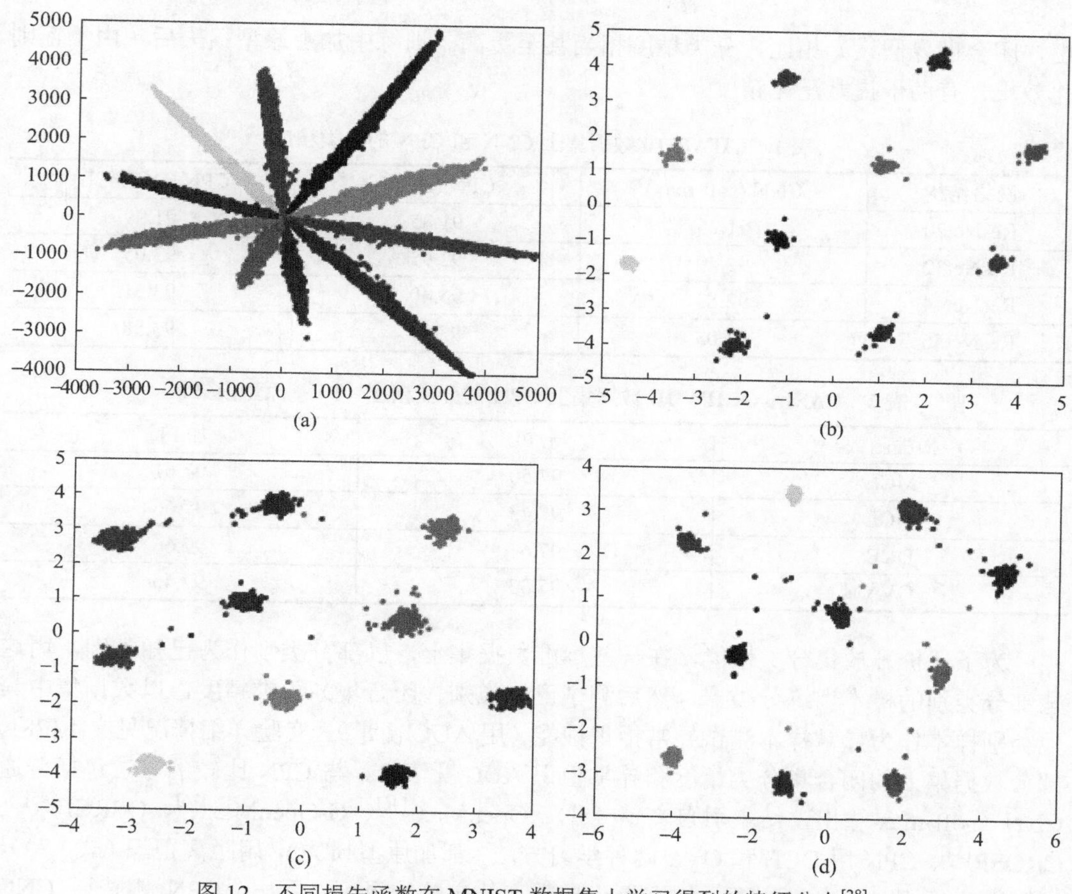

图 12　不同损失函数在 MNIST 数据集上学习得到的特征分布[28]

（a）DCE；（b）DCE+PL；（c）OVA；（d）OVA+PL

OVA 学习得到的特征空间中类内分布也比较紧凑，加上 ML 正则后变得更加紧凑。类内紧凑分布有助于原型分类器同时取得高分类精度和异常拒识性能。

由于 CPN 学习（加 ML 正则）得到的特征空间中每类样本围绕原型形成紧凑分布，一般来说每类只需一个原型就可得到足够高的分类精度。文献[28]的实验结果表明，增加原型数基本上不改变分类精度。

CPN 是为开放集分类设计的，因此我们同时评价它的已知类别（闭合集）分类精度和异常拒识性能（用 FPR-TPR 曲线的 AUC(area under ROC curve)衡量）。

表 1 和表 2 分别显示 CPN 和普通 CNN 在 CIFAR-10 数据集和联机手写汉字数据集 CASIA-OLHWDB（3755 类）上的分类精度。可以看出，CPN 跟普通 CNN（soft-max 输

出，闭合世界假设）相比，分类精度相当甚至更高。训练中加上原型损失后，由于正则化效果，有助于提升泛化精度。

表 1　CIFAR-10 数据集上 CNN 和 CPN 的分类精度

网络结构	CNN (soft-max)	CPN (DCE)	CPN (DCE+PL)
ResNet20	91.25	91.85	91.99
ResNet32	92.49	92.95	93.05
ResNet44	92.83	93.40	93.51
ResNet56	93.03	93.49	93.58

表 2　CASIA-OLHWDB 数据集上 CPN 的分类精度（PL：原型损失）

损失函数	无 PL	加 PL
MCE	97.50	97.61
MCL	97.43	97.66
DCE	97.61	97.66
OVA	97.33	97.45

为了评价开放集分类性能，在一些标准数据集上，以部分类别作为已知类别，用这些部分类别的样本训练分类器，然后测试已知类别（闭合集）分类精度，以数据集中其他类别样本作为异常样本测试异常拒识性能（用 AUC 衡量）。实验详细情况见文献[28]。部分数据集上的闭合集分类精度和异常拒识 AUC 见表 3。与 CPN 比较的方法包括普通 CNN（soft-max 输出，结果引自文献[45]）、OpenMax [46]、G-OpenMax[26]、OSRCI [45]、CROSR[34]。CPN 用 DCE 和 OVA 两种学习方法（都加原型损失），用距离拒识（式（7））、后验概率拒识（式（8））两种规则拒识异常。从实验结果可以看出，CPN 跟普通 CNN 和过去代表性的开放集分类方法相比，在闭合集分类精度相当或更高的同时，异常拒识性能在多数数据集上有明显优势。比较 DR 和 PR 两种拒识规则，CPN 上有时 DR 性能更好，有时 PR 性能更好。而普通 CNN 基于 soft-max 概率拒识也能得到较好的异常拒识效果。这是因为在这些数据集上，异常样本（数据集中部分类别样本）跟已知类别样本在外观上相似度较大，会被当作歧义模式拒识（PR 规则就是为歧义拒识设计的）。

表 3　标准数据集上开放集分类性能（DR：距离拒识；PR：后验概率拒识）

方法	闭合集精度			异常拒识 AUC		
	MNIST	SVHN	CIFAR-10	MNIST	SVHN	CIFAR-10
CNN (soft-max) [45]	0.995	0.947	0.801	0.978	0.886	0.677
OpenMax [46]	0.995	0.947	0.801	0.981	0.894	0.695

续表

方法	闭合集精度			异常拒识 AUC		
	MNIST	SVHN	CIFAR-10	MNIST	SVHN	CIFAR-10
G-OpenMax [26]	0.996	0.948	0.816	0.984	0.896	0.675
OSRCI [45]	0.996	0.951	0.821	0.988	0.910	0.699
CROSR [34]	0.992	0.945	**0.930**	**0.991**	0.899	—
CPN (DCE, DR)	**0.997**	0.966	0.927	0.989	0.919	0.754
CPN (DCE, PR)	**0.997**	0.966	0.927	0.987	0.923	0.815
CPN (OVA, DR)	**0.997**	**0.967**	0.929	0.990	**0.926**	0.771
CPN (OVA, PR)	**0.997**	**0.967**	0.929	0.987	0.924	**0.828**

表 4　噪声模式上的拒识性能

方法	高斯噪声		合成噪声	
	CIFAR-10	ImageNet-100	CIFAR-10	ImageNet-100
CNN (soft-max)	0.889	0.998	0.931	0.863
CPN (PR)	0.905	0.998	0.892	0.825
CPN (DR)	**0.916**	0.998	**0.938**	**0.893**

为进一步考察 CPN 对异常噪声的拒识性能，在 CIFAR-10 和 ImageNet（取 100 类数据作为已知类）上进行了实验。考虑了两种不同形式的噪声，即高斯噪声（Gaussian noise，GN）和合成噪声（Synthetic noise，SN）。GN 是在正常图像样加高斯噪声。SN 是将正常图像分成 $n \times n$ 的子块，然后对子块随机排列组合成新的图像作为异常样本（这样的图像看不去不像任何正常的物体图像）。实验结果见表 4。可以看出，PR 规则也可以较好地拒识噪声模式（当作歧义模式），而 CPN 上 DR 规则的拒识性能更好，因为这种情况下异样样本与已知类样本在外观上区别较大，在人看来是真正的异常图像，不像表 3 的实验中异常样本是正常物体图像（只是不属于已知类别）。

以上实验证实 CPN 具有这样的特点：联合特征表示和原型学习能同时产生高分类精度和对异常模式的拒识性能，因此非常适合开放集分类。CPN 模型也具有原理简单、容易实现、容易解释的特点。由于 CPN 的判别函数为欧氏距离，函数比较平滑，加上 ML 正则使分类器参数的变化受到约束，在训练样本较少时不易产生过学习，因此在小样本训练时泛化性能也有明显优势[43]。由于 CPN 是一种生成模型，以原型的形式存储了每个类别的样本分布，当分类器对新类别样本进行学习（增量学习）时，旧类别的原型有助于克服灾难性遗忘，因此 CPN 也非常适合增量学习[47]。

6 小结

过去模式分类器设计和学习一般假设输入模式来自规定的类别集，即符合闭合世界假设。然而，实际应用环境下，模式的类别集是不可控的，可能来自规定（已知）类别集，也可能是类别集之外（异常模式或新类别）。本文综合介绍了开放世界的模式分类和学习问题、分类决策规则、面向开放集的分类器设计与学习基本方法，并介绍了近来提出的一种面向开放集的深度神经网络模型：卷积原型网络（CPN）。CPN 利用了卷积神经网络的特征学习能力，同时判别函数为欧氏距离，可以产生闭合的决策区域，通过参数联合学习产生的特征空间中每类样本形成紧凑分布，因此可同时产生较高的分类精度和异常拒识性能。CPN 还具有优良的小样本泛化性能和增量学习能力。

除了开放集分类（识别），开放世界模式识别还有一些其他问题需要深入研究，如异常数据中新类别发现、混合样本（包括标记/无标记样本、噪声等）数据流连续学习、非静态数据的模型连续自适应、对噪声和对抗数据的鲁棒性、结构化数据理解和模型学习、多模态数据协同和连续自适应等。

致谢

本研究受到科技创新 2030——“新一代人工智能”重大项目（2018AAA0100400）和国家自然科学基金创新研究群体项目（61721004）资助。感谢我的同事和学生张煦尧、杨红明等对本课题研究的贡献。

参考文献

[1] FU K S. Sequential methods in pattern recognition and machine learning[M]//Mathematics in Science and Engineering Series. New York: Academic, 1968.

[2] BISHOP C M. Pattern recognition and machine learning[M]. Springer, 2006.

[3] CHOW C K. An optimum character recognition system using decision functions[J]. IRE Trans. Electronic Computers, 1957, 6(4): 247-254.

[4] FUKUNAGA K. Introduction to Statistical Pattern Recognition[M]. 1st edition. CRC Press, 1972.

[5] DUDA R O, HART P E. Pattern classification and scene analysis[M]. John Wiley & Sons, Inc., 1973.

[6] FU K S. Syntactic methods in pattern recognition[M]. New York: Academic, 1974.

[7] RUMELHART D E, HINTON G E, WILLIAMS R J. Learning internal representation by error propagation[M]. MIT Press, Cambridge, 1986: 318-362.

[8] CORTES C, VAPNIK V. Support vector networks[J]. Machine learning, 1995, 20: 273-297.
[9] SUEN C Y, NADAL C, MAI T A, et al. Recognition of totally unconstrained numerals based on the concept of multiple experts[C]//Int. Workshop on Frontiers in Handwriting Recognition. 1990: 131-143.
[10] HANSEN L K, SALAMON P. Neural network ensembles[J]. IEEE Trans. Pattern Analysis and Machine Intelligence, 1990, 12(10): 993-1001.
[11] SHAHSHAHANI B, LANDGREBE D. The effect of unlabeled samples in reducing the small sample size problem and mitigating the Hughes phenomenon[J]. IEEE Trans. Geoscience and Remote Sensing, 1994, 32(5): 1087-1095.
[12] MCCALLUM A. Multi-label text classification with a mixture model trained by EM[C]// Proceedings of AAAI'99 Workshop on Text Learning, 1999.
[13] CARUANA R. Multitask learning[J]. Machine Learning, 1997, 28(1): 41-75.
[14] DAI W, YANG Q, XUE G, et al. Boosting for Transfer Learning[C]// Proceedings of 24th Int. Conf. Machine Learning. 2007: 193-200.
[15] POLIKAR R, UDPA L, UDPA A S, et al. Learn++: an incremental learning algorithm for supervised neural networks[J]. IEEE Trans. SMC—Part C: Applications and Review, 2001, 31(4): 497-508.
[16] LECUN Y, BENGIO Y, HINTON G. Deep learning[J]. Nature, 521: 436-444, 2015.
[17] HINTON G E, OSINDERO S, THE Y W. A fast learning algorithm for deep belief nets[J]. Neural Computation, 2006, 18(7): 1527-1554.
[18] KRIZHEVSKY A, SUTSKEVER I, HINTON G. ImageNet classification with deep convolutional neural networks[C]//Advances in Neural Information Processing Systems. 2012, 25: 1090-1098.
[19] ZHANG X Y, LIU C L, SUEN C Y. Towards Robust Pattern Recognition: A review[J].Proceedings of the IEEE, 2020, 108(6): 894-922.
[20] SCHEIRER W J, ROCHA A R, SAPKOTA A, et al. Toward open set recognition[J]. IEEE Trans. Pattern Analysis and Machine Intelligence, 2013, 35(7): 1757-1772.
[21] GENG C X, HUANG S J, CHEN S C. Recent advances in open set recognition: a survey[J].IEEE Trans. Pattern Analysis and Machine Intelligence, 2021, 43(10): 3614-3631.
[22] CHOW C K. On optimal recognition error and reject tradeoff[J]. IEEE Trans. Information Theory, 1970, 16: 41-46.
[23] NGUYEN A, YOSINSKI J, CLUNE J. Deep neural networks are easily fooled: High confidence predictions for unrecognizable images[C]//IEEE Conference on Computer Vision and Pattern Recognition (CVPR). 2015: 27-436.
[24] DUBUISSON B, MASSON M. A statistical decision rule with incomplete knowledge about classes[J]. Pattern Recognition, 1993, 26(1): 155-165.
[25] LIU C L, SAKO H, FUJISAWA H. Discriminative learning quadratic discriminant function for handwriting recognition[J].IEEE Trans. Neural Networks, 2004, 15(2): 430-444.

[26] GE Z, DEMYANOV S, CHEN Z, et al. Generative OpenMax for multi-class open set classification[C]//Proceedings of British Machine Vision Conference (BMVC), 2017.

[27] NEAL L, OLSON M, FERN X, et al. Open set learning with counterfactual images[C]//Proceedings of European Conference on Computer Vision (ECCV), 2018.

[28] YANG H M, ZHANG X Y, YIN F, et al. Convolutional prototype network for open set recognition[J]. IEEE Tran. Pattern Analysis and Machine Intelligence, 2020.

[29] LIU C L, SAKO H, FUJISAWA H. Performance evaluation of pattern classifiers for handwritten character recognition[J]. Int. J. Document Analysis and Recognition, 4(3): 191-204, 2002.

[30] SCHUERMANN J. Pattern classification: A unified view of statistical and neural approaches[M]. New York: Wiley Interscience, 1996.

[31] LIU C L, NAKAGAWA M. Evaluation of prototype learning algorithms for nearest neighbor classifier in application to handwritten character recognition[J]. Pattern Recognition, 2001, 34(3): 601-615.

[32] JUANG B H, KATAGIRI S. Discriminative learning for minimization error classification[J]. IEEE Trans. Signal Processing, 1992, 40(12): 3043-3054.

[33] KIMURA F, TAKASHINA K, TSURUOKA S, et al. Modified quadratic discriminant functions and the application to Chinese character recognition[J]. IEEE Trans. Pattern Analysis and Machine Intelligence, 1987, 9(1): 149-153.

[34] YOSHIHASHI R, SHAO W, KAWAKAMI R. Classification-reconstruction learning for open-set recognition[J]. IEEE Conference on Computer Vision and Pattern Recognition (CVPR), 2019.

[35] OZA P, PATEL V M. C2AE: Class conditioned auto-encoder for open-set recognition[C]//Proceedings of IEEE Conference on Computer Vision and Pattern Recognition (CVPR), 2019.

[36] LIU C L, SAKO H, FUJISAWA H. Effects of classifier structures and training regimes on integrated segmentation and recognition of handwritten numeral strings[J]. IEEE Trans. Pattern Analysis and Machine Intelligence, 2004, 26(11): 1395-1407.

[37] POVEY D, WOOD P C. Minimum phone error and I-smoothing for improved discriminative training[C]//Proceedings of 2002 ICASSP. 2002, 1: 105-108.

[38] JIN X B, LIU C L, HOU X. Regularized margin-based conditional log-likelihood loss for prototype learning[J]. Pattern Recognition, 2010, 43(7): 2428-2438.

[39] PLATT J. Probabilistic outputs for support vector machines and comparisons to regularized likelihood methods[M]. MIT Press, 1999: 61-74.

[40] LIU C L. Classifier combination based on confidence transformation[J]. Pattern Recognition, 2005, 38(1): 11-28.

[41] BARNETT J A. Computational methods for a mathematical theory of evidence[C]//Proceedings of 7th International Joint Conference on Artificial Intelligence. Vancouver, Canada, 1981: 868-875.

[42] LIU C L. One-vs-all training of prototype classifiers for pattern classification and retrieval[C]// Proceedings of 20th ICPR. Istanbul, Turkey. 2010: 3328-3331.

[43] YANG H M, ZHANG X Y, YIN F, et al. Robust classification with convolutional prototype learning[C]// Proceedings of IEEE Conference on Computer Vision and Pattern Recognition (CVPR), 2018.

[44] GROSSMAN D, DOMINGOS P. Learning Bayesian network classifiers by maximizing conditional likelihood[C]//Proceedings of int. Conf. Machine Learning. 2004: 361-368.

[45] NEAL L, OLSON M, FERN X, et al. Open set learning with counterfactual images[C]//Proceedings of the European Conference on Computer Vision (ECCV). 2018: 613-628.

[46] Bendale A, Boult T E. Towards open set deep networks[C]//Proceedings of IEEE Conference on Computer Vision and Pattern Recognition. 2016: 1563-1572.

[47] ZHU F, ZHANG X Y, WANG C, et al. Prototype augmentation and self-supervision for incremental learning[C]//Proceedings of IEEE/CVF Conference on Computer Vision and Pattern Recognition (CVPR), 2021.

释放标记空间的威力：标记增强

耿　新　徐　宁　高永标　王秋锋

（东南大学计算机科学与工程学院，南京 210023）

1　引言

机器学习是人工智能的核心研究领域，其中监督学习又是机器学习中最受关注的方向之一。监督学习主要实现从示例空间到标记空间的映射，传统监督学习研究主要面向“单标记”或者“多标记”[1]样本，这里一个标记（label）通常对应一个概念类别，在图像识别[2]、文本分类[3]、视频分析[4]等领域获得了广泛应用。

在监督学习中，标记存在的形式往往为“硬标记”，即对于一个示例 $\boldsymbol{x}$，将 $l_x^y \in \{0,1\}$ 赋予每个可能的标记 y，用以表示该标记 y 是否描述了示例 $\boldsymbol{x}$。由于 l_x^y 表达了是与否的逻辑关系，所以本文称这种标记为逻辑标记。机器学习主要实现从示例空间到标记空间的映射，由于示例空间通常是欧式空间，因此在示例空间上使用特征提取、特征选择、流形嵌入、降维等技术。另一方面，由于通常使用的是逻辑标记，而逻辑标记空间中所有样本只能分布在单位超立方体的顶点上。受限于此，上述在示例空间的操作往往无法发挥作用。

对于一个示例 $\boldsymbol{x}$，可以将一个实数 d_x^y 赋予每个可能的标记 y，表示 y 描述 $\boldsymbol{x}$ 的程度。不失一般性，假设 $d_x^y \in [0,1]$，并进一步假设标记集合为完备集，即集合中的所有标记一定可以完整地描述一个示例，因此 $\sum_y d_x^y = 1$。由所有类别的连续值 d_x^y 构成的向量称为标记分布[5]。现实任务中应用标记分布则意味着更高的标注难度。一方面，对每个示例，为所有可能的标记赋予一个描述度使标注成本更高；另一方面，标记对示例的描述度也常常没有客观的量化标准。所以现实任务中大量的多义性数据仍然是以简单逻辑标记标注的。将所有标记划分为相关/无关两个子集的逻辑标记实际上是对多义性数据本质的一种简化，尽管如此，仍可假设这些数据的监督信息遵循某种更为本质的标记分布，这种标记分布虽然没有显式给出，却蕴含于训练样本中，如果能够通过某种方式将其自动恢复出来，由于这一恢复过程将每个示例原有的简单逻辑标记“增强”为包含更多类别监督信息的标记分布，这一过程称为“标记增强”（label enhancement，LE）[6-7]。标记增强将离散的逻辑空间转换为一个连续空间，使由机器学习连接的两个空间（示例空间和标记空间）对等，为释放标记空间的威力提供了基础，因此可以利用原本在示例空间才可以使用的工具和技术，为处理标记多义性、标记不确定性、标记相关性等提供更多的可能。而随着标记空间能力的释放，标记分布已经被成功应用于不同领域的实际问题，如计算机视觉[5-8]、自然语言处理[9-10]、问答系统[11-12]、情感计算[13-14]、医学诊断[15-16]等。

标记增强的目标是将训练样本中的原始逻辑标记转化为标记分布，这一过程依赖于对隐藏在训练样本中的标记相关信息的挖掘。利用隐含于数据中的标记间相关性标记增强，可以有效加强示例的监督信息，进而通过标记分布学习获得更好的预测效果。在标记增强这一概念提出之前，存在一些工作，尽管它们的应用背景和具体目标不尽相同，却可以用来实现（有些方法需要经过部分改造）标记增强功能。这些算法可分为三种类型，分别是面向模糊标记的标记增强、面向概率标记的标记增强和面向标记分布的标记增强。上述工作有些是直接为标记分布学习提出的，有些则是在其他领域提出但可以用来生成标记分布，不管哪种情况，它们都可以统一到标记增强概念之下。

本文首先介绍标记增强的应用背景，接着总结三类标记增强方法，给出标记增强的理论解释，最后详细介绍标记增强的应用。

2 标记增强方法

2.1 符号及形式化定义

本文主要的符号表示如下。示例用 $\boldsymbol{x}$ 表示，第 i 个示例用 $\boldsymbol{x}_i$ 表示；标记用 y 表示，第 j 个标记用 y_j 表示。$\boldsymbol{x}_i$ 的逻辑标记用 $\boldsymbol{L}_i=\left[l_{x_i}^{y_1},l_{x_i}^{y_2},\cdots,l_{x_i}^{y_c}\right]$ 表示，并且 $\boldsymbol{L}_i\in\{0,1\}^c$，$c$ 是可能的标记数目。y 对 $\boldsymbol{x}$ 的描述度用 d_x^y 表示，满足 $d_x^y\in[0,1]$ 且 $\sum_y d_x^y=1$。$\boldsymbol{x}_i$ 的标记分布用 $\boldsymbol{D}_i=\left[d_{x_i}^{y_1},d_{x_i}^{y_2},\cdots,d_{x_i}^{y_c}\right]$ 表示且 $\boldsymbol{D}_i\in[0,1]^c$。假设 $\mathbb{R}^q$ 表示示例的特征空间，$Y=\{y_1,y_2,\cdots,y_c\}$ 表示标记空间，则标记增强定义如下：给定训练集 $S=\{(\boldsymbol{x}_i,\boldsymbol{L}_i)\mid 1\leqslant i\leqslant n\}$，标记增强即根据 S 中蕴含的标记间相关性，将每个示例 $\boldsymbol{x}_i$ 的逻辑标记 $\boldsymbol{L}_i$ 转化为相应的标记分布 $\boldsymbol{D}_i$，从而得到标记分布训练集 $\mathcal{L}=\{(\boldsymbol{x}_i,\boldsymbol{D}_i)\mid 1\leqslant i\leqslant n\}$ 的过程。

2.2 面向模糊标记的标记增强

基于模糊方法的标记增强利用模糊数学的思想，通过模糊聚类、模糊运算和核隶属度等方法，挖掘出标记间相关信息，将逻辑标记转化为模糊隶属度。值得注意的是，这类方法提出的目的一般是为了将模糊性引入原本刚性的逻辑标记，而并非为了标记增强专门设计的算法。实际上，这些算法可以通过改造成为标记增强算法。本节介绍两种基于模糊方法的标记增强算法，分别是基于模糊聚类的标记增强算法和基于核隶属度的标记增强算法。

2.2.1 模糊聚类算法

基于模糊聚类的标记增强[17]通过模糊 C-均值聚类(fuzzy C-means algorithm, FCM)[18]和模糊运算，将训练集中每个示例的逻辑标记转化为相应的标记分布，从而得到标记分布训练集。FCM 是用隶属度确定每个数据点属于某个聚类的程度的一种聚类算法，该算法把 n 个样本分为 p 个模糊聚类，并求每个聚类的中心，使所有训练样本到聚类中心的加权（权值由样本点对相应聚类的隶属度决定）距离之和最小。假设 FCM 将训练集 S 分为 p 个聚类，$\boldsymbol{\mu}_k$ 表示第 k 个聚类中心，则可用如下公式计算示例 $\boldsymbol{x}_i$ 对每个聚类的隶属度 $\boldsymbol{m}_{\boldsymbol{x}_i}=\left[m_{\boldsymbol{x}_i}^1, m_{\boldsymbol{x}_i}^2, \cdots, m_{\boldsymbol{x}_i}^p\right]$：

$$m_{\boldsymbol{x}_i}^k=\frac{1}{\sum_{j=1}^p\left(\frac{\operatorname{Dist}\left(\boldsymbol{x}_i, \boldsymbol{\mu}_k\right)}{\operatorname{Dist}\left(\boldsymbol{x}_i, \boldsymbol{\mu}_j\right)}\right)^{\frac{1}{\beta-1}}} \tag{1}$$

其中，Dist 是任意的距离度量；β 是模糊因子，且满足 $\beta>1$。得到 $\boldsymbol{m}_{\boldsymbol{x}_i}$ 后，进一步构建一个关联矩阵 $\boldsymbol{A}$。首先初始化一个 $c\times p$ 的零矩阵 $\boldsymbol{A}$，然后用如下公式更新 $\boldsymbol{A}$ 的第 j 行 $\boldsymbol{A}_j$：

$$\boldsymbol{A}_j=\boldsymbol{A}_j+\boldsymbol{m}_{\boldsymbol{x}_i}, \text{如果} l_{\boldsymbol{x}_i}^{y_j}=1 \tag{2}$$

即 $\boldsymbol{A}_j$ 为所有属于第 j 个类的样本的隶属度向量之和。经过行归一化后得到的矩阵 $\boldsymbol{A}$ 可以被当作一个“模糊关系”矩阵，即 $\boldsymbol{A}$ 中的元素 a_{jk} 表示第 j 个类别（标记）与第 k 个聚类的关联强度。根据模糊逻辑推理机制[19]，将关联矩阵 $\boldsymbol{A}$ 与 $\boldsymbol{x}_i$ 对聚类的隶属度 $\boldsymbol{m}_{\boldsymbol{x}_i}$ 进行模糊合成运算 $\boldsymbol{v}_i=\boldsymbol{A}\cdot\boldsymbol{m}_{\boldsymbol{x}_i}$，从而将 $\boldsymbol{x}_i$ 对聚类的隶属度转化为对类别的隶属度。最后，对隶属度向量 $\boldsymbol{v}_i$ 进行归一化，使向量中元素的和为 1，即得到标记分布 $\boldsymbol{d}_i$。基于模糊聚类的标记增强算法利用模糊聚类过程中产生的示例对每个聚类的隶属度，通过类别和聚类的关联矩阵，将示例对聚类的隶属度转化为对类别的隶属度，从而生成标记分布。在这一过程中，模糊聚类反映了示例空间的拓扑关系，而通过关联矩阵，将这种关系转化到标记空间，从而有可能使简单的逻辑标记产生更丰富的语义，转变为标记分布。

2.2.2 核函数算法

基于核隶属度的标记增强方法源于一种模糊支持向量机中核隶属度的生成过程[20]，通过一个非线性映射函数将示例 $\boldsymbol{x}_i$ 映射到高维空间，利用核函数[21]计算该高维空间中正类的中心、半径和各示例 $\boldsymbol{x}_i$ 到正类中心的距离，进而通过隶属度函数计算示例 $\boldsymbol{x}_i$ 的标记分布。具体地，对于训练集 S 和某个标记 y_j，根据 y_j 的逻辑值，将 S 分为两个集合，其中 $\boldsymbol{x}_i$ 的逻辑标记 $l_{x_i}^{y_j}=1$ 的集合用 $C_+^{y_j}$ 表示。那么，正类集合在特征空间的中心为 $\boldsymbol{\Psi}_+^{y_j}=\frac{1}{n_+}\sum_{x_i\in C_+^{y_j}}\varphi(\boldsymbol{x}_i)$，这里 n_+ 表示该集合中示例的数量，$\varphi(\boldsymbol{x}_i)$ 是一个非线性映射函数，由核函数 $k(\boldsymbol{x}_i,\boldsymbol{x}_j)=\varphi(\boldsymbol{x}_i)\cdot\varphi(\boldsymbol{x}_j)$ 决定。该集合的半径定义为 $r_+=\max\|\boldsymbol{\Psi}_+^{y_j}-\varphi(\boldsymbol{x}_i)\|$，集合中的 $\boldsymbol{x}_i$ 到中心的距离是 $d_{i+}=\|\varphi(\boldsymbol{x}_i)-\boldsymbol{\Psi}_+^{y_j}\|$。那么，$\boldsymbol{x}_i$ 对于标记 y_j 的隶属度为

$$m_{x_i}^{y_j}=\begin{cases}1-\sqrt{\dfrac{d_{i+}^2}{\left(r_+^2+\delta\right)}}, & l_{x_i}^{y_j}=1\\ 0, & l_{x_i}^{y_j}=0\end{cases} \tag{3}$$

其中，$\delta>0$。涉及 $\varphi(\boldsymbol{x}_i)$ 的计算均可以由核函数 $k(\boldsymbol{x}_i,\boldsymbol{x}_j)$ 间接计算。最后，将 $m_{x_i}^{y_j}$ 归一化，即可得到 $\boldsymbol{x}_i$ 的标记分布 $\boldsymbol{d}_i$。基于核隶属度的标记增强算法利用核技巧在高维空间中计算示例对每个类别的隶属度，从而挖掘训练数据中类别标记间较为复杂的非线性关系。

2.3 面向概率标记的标记增强

概率标记面向单标记样本，表示该标记是正确标记的概率，是对标注置信度的度量。在训练神经网络时，使用概率标记作为输出，可以有效地缓解过拟合和增强模型对噪声标记的鲁棒性。除此之外，具有更多标记间信息的概率标记为模型的压缩提供了可能性。因此，本节介绍三种基于概率标记的标记增强算法，分别是基于端到端模型、基于标记平滑和基于知识蒸馏的标记增强算法。

2.3.1 基于端到端模型的算法

现实的任务中标记往往是带有噪声的，含有噪声的标记很容易导致过拟合，所以处理带有噪声的标记是一项具有挑战性的任务。Yi 等人[22]提出将带有噪声的标记转换为标记分布。PENCIL 模型定义训练集 $X=\{\boldsymbol{x}_1,\boldsymbol{x}_2,\cdots,\boldsymbol{x}_n\}$，每个样本 $\boldsymbol{x}_i$ 标注了含有噪声的标

记 $\hat{y}_i \in \mathcal{H}$，该方法将含有噪声的标记 $\hat{y}_i$ 转换成标记分布 $\boldsymbol{y}_i^d$，然后将 KL 散度作为损失函数，计算预测的标记分布和转换后的标记分布之间的损失，即

$$\begin{cases} \mathcal{L} = \dfrac{1}{n}\sum_{i=1}^{n} KL\left(f\left(\boldsymbol{x}_i;\theta\right) \| \boldsymbol{y}_i^d\right), \\ \mathrm{KL}\left(f\left(\boldsymbol{x}_i;\theta\right) \| \boldsymbol{y}_i^d\right) = \sum_{j=1}^{c} f_j\left(\boldsymbol{x}_i;\theta\right) \ln\left(\dfrac{f_j\left(\boldsymbol{x}_i;\theta\right)}{y_{ij}^d}\right) \end{cases} \tag{4}$$

其中，$f\left(\boldsymbol{x}_i;\theta\right)$ 表示模型输出；θ 表示模型参数；y_{ij}^d 表示含有噪声的标记转换成的标记分布中第 i 个样本的第 j 个描述度。该模型提出了同时更新网络参数和标记分布的方法，且噪声标记转化成的标记分布同样由梯度下降和反向传播来更新，为了达到这一目的，首先根据含有噪声的标记进行初始化概率形式的标记分布，即

$$\tilde{\boldsymbol{y}} = K\hat{y} \tag{5}$$

其中，$\tilde{\boldsymbol{y}}$ 表示初始化的标记分布；$\hat{y}$ 表示含有噪声的标记；K 表示一个常数，对噪声标记进行尺度扩充。为了满足标记分布的所有概率值和为 1，对转换后的标记分布进行归一化操作，即 $\boldsymbol{y}^d = \mathrm{softmax}(\tilde{\boldsymbol{y}})$，其中 $\boldsymbol{y}^d$ 表示归一化后的标记分布。为了同时更新网络参数和带有噪声的标记分布，如图 1 所示，PENCIL 采用三个损失函数进行模型学习，损失函数如下：

（1）兼容性损失（compatibility loss）

$$\mathcal{L}_{\mathrm{o}}\left(\hat{\boldsymbol{Y}}, \boldsymbol{Y}^d\right) = -\frac{1}{n}\sum_{i=1}^{n}\sum_{j=1}^{c}\hat{y}_{ij} \ln y_{ij}^d \tag{6}$$

其中，$\boldsymbol{Y}^d$ 表示需要评估的标记分布；$\hat{\boldsymbol{Y}}$ 表示含有噪声的标记分布。由于数据集中并不是

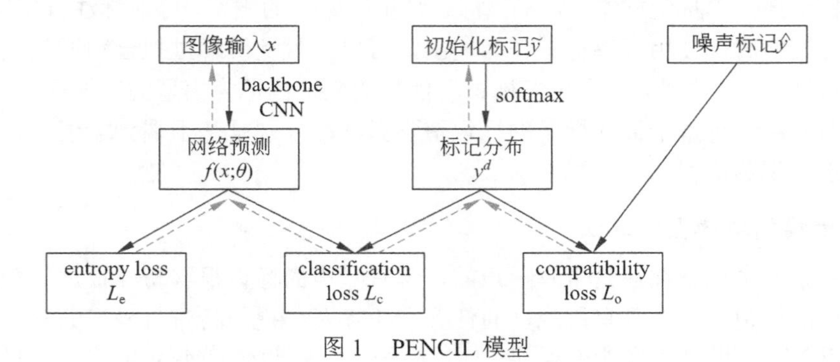

图 1　PENCIL 模型

所有样本都包含噪声，且含有噪声的标记中有大量真实信息，不能偏差太远，所以需要保持学习得到的标记分布和含有噪声标记的标记分布具有一定的相似性。

（2）分类损失（classification loss）

$$\frac{\partial \mathcal{L}_{\mathrm{c}}}{\partial y_{ij}^{d}}=-\sum_{j=1}^{c}\frac{f_j\left(\boldsymbol{x}_i;\boldsymbol{\theta}\right)}{y_{ij}^{d}} \tag{7}$$

该损失函数用于计算分类损失。

（3）熵损失（entropy loss）

$$\mathcal{L}_{\mathrm{e}}(f(\boldsymbol{x};\theta))=-\frac{1}{n}\sum_{i=1}^{n}\sum_{j=1}^{c}f_j(\boldsymbol{x};\theta)\ln f_j(\boldsymbol{x};\theta) \tag{8}$$

模型将标记分布作为监督信息，很容易导致所有标记的概率相同，为了解决这个问题，增加了一个熵损失，让分布更离散。

最后把三个损失函数组合在一起进行梯度更新，即

$$\mathcal{L}=\frac{1}{c}\mathcal{L}_c\left(f(\boldsymbol{x};\theta),\boldsymbol{Y}^d\right)+\alpha\mathcal{L}_o\left(\hat{\boldsymbol{Y}},\boldsymbol{Y}^d\right)+\frac{\beta}{c}\mathcal{L}_e(f(\boldsymbol{x};\theta)) \tag{9}$$

其中，α 和 β 表示超参，用于平衡 compatibility loss 和 entropy loss。

2.3.2 基于标记平滑的算法

在训练神经网络时，通过最小化预测标记概率和真实标记概率之间的交叉熵，从而得到最优的预测标记概率分布。神经网络会促使自身往正确标记和错误标记差值最大的方向学习。在训练数据较少、不足以表征所有的样本特征的情况下，会导致网络过拟合。

标记平滑[23]可以解决上述问题，这是一种正则化策略，主要通过软标记替代独热编码（one-hot）标记，减少真实样本标记在计算损失函数时的权重，最终起到抑制过拟合的效果。效果如图 2 所示，神经网络会输出一个当前数据对应于各个标记的置信度分数，将这些分数通过 softmax 进行归一化处理，最终得到当前数据属于每个标记的概率。

$$q_i=\frac{\exp\left(z_i\right)}{\sum_{j=1}^{K}\exp\left(z_j\right)} \tag{10}$$

然后计算加入标记平滑的交叉熵损失函数：

$$\mathrm{Loss}_i=\begin{cases}(1-\varepsilon)\times\mathrm{Loss}, & i=y\\ \varepsilon\times\mathrm{Loss}, & i\neq y\end{cases} \tag{11}$$

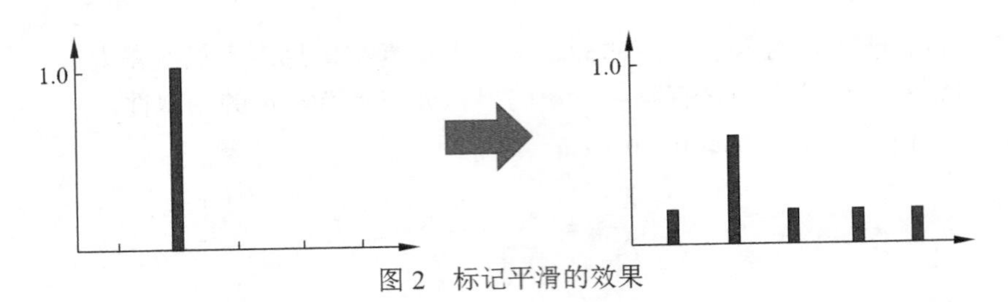

图 2　标记平滑的效果

其中，ε 为平滑权重，真实的标记概率分布也变为

$$p_i = \begin{cases} 1-\varepsilon, & i = y \\ \dfrac{\varepsilon}{K-1}, & i \neq y \end{cases} \tag{12}$$

2.3.3 基于知识蒸馏的算法

简单地说，知识蒸馏[24]就是将大模型对样本输出的概率向量作为软标记，使小模型的输出尽可能和这个增强的软标记靠近（原来是往 one-hot 标记上靠近）。知识蒸馏过程所用的训练样本可以和训练大模型用的训练样本一样，或者使用新的迁移数据集。因为软标记比 one-hot 标记携带的信息更丰富，所以在训练小模型时可以用比训练大模型时更少的训练集和更大的学习率。如果直接使用 softmax 层的输出值作为软标记，这又会带来一个问题:当 softmax 输出的概率分布熵相对较小时，负标签的值都很接近 0，对损失函数的贡献非常小，小到可以忽略不计。因此，知识蒸馏引入了“温度”这个变量。下面的公式为加了温度这个变量之后的 softmax 函数：

$$q_i = \frac{\exp\left(\dfrac{z_i}{T}\right)}{\sum_j \exp\left(\dfrac{z_j}{T}\right)} \tag{13}$$

其中，T 就是温度。大模型的输出如图 3 所示，原来的 softmax 函数是 $T=1$ 的特例。T 越高，softmax 的输出概率标记分布越趋于平滑，其分布的熵越大，负标记携带的信息会被相对地放大，模型训练将更加关注负标签。

2.4 面向标记分布的标记增强

标记分布通过连续的描述度来显式表达各标记对于数据对象的相关程度并覆盖所有可能的标记，从而使标记与对象的关系及标记间的关系都能通过描述度值及排序明确表

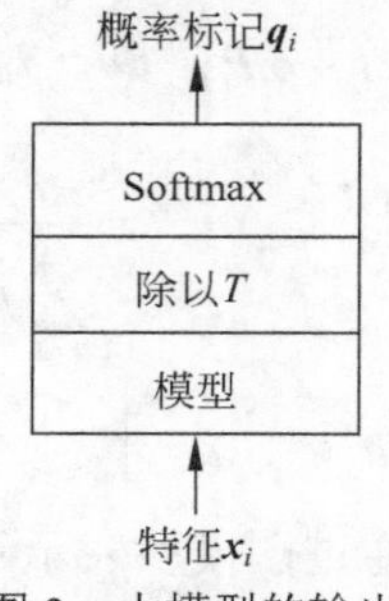

图 3　大模型的输出

达和利用。本节介绍三种将逻辑标记增强为标记分布的标记增强算法，分别是基于标记传播的标记增强算法、基于流形学习的标记增强算法和基于图的拉普拉斯的标记增强算法。

2.4.1　基于标记传播的算法

基于标记传播的标记增强[25]将半监督学习[26]中的标记传播技术应用于标记增强。该方法首先根据示例间相似度构建一个图，然后根据图中的拓扑关系在示例间传播标记。由于标记的传播会受路径上权值的影响，会自然形成不同标记的描述度差异。当标记传播收敛时，每个示例的原有逻辑标记即可增强为标记分布。

具体地，假设多标记训练集$S=\{(\boldsymbol{x}_i,\boldsymbol{L}_i)\,|\,1\leqslant i\leqslant n\}$，$G=(V,E,W)$表示以$S$中的示例为顶点的全连通图，其中$V=\{\boldsymbol{x}_i\,|\,1\leqslant i\leqslant n\}$表示顶点，$E$表示顶点两两之间的边，$\boldsymbol{x}_i$与$\boldsymbol{x}_j$之间的边上的权值为它们之间的相似度：

$$\forall_{i,j=1}^{n}: w_{ij}=\begin{cases}\exp\left(-\dfrac{\|\boldsymbol{x}_i-\boldsymbol{x}_j\|_2^2}{2}\right), & i\neq j\\ 0, & i=j\end{cases}\tag{14}$$

所有边的权值构成相似度矩阵$\boldsymbol{W}=[w_{ij}]_{n\times n}$。标记传播矩阵$\boldsymbol{P}$由相似度矩阵$\boldsymbol{W}$计算而来：$\boldsymbol{P}=\boldsymbol{Q}^{-\frac{1}{2}}\boldsymbol{W}\boldsymbol{Q}^{-\frac{1}{2}}$，这里$\boldsymbol{Q}=\mathrm{diag}[\boldsymbol{d}_1,\boldsymbol{d}_2,\cdots,\boldsymbol{d}_m]$，其中$\boldsymbol{d}_i=\sum_{j}^{n}w_{ij}$。假设所有标记对所有示例的描述度构成一个描述度矩阵$\boldsymbol{F}$，该算法使用迭代方法不断更新$\boldsymbol{F}$。$\boldsymbol{F}$的初始值$\boldsymbol{F}^{(0)}=\boldsymbol{\Phi}=[\phi_{ij}]_{n\times c}$由示例$\boldsymbol{x}$的逻辑标记构成，即$\forall_{i=1}^{n}\forall_{j=1}^{c}:\phi_{ij}=l_{x_i}^{y_j}$。在此基础上，使用标记传播对描述度矩阵$\boldsymbol{F}$进行更新：

$$\boldsymbol{F}^{(t)}=\alpha\boldsymbol{P}\boldsymbol{F}^{(t-1)}+(1-\alpha)\boldsymbol{\Phi}\tag{15}$$

其中，α是平衡参数，控制了初始的逻辑标记和标记传播对最终描述度的影响程度。经过迭代，最终 $\boldsymbol{F}$ 收敛到 $\boldsymbol{F}^{*}=(1-\alpha)(I-\alpha\boldsymbol{P})^{-1}\boldsymbol{\Phi}$ 。对 $\boldsymbol{F}^{*}$ 做归一化处理：

$$\forall_{i=1}^{n}\forall_{j=1}^{c}:d_{x_i}^{y_j}=\frac{f_{ij}^{*}}{\sum_{k=1}^{c}f_{ik}^{*}} \tag{16}$$

即得到示例 $\boldsymbol{x}_i$ 的标记分布 $\boldsymbol{D}_i=\left[d_{x_i}^{y_1},d_{x_i}^{y_2},\cdots,d_{x_i}^{y_c}\right]$ 。

基于标记传播的标记增强算法通过图模型表示示例间的拓扑结构，构造了基于示例间相关性的标记传播矩阵，利用传播过程中路径权值的不同使不同标记的描述度自然产生差异，从而反映出蕴含在训练数据中的标记间关系。

2.4.2 基于流形学习的算法

基于流形的标记增强算法[27]假设数据在特征空间和标记空间均分布在某种流形上，并利用平滑假设将两个空间的流形联系起来，从而利用特征空间流形的拓扑关系指导标记空间流形的构建，在此基础上将示例的逻辑标记增强为标记分布。具体地，该算法用图 $\boldsymbol{G}=\langle V,E,\boldsymbol{W}\rangle$ 表示多标记训练集 S 的特征空间的拓扑结构，其中 V 是由示例构成的顶点集合，E 是边的集合，$\boldsymbol{W}$ 是图的边权重矩阵。首先，在特征空间中，假设示例分布的流形满足局部线性，即任意示例 $\boldsymbol{x}_i$ 可以由它的 k-近邻的线性组合重构，重构权值矩阵 $\boldsymbol{W}$ 可通过最小化下式得到:

$$\Omega(\boldsymbol{W})=\sum_{i=1}^{n}\boldsymbol{x}_i-\sum_{j\neq i}w_{ij}\boldsymbol{x}_j{}^{2} \tag{17}$$

其中，$\sum_{j=1}^{n}w_{ij}=1$ 。如果 $\boldsymbol{x}_j$ 不是 $\boldsymbol{x}_i$ 的 k-近邻，那么 $w_{ij}=0$ 。通过平滑假设[28]，即特征相似的示例的标记也很可能相似，可将特征空间的拓扑结构迁移到标记空间中，即共享同样的局部线性重构权值矩阵 $\boldsymbol{W}$ 。这样，标记空间的标记分布可由最小化下式得到:

$$\Psi\left(\hat{\boldsymbol{d}}\right)=\sum_{i=1}^{n}\hat{\boldsymbol{d}}_i-\sum_{j\neq i}w_{ij}\hat{\boldsymbol{d}}_j{}^{2} \tag{18}$$

$$\text{s.t.}\quad d_{x_i}^{y_l}l_{x_i}^{y_l}>\lambda,\forall 1\leqslant i\leqslant n,1\leqslant j\leqslant c$$

其中，$\lambda>0$ 是预先设定的参数。值得指出的是，为了方便构建上述约束条件，文献[27]中定义逻辑标记 $l_i\in\{-1,1\}^c$ ，而不是其他方法中常用的 $l_i\in\{0,1\}^c$ ，但两者本质上并没有区别。这样，约束条件 $d_{x_i}^{y_l}l_{x_i}^{y_l}>\lambda$ 可以确保 d_{x_i} 与 $l_{x_i}^{y_l}$ 同号。通过求解上述二次规划问题确定 $\hat{\boldsymbol{d}}_i$ 后，经过归一化即可得到标记分布 $\boldsymbol{d}_i$ ，进而得到标记分布训练集。基于流形的方法通过重构特征空间和标记空间的流形，利用平滑假设，将特征空间的拓扑关系迁移到标记空间中，建立示例间相关性与标记间相关性之间的关系，从而将逻辑标记增强为标记分布。

2.4.3 基于图的拉普拉斯的算法

本节介绍基于图的拉普拉斯的标记增强 GLLE（graph laplacian label enhancement)，该算法应能够在不依赖具体应用的先验知识前提下，充分挖掘特征空间的拓扑结构信息、标记间相关性信息，产生标记之间的重要程度差异，进而恢复出标记分布。

给定一个训练集 S，建立特征矩阵 $\boldsymbol{X}=\left[\boldsymbol{x}_1,\boldsymbol{x}_2,\cdots,\boldsymbol{x}_n\right]$ 和逻辑标记矩阵 $\boldsymbol{L}=\left[\boldsymbol{l}_1,\boldsymbol{l}_2,\cdots,\boldsymbol{l}_n\right]$。我们的目标是从逻辑标记矩阵 $\boldsymbol{L}$ 中恢复出标记分布矩阵 $\boldsymbol{D}=\left[\boldsymbol{d}_1,\boldsymbol{d}_2,\cdots,\boldsymbol{d}_n\right]$。为了解决该问题，考虑如下的参数模型：

$$\boldsymbol{d}_i=\boldsymbol{W}^{\mathrm{T}}\varphi\left(\boldsymbol{x}_i\right)+\boldsymbol{b}=\hat{\boldsymbol{W}}\phi_i \tag{19}$$

其中，$\boldsymbol{W}=\left[\boldsymbol{w}^1,\cdots,\boldsymbol{w}^c\right]$ 是一个权重矩阵；$\boldsymbol{b}\in\mathbb{R}^c$ 是偏置向量；$\varphi\left(\boldsymbol{x}\right)$ 是一个非线性映射，将 $\boldsymbol{x}$ 映射到一个更高维度的特征空间。为了方便描述，记 $\hat{\boldsymbol{W}}=\left[\boldsymbol{W}^{\mathrm{T}},\boldsymbol{b}\right]$ 且 $\phi_i=\left[\varphi\left(\boldsymbol{x}_i\right);1\right]$。因此，目标是确定最优参数 $\hat{\boldsymbol{W}}^*$，使该模型在给定示例 $\boldsymbol{x}_i$ 时可以产生合理的标记分布 $\boldsymbol{d}_i$。为了得到最优参数 $\hat{\boldsymbol{W}}^*$，需要最小化目标函数：

$$\hat{\boldsymbol{W}}^*=\arg\min_{\hat{\boldsymbol{W}}} l\left(\hat{\boldsymbol{W}}\right)+\lambda_1\Omega\left(\hat{\boldsymbol{W}}\right)+\lambda_2 z(\hat{\boldsymbol{W}}) \tag{20}$$

其中，l 是一个损失函数；Ω 是一个利用特征空间拓扑结构的函数；z 是一个利用标记间相关性的函数。值得注意的是，标记增强本质上是一个对训练集的预处理，并不需要考虑泛化能力，这与传统的监督学习不一样。因此，该优化框架不需要考虑过拟合问题。

由于标记分布中的一些信息应该继承于逻辑标记，因此选择最小二乘损失函数(least squares)作为式（20）中的损失函数：

$$l(\hat{\boldsymbol{W}})=\sum_{i=1}^{n}\|\hat{\boldsymbol{W}}\phi_i-\boldsymbol{l}_i\|^2=\operatorname{tr}\left[(\hat{\boldsymbol{W}}\boldsymbol{\Phi}-\boldsymbol{L})^{\top}(\hat{\boldsymbol{W}}\boldsymbol{\Phi}-\boldsymbol{L})\right] \tag{21}$$

其中，$\boldsymbol{\Phi}=[\phi_1,\phi_2,\cdots,\phi_n]$。

为了挖掘训练集中隐藏的标记重要程度信息，可以利用特征空间的拓扑结构信息。因此，定义如下的相似度矩阵 $\boldsymbol{A}$：

$$a_{ij}=\exp\left(-\frac{\|\boldsymbol{x}_i-\boldsymbol{x}_j\|^2}{2\sigma^2}\right) \tag{22}$$

其中，$\sigma>0$ 是相似度计算的宽度参数，一般设为 1。根据平滑假设[28]，两个相互靠近的示例极有可能具有相同的标记。直观地，如果 $\boldsymbol{x}_i$ 和 $\boldsymbol{x}_j$ 具有很高的相似度(可用 a_{ij} 度量)，那么 $\boldsymbol{d}_i$ 和 $\boldsymbol{d}_j$ 应该彼此接近。这个直觉引出如下函数 $\Omega(\hat{\boldsymbol{W}})$：

$$\Omega(\hat{\boldsymbol{W}})=\sum_{i,j}a_{ij}\|\boldsymbol{d}_i-\boldsymbol{d}_j\|^2=\mathrm{tr}\left(\boldsymbol{DGD}^{\mathrm{T}}\right)=\mathrm{tr}\left(\hat{\boldsymbol{W}}\boldsymbol{\Phi}\boldsymbol{G}\boldsymbol{\Phi}^{\mathrm{T}}\hat{\boldsymbol{W}}^{\mathrm{T}}\right) \tag{23}$$

其中，$\boldsymbol{G}=\hat{\boldsymbol{A}}-\boldsymbol{A}$ 是图的拉普拉斯(graph Laplacian)矩阵，且 $\hat{\boldsymbol{A}}$ 是对角矩阵，该矩阵的元素是 $\hat{a}_{ii}=\sum_{j=1}^{n}a_{ij}$。

标记相关性[29]可以提供额外的信息，将训练集的逻辑标记恢复为标记分布。特别地，两个标记越相关，标记对应的描述度也越接近。换言之，如果第 i 个标记和第 j 个标记更为相关，则 $\boldsymbol{d}^i$ 应该与 $\boldsymbol{d}^j$ 更相似。其中，$\boldsymbol{d}^i$ 是由所有的第 i 个标记对应的描述度组成的向量，也就是 $\boldsymbol{d}^i=[d_{x_1}^{y_i},d_{x_2}^{y_i},\cdots,d_{x_n}^{y_i}]$。在实际任务中，标记相关性往往是局部的，即标记相关性自然地存在于示例的子集中，而非存在于所有示例中[30]。假设训练集可以被分为 m 个集合 $\{G_1,G_2,\cdots,G_m\}$，其中相同集合中的示例共享相同的标记相关性。而该集合恰好可以通过聚类手段得到[31]，因此得到以下的函数替代式：

$$z(\hat{\boldsymbol{W}})=\sum_{i=1}^{m}\mathrm{tr}\left(\boldsymbol{D}_i^{\mathrm{T}}\boldsymbol{C}_i\boldsymbol{D}_i\right)=\sum_{i=1}^{m}\mathrm{tr}\left(\boldsymbol{\Phi}_i^{\mathrm{T}}\hat{\boldsymbol{W}}^{\mathrm{T}}\boldsymbol{C}_i\hat{\boldsymbol{W}}\boldsymbol{\Phi}_i\right) \tag{24}$$

其中，$\boldsymbol{D}_i$ 是 G_i 中所有示例的标记分布组成的矩阵；$\boldsymbol{C}_i$ 是对应的拉普拉斯矩阵，代表 G_i 的局部标记相关性；$\boldsymbol{\Phi}_i$ 表示 G_i 中示例的特征矩阵。将标记增强问题形式化为如下的优化问题：

$$\min_{\hat{\boldsymbol{W}}}\ \mathrm{tr}\left[(\hat{\boldsymbol{W}}\boldsymbol{\Phi}-\boldsymbol{L})^{\mathrm{T}}(\hat{\boldsymbol{W}}\boldsymbol{\Phi}-\boldsymbol{L})\right]+\lambda_1\mathrm{tr}\left(\hat{\boldsymbol{W}}\boldsymbol{\Phi}\boldsymbol{G}\boldsymbol{\Phi}^{\mathrm{T}}\hat{\boldsymbol{W}}^{\mathrm{T}}\right)+\lambda_2\sum_{i=1}^{m}\mathrm{tr}\left(\boldsymbol{\Phi}_i^{\mathrm{T}}\hat{\boldsymbol{W}}^{\mathrm{T}}\boldsymbol{C}_i\hat{\boldsymbol{W}}\boldsymbol{\Phi}_i\right) \tag{25}$$

3 标记增强理论解释

如前所述，标记分布可看作是对多义性对象类别监督信息更为本质的表达，但由于标注代价高昂、量化困难等原因，标记分布在训练集中常常并未显式给出，而代之以更为简单的逻辑标记。此时，可以将标记分布视为一种隐变量，而逻辑标记是由标注者通过观测标记分布产生的观测变量。假设标记分布用 $\boldsymbol{d}$ 表示，逻辑标记用 $\boldsymbol{l}$ 表示，则观测变量 $\boldsymbol{l}$ 产生的条件概率为 $P(\boldsymbol{l}\mid\boldsymbol{d})$，后验概率 $P(\boldsymbol{d}\mid\boldsymbol{l})$ 却无法直接得到。因此，本文采用图 4 所示

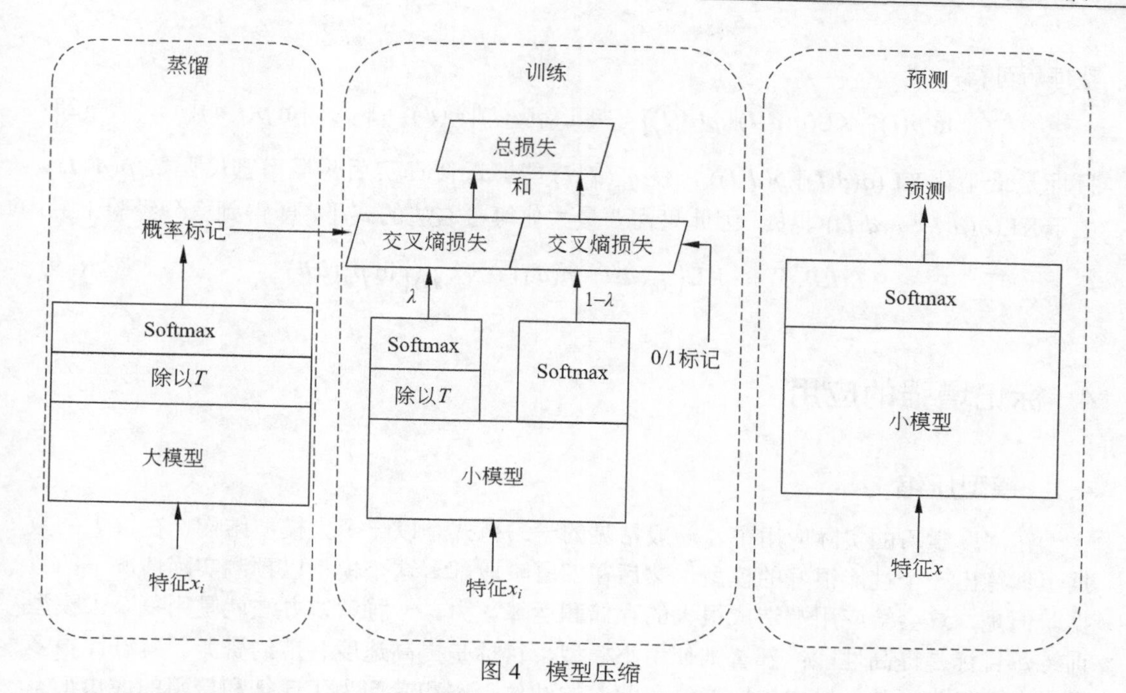

图4　模型压缩

的模型来描述标记分布的生成机制。假设$\boldsymbol{d}$的先验概率密度为$p(\boldsymbol{d})$，根据上述模型，观测到变量$\boldsymbol{l}$的概率密度为$p(\boldsymbol{d})P(\boldsymbol{l}\mid\boldsymbol{d})$。可以采用一个多层感知机对$p_{\theta}(\boldsymbol{l}\mid\boldsymbol{d})$建模，但作为模型输入的$\boldsymbol{d}$是潜在的隐变量，无法直接得到其后验概率密度$p(\boldsymbol{d}\mid\boldsymbol{l})$。因此，拟构造参数为$\phi$的另一个多层感知机$q_{\phi}(\boldsymbol{d}\mid\boldsymbol{l})$，用其近似$p(\boldsymbol{d}\mid\boldsymbol{l})$，进而依据该后验概率密度函数采样生成$\boldsymbol{d}$，然后作为$p_{\theta}(\boldsymbol{l}\mid\boldsymbol{d})$的输入。通过变分推断(variational inference)[32]可对上述模型进行优化，最终使$q_{\phi}(\boldsymbol{d}\mid\boldsymbol{l})$能够近似收敛到真实后验概率密度函数$p(\boldsymbol{d}\mid\boldsymbol{l})$，即该模型推断出了$\boldsymbol{d}$的后验概率密度函数，通过对该后验概率密度函数多次采样取平均得到$\boldsymbol{d}$，至此即对标记分布的产生机制给出理论解释。

具体地，为了推断出$\boldsymbol{d}$的后验概率密度函数，需要推导出变分经验下界(empirical lower bound) [33]。首先从$p(\boldsymbol{d}\mid\boldsymbol{l})$与$q(\boldsymbol{d}\mid\boldsymbol{l})$的KL散度（Kullback-Leibler divergence）定义出发：

$$\mathrm{KL}\left(q(\boldsymbol{d}\mid\boldsymbol{l})\| p(\boldsymbol{d}\mid\boldsymbol{l})\right)=\mathbb{E}_{q(\boldsymbol{d}\mid\boldsymbol{l})}[\ln q(\boldsymbol{d}\mid\boldsymbol{l})-\ln p(\boldsymbol{d}\mid\boldsymbol{l})] \tag{26}$$

利用贝叶斯公式将其展开：

$$\mathrm{KL}(q(\boldsymbol{d}\mid\boldsymbol{l})\| p(\boldsymbol{d}\mid\boldsymbol{l}))=\mathbb{E}_{q(\boldsymbol{d}\mid\boldsymbol{l})}[\ln q(\boldsymbol{d}\mid\boldsymbol{l})-\ln p(\boldsymbol{l}\mid\boldsymbol{d})-\ln p(\boldsymbol{d})]+\ln p(\boldsymbol{l}) \tag{27}$$

整理后可得：

$$\ln p(\boldsymbol{l})-\mathrm{KL}(q(\boldsymbol{d}|\boldsymbol{l})\|p(\boldsymbol{d}|\boldsymbol{l}))=-\mathrm{KL}(q(\boldsymbol{d}|\boldsymbol{l})\|p(\boldsymbol{d}))+\mathbb{E}_{q(\boldsymbol{d}|\boldsymbol{l})}[\ln p(\boldsymbol{l}|\boldsymbol{d})] \tag{28}$$

目标是最小化 $\mathrm{KL}(q(\boldsymbol{d}|\boldsymbol{l})\|p(\boldsymbol{d}|\boldsymbol{l}))$，使 $q_{\phi}(\boldsymbol{d}|\boldsymbol{l})$ 能够近似真实后验概率密度函数 $p(\boldsymbol{d}|\boldsymbol{l})$。由于 $\mathrm{KL}(q(\boldsymbol{d}|\boldsymbol{l})\|p(\boldsymbol{d}|\boldsymbol{l}))$ 非负，因此只需要最大化等号右边的式子，即得到变分经验下界：

$$L(\boldsymbol{l};\boldsymbol{\theta},\phi)=-\mathrm{KL}\left(q_{\phi}(\boldsymbol{d}|\boldsymbol{l})\|p(\boldsymbol{d})\right)+\mathbb{E}_{q_{\phi}(\boldsymbol{d}|\boldsymbol{l})}\left[\ln p_{\theta}(\boldsymbol{l}|\boldsymbol{d})\right] \tag{29}$$

4　标记增强的应用

4.1　模型压缩

在深度学习的实际应用中，一般常见的学习范式是以一个大模型结构，在很大的数据上训练出一个性能很好的模型，然后在部署阶段部署这个模型以预测实际情况下的数据。但是，这会给应用端造成很大的存储和运算压力。一种解决办法就是用复杂大模型训练，目标是提高性能；部署则使用小模型，目标是提高速度和节约资源。就好比很多昆虫有专门为了从环境中提取能量和营养的幼体形态和专门为了迁徙和繁殖的成虫形态一样。Hinton 等人[24]提出用大模型来学习抽取大规模、强冗余的训练集信息，然后利用新的训练方法（即前文提到的知识蒸馏）把大模型学到的知识转移到一个更适合部署的小模型上，大模型可称作教师模型，小模型可称作学生模型。

如图 5 所示，首先是一个训练教师模型，使它在训练集上性能良好（有一个好的教师模型是知识蒸馏的前提），然后用这个教师模型来训练学生模型。在训练时，对样本 x_i，教师模型的倒数第二层先除以一个温度 T，然后通过 Softmax 预测一个软标记。学生模型也一样，倒数第二层除以同样的温度 T，然后通过 Softmax 预测一个结果，再把这个

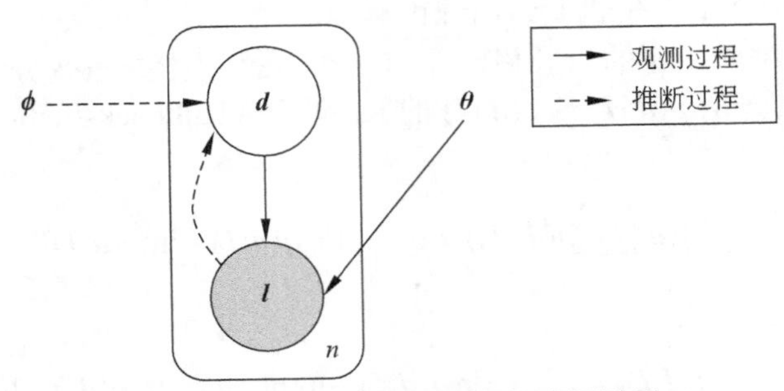

图 5　标记分布生成机制的概率图模型

结果和软标记的交叉熵作为训练的一部分损失函数：

$$L_{\text{soft}} = -\sum_{j=1}^{N} \boldsymbol{p}_j^T \ln\left(\boldsymbol{q}_j\right) \tag{30}$$

其中，$\boldsymbol{p}_i^T = \frac{\exp\left(\frac{\boldsymbol{v}_i}{T}\right)}{\sum_{k=1}^{N} \exp\left(\frac{\boldsymbol{v}_{ik}}{T}\right)}$；$q_i^T = \frac{\exp\left(\frac{\boldsymbol{v}_i}{T}\right)}{\sum_{k=1}^{N} \exp\left(\frac{\boldsymbol{z}_{ik}}{T}\right)}$；$v_i$ 是教师模型的 logits；z_i 是学生模型的 logits。

训练的损失函数还包括正常的输出和真值标记的交叉熵：

$$L_{\text{hard}} = -\sum_{j=1}^{N} \boldsymbol{c}_j \ln\left(\boldsymbol{q}_j\right) \tag{31}$$

其中，$\boldsymbol{q}_i = \frac{\exp\left(z_i\right)}{\sum_{k=1}^{N} \exp\left(z_k\right)}$。

最后把这两个损失函数加权起来作为训练学生模型的最后的损失函数：

$$L = \alpha L_{\text{soft}} + \beta L_{\text{hard}} \tag{32}$$

当学生模型训练完成、预测的时候，就不需要再有温度 T，直接使用常规的 Softmax 输出预测结果。

4.2 线性判别分析

处理高维数据一直是机器学习和模式识别领域一项具有挑战性的任务，线性判别分析[34]是维度压缩的一种常用方法。但是线性判别分析往往只能处理带有标记的数据，实际任务中的大多数数据是无标记标注，所以 Zhao[35]等人提出利用标记传播[25]的方法实现对未标注数据的标记增强，进而实现基于软标记的线性判别分析。

该方法用 $\boldsymbol{X} = \left[\boldsymbol{X}_l, \boldsymbol{X}_u\right] \in \mathbb{R}^{D\times(l+u)}$ 表示已标注和未标注的数据集，基于标记传播的标记增强首先定义相似矩阵判断任意两个样本之间的相似度，该方法采用基于样本重建的方法计算相似矩阵的权重，即

$$\begin{aligned} &\min \quad \| \boldsymbol{x}_i - \sum_{j\in N_k(x_i)} w_{ij}\boldsymbol{x}_j \|_F^2 \\ &\text{s.t.} \quad w_{ij} \geqslant 0, \sum_{j\in N_k(\boldsymbol{x}_i)} w_{ij} = 1 \end{aligned} \tag{33}$$

表示样本 $\boldsymbol{x}_i$ 可以由其邻近样本 $\boldsymbol{x}_j$ 重建得到。$\boldsymbol{Y}=\left[\boldsymbol{y}_1,\boldsymbol{y}_2,\cdots,\boldsymbol{y}_{l+u}\right]\in\mathbb{R}^{(C+1)(l+u)}$ 表示所有样本的原始标记，对于有标记的数据，如果样本 $\boldsymbol{x}_j$ 属于第 i 个类别，则 $y_{ij}=1$，否则等于 0。对于未标记数据，如果 x_j 属于第 c+1 个类别，则 $y_{ij}=1$，否则等于 0，该方法采用一个额外的类别 $c+1$ 用于检测异常值。用 $\boldsymbol{F}=\left[\boldsymbol{f}_1,\boldsymbol{f}_2,\cdots,\boldsymbol{f}_{l+u}\right]\in\mathbb{R}^{(C+1)(l+u)}$ 表示所有样本的标记，划分矩阵 $\hat{\boldsymbol{W}}$，$\boldsymbol{Y}$ 和 $\boldsymbol{F}$ 如下：

$$\hat{\boldsymbol{W}}=\begin{bmatrix}\hat{\boldsymbol{W}}^{ll}\hat{\boldsymbol{W}}^{lu}\\ \hat{\boldsymbol{W}}^{ul}\hat{\boldsymbol{W}}^{mu}\end{bmatrix},\boldsymbol{Y}=\left[\boldsymbol{Y}^l,\boldsymbol{Y}^u\right],\boldsymbol{F}=\left[\boldsymbol{F}^l,\boldsymbol{F}^u\right] \tag{34}$$

每一次标记传播的迭代希望保持已标注样本的标记不变，标记样本的标记可以从其邻近的样本标记中获取，即，

$$f_{ij}^u(t+1)=(1-\lambda)\sum_{k=1}^{l+u}f_{ik}(t)\hat{w}_{kj}+\lambda y_{ij}^u \tag{35}$$

用矩阵形式表示为如下形式：

$$F^u(t+1)=(1-\lambda)(F^l(t)\hat{\boldsymbol{W}}^{lu}+F^u(t)\hat{\boldsymbol{W}}^{uu})+\lambda\boldsymbol{Y}^u \tag{36}$$

根据式（36），令 $F^u(0)=Y^u$，则可以得到：

$$\begin{aligned}F^u(t+1)=&(1-\lambda)\boldsymbol{Y}^l\hat{\boldsymbol{W}}^{lu}\sum_{m=0}^{t}[(1-\lambda)\hat{\boldsymbol{W}}^{uu}]^m+\\ &\lambda\boldsymbol{Y}^u\sum_{m=0}^{t}[(1-\lambda)\hat{\boldsymbol{W}}^{uu}]^m+\boldsymbol{Y}^u[(1-\lambda)\hat{\boldsymbol{W}}^{uu}]^{t+1}\end{aligned} \tag{37}$$

根据矩阵的两个性质，即

$$\begin{gathered}\lim_{t\to\infty}[(1-\lambda)\hat{\boldsymbol{W}}^{uu}]^{t+1}=0\\ \lim_{t\to\infty}\sum_{t}^{m=0}[(1-\lambda)\hat{\boldsymbol{W}}^{uu}]^m=[\boldsymbol{I}-(1-\lambda)\hat{\boldsymbol{W}}^{uu}]^{-1}\end{gathered} \tag{38}$$

迭代过程可以收敛到如下形式：

$$\boldsymbol{F}^u=\lim_{t\to\infty}\boldsymbol{F}^u(t)=\left[(1-\lambda)\boldsymbol{Y}^l\hat{\boldsymbol{W}}^{lu}+\lambda\boldsymbol{Y}^u\right]\left[\boldsymbol{I}-(1-\lambda)\hat{\boldsymbol{W}}^{uu}\right]^{-1} \tag{39}$$

由于 $\boldsymbol{I}-(1-\lambda)\hat{\boldsymbol{W}}^{uu}$ 是满秩矩阵，所以 $\boldsymbol{F}^u$ 一定存在。通过以上标记传播标记增强方法，可以对未标注的样本进行标记增强，然后利用线性判别分析实现维度压缩。

线性判别分析的目标是最大化类间矩阵和最小化类内矩阵，即

$$J(V)=\max_{V}\operatorname{tr}\left[\left(\boldsymbol{V}^{T}\boldsymbol{S}_{w}\boldsymbol{V}\right)^{-1}\boldsymbol{V}\boldsymbol{S}_{b}\boldsymbol{V}\right] \tag{40}$$

通过标记传播增强得到的标记是软标记 $f_{ij}^{u}\left(0\leqslant f_{ij}^{u}\leqslant 1\right)$，标记增强之后得到的软标记包含重要的概率判别信息，分为三类：① f_{ij}^{u} 接近 1，表示样本 $\boldsymbol{x}_j$ 是类别 i 中新添加的样本；② f_{ij}^{u} 接近 1，并且 $i=c+1$，表示样本 $\boldsymbol{x}_j$ 是异常点；③ 对于其他情况 $f_{ij}^{u}\left(0\leqslant f_{ij}^{u}\leqslant 1\right)$，表示样本 x_j 具有概率判别信息。基于软标记的 total-class，within-class 和 between-class 矩阵定义为

$$\begin{cases}\boldsymbol{S}_{t}=\sum\limits_{i=1}^{c}\sum\limits_{j=1}^{l+u}f_{ij}\left(\boldsymbol{x}_{j}-\tilde{\boldsymbol{\mu}}\right)\left(\boldsymbol{x}_{j}-\tilde{\boldsymbol{\mu}}\right)^{\mathrm{T}}=\boldsymbol{X}\left(\boldsymbol{E}-\boldsymbol{E}\boldsymbol{e}\boldsymbol{e}^{\mathrm{T}}\boldsymbol{E}/\boldsymbol{e}\boldsymbol{E}\boldsymbol{e}^{\mathrm{T}}\right)\boldsymbol{X}^{\mathrm{T}}\\ \boldsymbol{S}_{w}=\sum\limits_{i=1}^{c}\sum\limits_{j=1}^{l+u}f_{ij}\left(\boldsymbol{x}_{j}-\boldsymbol{\mu}_{i}\right)\left(\boldsymbol{x}-\boldsymbol{\mu}_{i}\right)^{\mathrm{T}}=\boldsymbol{X}\left(\boldsymbol{E}-\boldsymbol{F}_{c}^{\mathrm{T}}\boldsymbol{G}^{-1}\boldsymbol{F}_{c}\right)\boldsymbol{X}^{\mathrm{T}}\\ \boldsymbol{S}_{b}=\sum\limits_{i=1}^{c}\sum\limits_{j=1}^{l+u}f_{ij}\left(\boldsymbol{\mu}_{i}-\tilde{\boldsymbol{\mu}}\right)\left(\boldsymbol{\mu}_{i}-\tilde{\boldsymbol{\mu}}\right)^{\mathrm{T}}=\boldsymbol{X}\left(\boldsymbol{F}_{c}^{\mathrm{T}}\boldsymbol{G}^{-1}\boldsymbol{F}_{c}-\boldsymbol{E}\boldsymbol{e}\boldsymbol{e}^{\mathrm{T}}\boldsymbol{E}/\boldsymbol{e}\boldsymbol{E}\boldsymbol{e}^{\mathrm{T}}\right)\boldsymbol{X}^{\mathrm{T}}\end{cases} \tag{41}$$

其中，

$$\begin{cases}\boldsymbol{\mu}_{i}=\sum\limits_{j=1}^{l+u}f_{ij}\boldsymbol{x}_{j}\Big/\sum\limits_{j=1}^{l+u}f_{ij}\\ \tilde{\boldsymbol{\mu}}=\sum\limits_{i=1}^{c}\sum\limits_{j=1}^{l+u}f_{ij}\boldsymbol{x}_{j}\Big/\sum\limits_{i=1}^{c}\sum\limits_{j=1}^{l+u}f_{ij}\end{cases} \tag{42}$$

则基于软标记的线性判别分析目标函数为

$$J(\boldsymbol{V})=\max_{\boldsymbol{V}}\operatorname{tr}\left[\left(\boldsymbol{V}^{T}\widetilde{\boldsymbol{S}}_{w}\boldsymbol{V}\right)^{-1}\boldsymbol{V}\widetilde{\boldsymbol{S}}_{b}\boldsymbol{V}\right] \tag{43}$$

最大化 between-class 矩阵、最小化 within-class 矩阵以实现目标函数的优化。

4.3 神经网络正则化

尽管使用了大量数据集，大型神经网络仍然经常出现过拟合现象，具体表现为：模型在给目标分类时，给某个类过大的信任导致分类出错。针对上述问题，Pereyra 等人[36]提出使用基于最大熵的惩罚机制和标记平滑对输出分布进行正则化。

$p_{\theta}(y\mid x)$ 为神经网络对于给定输入值 x 产生的条件分布，y 为相应的标记，该条件

概率下的熵为

$$H(p_\theta(y|x)) = -\sum_i p_\theta(y_i|x)\ln(p_\theta(y_i|x)) \tag{44}$$

假设 z_i 为第 i 个 logit，熵值对于 z_i 的梯度为

$$\frac{\partial H(p_\theta)}{\partial z_i} = p_\theta(y_i|x)[-\ln p_\theta(y_i|x) - H(p_\theta)] \tag{45}$$

为了惩罚输出分布，将熵值加到似然函数中，得到：

$$L(\theta) = -\sum \ln p_\theta(y|x) - \beta H(p_\theta(y|x)) \tag{46}$$

其中，β 为惩罚力度参数。

在强化学习中，惩罚低熵分布会阻止网络提前收敛并鼓励模型继续训练，然而在有监督学习中，我们希望模型在训练开始时可以快速拟合并且避免在训练后期出现过拟合现象，这需要在训练初期给予较小的惩罚力度，在模型逼近收敛时增大惩罚参数。在上述公式中添加阈值可以使初期模型不受惩罚机制的影响：

$$L(\theta) = -\sum \ln p_\theta(y|x) - \beta \max(0, \Gamma - H(p_\theta(y|x)) \tag{47}$$

Γ 即为阈值，使用加入阈值的损失函数可以在训练过程中只惩罚低于某个阈值的输出分布，提升了训练初期模型的拟合速度。

本文进一步使用了标记平滑，假设预设标记是均匀分布的，那么标记平滑可以简化为在均匀分布 u 和网络预测分布 p_θ 之间的 KL 散度，即

$$L(\theta) = -\sum \ln p_\theta(y|x) - D_{KL}(u \| p_\theta(y|x)) \tag{48}$$

使用翻转后的相对熵 $D_{KL}(p_\theta(y|x) \| u)$ 即可恢复惩罚制度。

综上所述，使用惩罚机制和标记平滑进行神经网络正则化在强化学习中可以提高模型的探索性，在有监督学习中也可以起到很强的正则化作用，在图像分类、语言建模、机器翻译等领域已广泛试用，可以在不改变现有参数的前提下提升模型的性能。

4.4 标记嵌入

现有的神经网络大多使用 one-hot 方法表示标记，但使用这种表示方法后，每个标记与其他标记都处于不同的维度分布，且每个维度上的值均为 0 或 1，不存在中间值。这些限制一方面使测量标记之间的相关性变得困难，可能导致数据稀疏问题；另一方面，0/1 编码使得在训练过程中无法对参数进行进一步修改，容易引起过拟合问题。Sun 等

人[37]提出标记嵌入网络的方法，将 one-hot 损失函数转换为 softmax 分布函数，使原来不相关的标记在训练过程中具有连续的交互作用。基于标记嵌入网络，通过反向传播自适应地学习标记嵌入，可以在深层网络的训练过程中自动学习标记的表示。

神经网络通常由几个隐藏层和一个输出层组成，隐藏层负责将输入映射到隐藏表示。假设神经网络中最后一层的隐藏表示为 $\boldsymbol{h}$，那么

$$\boldsymbol{h} = f(\boldsymbol{x}) \tag{49}$$

其中，x 为神经网络的输入；f 为输入到隐藏表示的映射。输出层将隐藏表示映射到输出，通常使用线性变化，即

$$z_1 = o_1(\boldsymbol{h}),\quad z_2 = o_2(\boldsymbol{h}) \tag{50}$$

其中，o 表示线性映射；z_1 表示用来预测标记的输出层；z_2 为负责学习隐藏表示相似性的输出层。两个输出层共享相同的隐藏表示，但各自具有独立的参数。之后使用 softmax 函数输出层进行归一化，得到标记的概率分布：

$$z_1' = \text{softmax}(z_1),\quad z_2' = \text{softmax}(z_2) \tag{51}$$

其中，m 为标记的个数。损失函数表示为

$$\text{Loss}(z_1', \boldsymbol{y}) = -\sum_i y_i \ln(z_1')_i,\quad i = 1,2,\cdots,m \tag{52}$$

$$\text{Loss}(z_2', \boldsymbol{y}) = -\sum_i y_i \ln(z_2')_i,\quad i = 1,2,\cdots,m \tag{53}$$

在反向传播过程中，从 z_2 开始的梯度不会传播到 h，因此 o_2 的学习不会影响隐藏表示，标记嵌入获得了更稳定的学习目标。

在标记个数为 m 时，嵌入表示为

$$\boldsymbol{E} \in \mathbb{R}^{m\times m} \tag{54}$$

真实标记 y 的嵌入向量为

$$\boldsymbol{e} = \boldsymbol{E}_y \tag{55}$$

普通方法在训练中使用 $z_2' = \text{softmax}(z_2)$ 计算交叉熵损失以学习标记嵌入，但是 z_2' 太接近 one-shot 分布，不能捕获标记之间的相似性，所以使用带温度 τ 的 softmax 函数对输出进行归一化：

$$(z'')_i = \frac{\exp((z_2)_i / \tau)}{\sum_{j=1}^{m}\exp((z_2)_j / \tau)},\quad i = 1,2,\cdots,m \tag{56}$$

损失函数为

$$\mathrm{Loss}(\boldsymbol{e}', \boldsymbol{z}_2'') = -\sum_i (z_2'')_i \ln e_i', \quad i = 1, 2, \cdots, m \tag{57}$$

其中，$e' = \mathrm{softmax}(e)$。通过调整温度参数τ，标记嵌入获得了输出分布的更多细节，但是这也使错误标记之间的差异更接近，为了解决上述问题，对归一化输出进行正则化，使分布的最大值不至于太高，同时保持标记之间的差异：

$$\mathrm{Loss}(\boldsymbol{z}_2') = \| \max(0, (z_2')_y - \alpha \|_p \tag{58}$$

在训练中通过交叉熵损失函数使嵌入逼近输出：

$$\mathrm{Loss}(\boldsymbol{z}_1', \boldsymbol{e}') = -\sum_i e_i' \ln(z_1')_i, \quad i = 1, 2, \cdots, m \tag{59}$$

综上所述，该方法的最终目标为

$$\begin{aligned}\mathrm{Loss}(x, y; \boldsymbol{\theta}) = &\ \mathrm{Loss}(\boldsymbol{z}_1', y) + \mathrm{Loss}(\boldsymbol{z}_1', \boldsymbol{e}') + \mathrm{Loss}(\boldsymbol{z}_2', y) + \\ &\| \max(0, (z_2')_y - \alpha \|_p + \mathrm{Loss}(\boldsymbol{e}', \boldsymbol{z}_2'')\end{aligned} \tag{60}$$

总体架构如图 6 所示，图中圆代表矢量，正方形代表带参数的图层，虚线表示交叉熵运算，正方形代表神经网络。使用标记嵌入方法学习了真实标记的细粒度分布，收敛速度更快，同时降低了过拟合的风险。

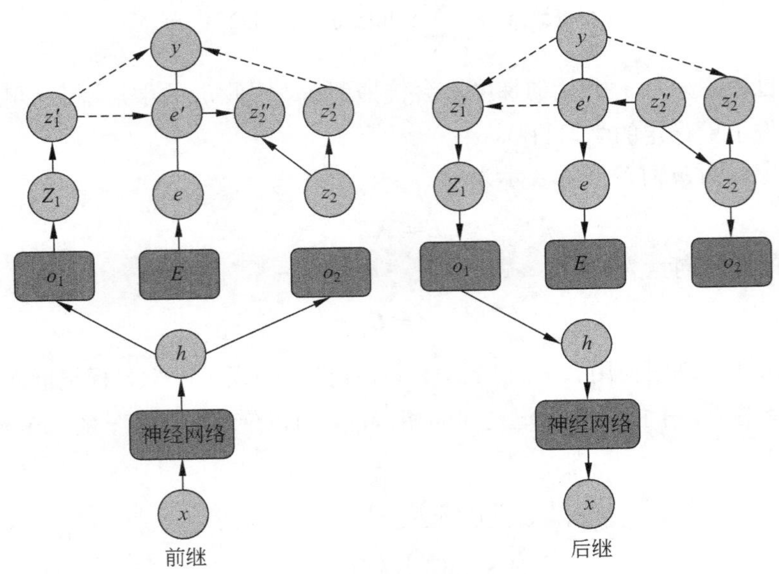

图 6　标记嵌入

4.5 标记分布学习

标记分布学习[38]是一种更加泛化的机器学习范式，然而标记分布的获取比较困难。标记增强可以根据训练集$\boldsymbol{S}=\{(\boldsymbol{x}_i,\boldsymbol{L}_i)\,|\,1\leqslant i\leqslant n\}$中蕴含的标记间相关性，将每个示例$\boldsymbol{x}_i$的逻辑标记$\boldsymbol{L}_i$转化为相应的标记分布$\boldsymbol{D}_i$，从而得到标记分布训练集$\mathcal{E}=\{(\boldsymbol{x}_i,\boldsymbol{D}_i)\,|\,1\leqslant i\leqslant n\}$，所以标记增强无疑给标记分布学习提供了良好的数据预处理支持，对于标记分布学习拓展具有很强的促进作用。

假设$p(y\,|\,\boldsymbol{x})$的参数模型表示为$p(y\,|\,\boldsymbol{x};\boldsymbol{\theta})$，其中$\boldsymbol{\theta}$是参数向量。给定训练集$S$，标记分布学习的目标是找到一个$\boldsymbol{\theta}$，使得对于给定示例$\boldsymbol{x}_i$，$p(y\,|\,\boldsymbol{x};\boldsymbol{\theta})$能产生与$\boldsymbol{x}_i$的真实标记分布$\boldsymbol{D}_i$尽可能相似的标记分布。如果使用 Kullback-Leibler 散度来度量两个分布之间距离，那么最佳的参数$\boldsymbol{\theta}$为

$$\boldsymbol{\theta}^*=\underset{\boldsymbol{\theta}}{\operatorname{argmin}}\sum_i\sum_j\left(d_{\boldsymbol{x}_i}^{y_j}\ln\frac{d_{\boldsymbol{x}_i}^{y_j}}{p\left(y_j|\boldsymbol{x}_i;\boldsymbol{\theta}\right)}\right)=\underset{\boldsymbol{\theta}}{\operatorname{argmax}}\sum_i\sum_j d_{\boldsymbol{x}_i}^{y_j}\ln p\left(y_j|\boldsymbol{x}_i;\boldsymbol{\theta}\right) \tag{61}$$

有了式（61）中的优化目标，首先可以回望一下传统的单标记和多标记学习这两个特例在这一优化目标下会得到什么结果。对于单标记学习，$d_x^y=\mathrm{Kr}(y_j,y(\boldsymbol{x}_i))$，这里$\mathrm{Kr}(\cdot,\cdot)$是 Kronecker delta 函数，$y(\boldsymbol{x}_i)$是$\boldsymbol{x}_i$的单标记。这时，式（61）可以简化为

$$\boldsymbol{\theta}^*=\underset{\boldsymbol{\theta}}{\operatorname{argmax}}\sum_i\ln p(y(\boldsymbol{x}_i)\,|\,\boldsymbol{x}_i;\boldsymbol{\theta}) \tag{62}$$

这实际上是$\boldsymbol{\theta}$的极大似然估计（maximum likelihood，ML），而后面使用$p(y\,|\,\boldsymbol{x}_i;\boldsymbol{\theta})$进行分类等价于最大后验决策（maximum a posteriori，MAP）。对于多标记学习，每个示例$\boldsymbol{x}_i$使用一个标记集合Y_i来标注，因此，$d_{\boldsymbol{x}_i}^{y_j}=\begin{cases}\dfrac{1}{|Y_\mathrm{i}|}, & y_j\in Y_i\\ 0, & y_j\notin Y_i\end{cases}$。此时，KL 散度变化为

$$\boldsymbol{\theta}^*=\underset{\boldsymbol{\theta}}{\operatorname{argmax}}\sum_i\frac{1}{|Y_i|}\sum_{y\in Y_i}\ln p(y\,|\,\boldsymbol{x}_i;\boldsymbol{\theta}) \tag{63}$$

式（63）可以看作是使用示例的相关标记集合的势的倒数进行加权的极大似然估计。实际上，这等价于首先采用一种基于熵的标记分配方法（entropy-based label assignment，ELA）[1]将多标记数据转化为加权的单标记数据，然后再用极大似然估计来估计参数$\boldsymbol{\theta}$。

通过以上分析可以看出，在适当的约束条件下，标记分布学习模型可以转化为常见的单标记或多标记学习方法。因此，标记分布学习可以看作一个更加通用的学习框架，包

含了作为特例的单标记和多标记学习。标记分布学习已经在美学感知[39]、图像情感[40-41]、电影评分预测[42]等应用中取得了良好的效果。

5 结束语

标记分布通过连续的描述度来显式表达每个标记与数据对象的关联强度，构成了连续的标记空间，使由机器学习连接的示例空间和标记空间对等。因此，可以利用原本在示例空间才可以使用的工具和技术为处理标记多义性、标记不确定性、标记相关性等提供更多的可能。由于描述度的标注成本更高且常常没有客观的量化标准，现实任务中大量的多义性数据仍然是以简单逻辑标记标注的，而标记增强可以在不增加额外数据标注负担的前提下，挖掘训练样本中蕴含的标记重要性差异信息，将离散的逻辑标记空间转换为连续的标记分布空间，释放标记空间的威力。因此，对标记增强的研究对于探索标记信息的本质具有重要意义，有望为传统机器学习研究中的焦点问题提供新的解决思路。

参考文献

[1] TSOUMAKAS G, KATAKIS I. Multi-label classification: An overview[J]. International Journal of Data Warehousing and Mining, 2007, 3(3): 1-13.

[2] CABRAL R S, TORRE F, COSTEIRA J P, et al. Matrix completion for multi-label image classification[C]//Proceedings of the 25th Annual Conference on Neural Information Processing Systems. Granada, Spain, 2011: 190-198.

[3] RUBIN T N, CHAMBERS A, SMYTH P, et al. Statistical topic models for multi-label document classification[J]. Machine Learning, 2012, 88(1-2): 157-208.

[4] WANG J, ZHAO Y, WU X, et al. A transductive multi-label learning approach for video concept detection[J]. Pattern Recognition, 2011, 44(10-11): 2274-2286.

[5] GAO B B, ZHOU H Y, WU J, et al. Age Estimation Using Expectation of Label Distribution Learning[C]//Proceedings of the 27th International Joint Conference onArtificial Intelligence. Stockholm, Sweden, 2018: 712-718.

[6] XU N, TAO A, GENG X. Label enhancement for label distribution learning[C]//Proceedings of the 27th International Joint Conference on Artificial Intelligence. Stockholm, Sweden, 2018: 2926-2932.

[7] XU N, LIU Y P, GENG X. Label enhancement for label distribution learning[J]. IEEE Transactions on Knowledge and Data Engineering, 2021, 33(4): 1632-1643.

[8] SU K, GENG X. Soft facial landmark detection by label distribution learning[C]//Proceedings of the 33rd AAAI Conference on Artificial Intelligence. Honolulu, Hawaii, 2019: 5008-5015.

[9] SHIRANI A, DERNONCOURT F, ASENTE P, et al. Learning emphasis selection for written text in visual media from crowd-sourced label distributions[C]//Proceedings of the 57th Annual Meeting of the Association for Computational Linguistics. Florence, Italy, 2019: 1167-1172.

[10] ZHAO Z, MA X. Text emotion distribution learning from small sample: A meta-learning approach[C]//Proceedings of the 2019 Conference on Empirical Methods in Natural Language. Hong Kong, China, 2019: 3948-3958.

[11] CHUNG J J Y, SONG J Y, KUTTY S, et al. Efficient elicitation approaches to estimate collective crowd answers[J]. Proceedings of the ACM on Human-Computer Interaction, 2019, 3(CSCW): 1-25.

[12] YU Z, YU J, XIANG C, et al. Beyond bilinear: Generalized multimodal factorized high-order pooling for visual question answering[J]. IEEE Transactions on Neural Networks and Learning Systems, 2018, 29(12): 5947-5959.

[13] ZHOU Y, XUE H, GENG X. Emotion distribution recognition from facial expressions[C]//Proceedings of the 23rd ACM International Conference on Multimedia. Brisbane, Australia, 2015: 1247-1250.

[14] ZHOU D, ZHANG X, ZHOU Y, et al. Emotion distribution learning from texts[C]// Proceedings of the 2016 Conference on Empirical Methods in Natural Language Processing. Austin, Texas, 2016: 638-647.

[15] LIU C, WANG W, LI Z, et al. Biological age estimated from retinal imaging: A novel biomarker of aging[C]//Proceedings of the 22nd International Conference on Medical Image Computing and Computer-Assisted Intervention. Shenzhen, China, 2019: 138-146.

[16] WU X, WEN N, LIANG J, et al. Joint acne image grading and counting via label distribution learning[C]//Proceedings of the 17th IEEE International Conference on Computer Vision. Seoul, Korea (South), 2019: 10642-10651.

[17] EL GAYAR N, SCHWENKER F, PALM G. A study of the robustness of KNN classifiers trained using soft labels[C]//Proceedings of 2nd International Association for Pattern Recognition Workshop on Artificial Neural Networks in Pattern Recognition. Ulm, Germany, 2006: 67-80.

[18] MELIN P, CASTILLO O. Hybrid intelligent systems for pattern recognition using soft computing: an evolutionary approach for neural networks and fuzzy systems[M]. Berlin: Springer Science & Business Media, 2005.

[19] KLIR G, YUAN B. Fuzzy sets and fuzzy logic[M]. New Jersey: Prentice hall, 1995.

[20] JIANG X, YI Z, LV J C. Fuzzy SVM with a new fuzzy membership function[J]. Neural Computing & Applications, 2006, 15(3): 268-276.

[21] SMOLA A J. Learning with kernels[D]. Technischen Universitat Berlin, 1998.

[22] YI K, WU J. Probabilistic end-to-end noise correction for learning with noisy labels[C]//Proceedings of the 32nd IEEE Conference on Computer Vision and Pattern Recognition. Long Beach, California, 2019: 7017-7025.

[23] SZEGEDY C, VANHOUCKE V, IOFFE S, et al. Rethinking the inception architecture for computer vision[C]//Proceedings of the 29th IEEE Conference on Computer Vision and Pattern Recognition.

Las Vegas, Nevada, 2016: 2818-2826.

[24] HINTON G, VINYALS O, DEAN J. Distilling the knowledge in a neural network[J/OL]. https://arxiv.org/abs/1503.02531.

[25] LI Y K, ZHANG M L, GENG X. Leveraging implicit relative labeling-importance information for effective multi-label learning[C]//Proceedings of 15th IEEE International Conference on Data Mining. Atlantic City, New Jersey, 2015: 251-260.

[26] ZHU X, GOLDBERG A B. Introduction to semi-supervised learning[J]. Synthesis Lectures on Artificial Intelligence and Machine Learning, 2009, 3(1): 1-130.

[27] HOU P, GENG X, ZHANG M L. Multi-label manifold learning[C]//Proceedings of the 30th AAAI Conference on Artificial Intelligence. Phoenix, Arizona,2016: 1680-1686.

[28] ZHU X, LAFFERTY J, ROSENFELD R. Semi-supervised learning with graphs[D]. Carnegie Mellon University, 2005.

[29] TSOUMAKAS G, DIMOU A, SPYROMITROS E, et al. Correlation-based pruning of stacked binary relevance models for multi-label learning[C]//Proceedings of the 1st International Workshop on Learning from Multi-Label Data. Bled, Slovenia, 2009: 101-116.

[30] HUANG S J, ZHOU Z H. Multi-label learning by exploiting label correlations locally[C]//Proceedings of the 26th AAAI Conference on Artificial Intelligence. Toronto, Ontario, 2012: 949-955.

[31] ZHOU Z H, ZHANG M L, HUANG S J, et al. Multi-instance multi-label learning[J]. Artificial Intelligence, 2012, 176(1): 2291-2320.

[32] KINGMA D P, WELLING M. Auto-encoding variational bayes[C]//Proceedings of the 2nd International Conference on Learning Representations. Banff, Canada, 2014.

[33] REZENDE D J, MOHAMED S, WIERSTRA D. Stochastic backpropagation and approximate inference in deep generative models[C]//Proceedings of the 31st International Conference on Machine Learning. Beijing, China, 2014: 1278-1286.

[34] DEVROYE L. Random variate generation in one line of code[C]//Proceedings of the 28th Conference on Winter Simulation. Coronado, California, 1996: 265-272.

[35] ZHAO M, ZHANG Z, ZHANG H. A soft label based linear discriminant analysis for semi-supervised dimensionality reduction[C]//Proceedings of the 30th International Joint Conference on Neural Networks. Dallas,Texas, 2013: 1-8.

[36] PEREYRA G, TUCKER G, CHOROWSKI J, et al. Regularizing neural networks by penalizing confident output distributions[C]//Proceedings of the 5th International Conference on Learning Representations. Toulon, France, 2017.

[37] SUN X, WEI B, REN X, et al. Label embedding network: Learning label representation for soft training of deep networks[J/OL]. https://arxiv.org/abs/1710.10393.

[38] GENG X. Label distribution learning[J]. IEEE Transactions on Knowledge and Data Engineering, 2016, 28(7): 1734-1748.

[39] REN Y, GENG X. Sense beauty by label distribution learning[C]//Proceedings of the 26th International Joint Conference on Artificial Intelligence. Melbourne, Australia,2017: 2648-2654.

[40] YANG J, SHE D, SUN M. Joint Image Emotion Classification and Distribution Learning via Deep Convolutional Neural Network[C]//Proceedings of the 26th International Joint Conference onArtificial Intelligence. Melbourne, Australia, 2017: 3266-3272.

[41] YANG J, SUN M, SUN X. Learning visual sentiment distributions via augmented conditional probability neural network[C]//Proceedings of the 31st AAAI Conference on Artificial Intelligence. San Francisco, California,2017: 224-230.

[42] GENG X, HOU P. Pre-release Prediction of Crowd Opinion on Movies by Label Distribution Learning[C]//Proceedings of the 24th International Joint Conference onArtificial Intelligence. Buenos Aires, Argentina, 2015: 3511-3517.

因果推断与因果性学习

陈　薇[*]　蔡瑞初[*]　郝志峰[1,2]　张　坤[3]

[1]广东工业大学计算机学院，广州 中国 510006;
[2]佛山科学技术学院数学与大数据学院，佛山 中国 52800;
[3]卡内基梅隆大学哲学系，匹兹堡 美国 15213

1　引言

因果关系一直是人类认识世界的基本方式和现代科学的基石。通俗地讲，科学研究中的因果关系与我们生活中所说的因果关系是一回事。比如我们平时所说的天下雨导致地上湿就是一个典型的因果关系的例子。因果关系抽象出来的定义是这样的——如果保持系统中其他变量都不变，只改变其中一个变量（比如通过人工降雨来让天下雨），然后发现有另一个变量也随之改变（比如我们发现地上由干变湿了），那么我们就说前面一个变量（是否下雨）是后一个变量（地上是否湿）的一个原因。一个典型的因果模型可以通过直观的图模型来描述，或者通过从因到果的数学函数（结构方程模型）来刻画。

与相关关系相比，因果关系严格区分了“原因”变量和“结果”变量，在揭示事物发生机制、指导干预行为等方面有相关关系不能替代的重要作用。以图 1 为例，吸烟、黄牙都与肺癌具有较强的相关关系，然而吸烟才是肺癌的原因，相应地，戒烟才能降低肺癌的发病概率，而把牙齿洗白不能降低肺癌的发病概率。

经典的例子还有辛普森悖论。在治疗肾结石的场景中，单独从结石的大小角度看，治疗方案 A 比较有效；但是从整体数据来看，治疗方案 B 比较有效。这是因为其背后的因果机制是存在一个隐含的混淆因子——石头的大小，影响治疗方案 A 或 B 的选择及康复效果。

---

[*] 同等贡献。

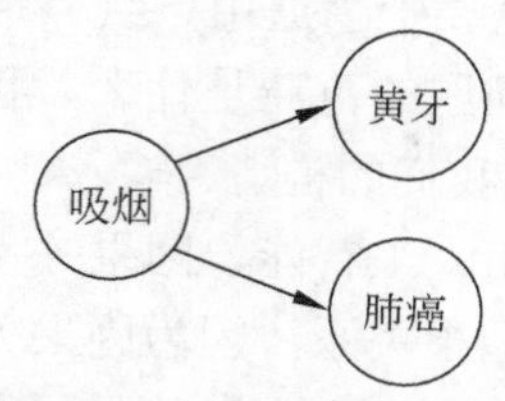

图 1 吸烟与黄牙、肺癌之间的因果关系发现

传统机器学习中对训练数据存在过度依赖，逐渐突显出各种弊端。例如，在谷歌图片的照片分类问题中，谷歌错误地把黑人分类为黑猩猩；在对抗学习的过程中，将一张加了噪声的熊猫图片识别为长臂猿。造成这类错误的主要原因是传统的机器学习方法忽视了事物背后的因果机制。近年来，研究者越来越关注更实际的机器学习或者人工智能问题，比如如何在变化的环境中做迁移学习，在这类问题中研究者需要找到并且利用数据产生过程的性质；而因果关系对数据背后的过程提供了一个很自然的描述。

人类为何自发地习惯于因果的思维方式？我们有时需要有的放矢地改变这个世界，这当然需要因果的认识；更多情形下，因果的思维方式帮助我们把眼前复杂的世界分解成各个相互没有联系的模块，从而简单地分而治之，否则，沉醉于万物皆有联系、牵一发而动全局的想法，就很难主动地去做好预测或干预。在机器学习领域亦是如此，只要机器学习的问题脱离了传统的数据独立同分布假设，因果的思维方式就可以很自然地帮助我们理解、应用数据背后的规律来解决问题。这可以改善传统机器学习对训练数据过度依赖的问题。这类问题包括迁移学习、半监督学习、有标签噪声的学习、强化学习，以及如何处理对抗攻击。在这类问题中，不能以简单的方式完全依赖训练数据，而因果的思维方式可以让我们理解数据背后的产生规律并为之建模，从而针对待分析数据或待解决的特定问题，得到更有适应性或者更恰当的模型。

我们提出的因果性学习体现了因果推断对于机器学习算法设计的指导作用。随着人工智能的发展，越来越多学者开始认识到因果推断对于克服现有人工智能在抽象、推理、可解释性等方面的不足具有重要意义。正如图灵奖得主 Judea Pearl 在新作《The Book of Why》一书中提出的“因果关系之梯”，他把因果推断分成三个层面：第一层是“关联”，第二层是“干预”，第三层是“反事实推理”。他特别指出，当前的机器学习只处于第一层，只是“弱人工智能”，要实现“强人工智能”还需要干预和反事实推理。目前实现通用人工智能的困难至少有两点。第一点是如何让智能系统学习出整个环境的性质、它的行为如何与环境交互，并且有能力恰当使用这些学习出来的规律。环境包括所有需要处理的问题及所有可能见到的场景，而学习出来的规律需要把这些问题、场景和行为内部及之间的联系表述出来。第二点是需要把“通用”给界定出来——我们希望人工智能系统处于什么环境？处理哪些问题？问题的范畴是什么？我们对身处的世界及其中要解决的问题习以为常，但人工智能系统跟我们的构造、输入、产生方式都不一样，我们需要为智能系统提供一个环境。为实现通用人工智能，因果关系至少有两方面的用途：第一，上面提到的学习出来的规律往往是因果关系，因为因果关系一般为环境提供了一个紧凑的描述；第二，智能系统要处理的很多问题本身就是因果问题，比如要做什么才能把一瓶水打开，要解决这些问题就必须认识到事物之间的因果联系。同时，Pearl 假设我们在意的因果变量已经被定义出来了，但现实中，如何发现这些有意义的因果变量也是一个大问题。

因此，当前因果关系的研究问题主要有两个方向。第一个方向是因果推断，即如何从观察数据中找到背后的因果关系，这就需要发展数据分析或者机器学习的方法来解决。第二个研究方向就是因果性学习，即如何利用因果的思维方式解决更复杂更实用的机器学习问题。接下来将从经典方法、隐变量场景和非稳态/异质数据场景三个方面介绍现有的因果推断方法和因果性学习方法。

2 经典因果推断方法

从观察数据中进行因果推断的经典方法有基于约束的方法(constraint-based methods)和基于函数因果模型的方法(functional causal model-based methods)。

2.1 基于约束的方法

基于约束的方法是通过数据中变量间的条件独立性来判断特定结构的存在性。这种测试通常用统计或信息论的度量来实现。因此，基于约束的方法也被称为基于条件独立性的方法。经典的基于条件独立约束的算法包括 PC(Peter-Clark)算法和快速因果推断(fast causal inference, FCI)算法[1]。

PC 算法假定没有混淆因子（confounder，即两个观察变量隐的直接的共同原因），且其发现的因果信息是渐近正确的。该算法主要分为无向图学习和方向学习两个阶段。在无向图学习阶段，从完全连通图出发，基于（条件）独立性假设检验等统计方法给出的变量之间的独立性而砍掉相应的边，从而获得变量间的无向图；在方向学习阶段，则依赖于 V 结构(V-Structure)（定义如图 2（b）所示）定向规则确定部分边的方向。PC 算法应用于观察数据（如图 3（a）所示）的例子如图 3（c）所示。

将结合基于核的条件独立性检验(Kernel-based conditional independence test, KCI)[2]

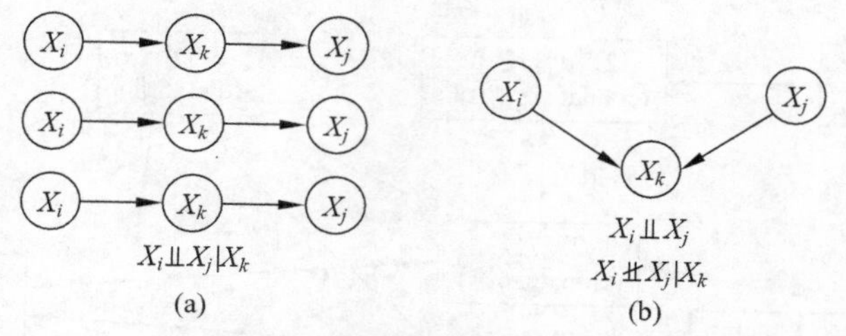

图 2　三元变量之间的因果关系

（a）马尔可夫等价类（三元变量之间的关系均满足在变量 X_k 的条件下，变量 X_i 和变量 X_j 互相独立）；（b）V 结构（三元变量之间的关系满足变量 X_i 和变量 X_j 互相独立，但在变量 X_k 的条件下，变量 X_i 和变量 X_j 不独立）

X_1	X_2	X_3	X_4	X_5
−1.1	1	1.3	0.2	−0.7
2.1	2	3.1	−1.3	−1.6
3.1	4.2	−2.6	0.6	2.1
2.3	−0.6	−3.5	0.8	2.3
1.3	−1.7	0.9	2.4	−1.4
−1.8	0.9	−1.3	0.9	0.7
…	…	…	…	…

(a)

$X_1 \perp\!\!\!\perp X_5 | X_3$

$X_2 \perp\!\!\!\perp X_4 | X_1$

$X_2 \perp\!\!\!\perp X_5 | X_3$

$X_4 \perp\!\!\!\perp X_5 | X_3$

$X_1 \perp\!\!\!\perp X_3 | \{X_2, X_4\}$

(b)

(c)

图 3　PC 算法应用于观察数据的例子

（a）数据；（b）条件独立性；（c）PC 算法流程

的 PC 算法应用到考古数据上，结果如图 4 所示，准确识别了因果骨架和因果方向。通过数据发现出来的因果关系与领域知识一致，但可能更深一层——比如，我们发现饮食(diet)对颅骨形态分化(cranial shape differentiation)有因果影响，但是这个影响不是直接的因果效应，而是通过牙齿磨损程度(level of attribution in teeth)及咀嚼行为(paramasticatory behavior)发生的。

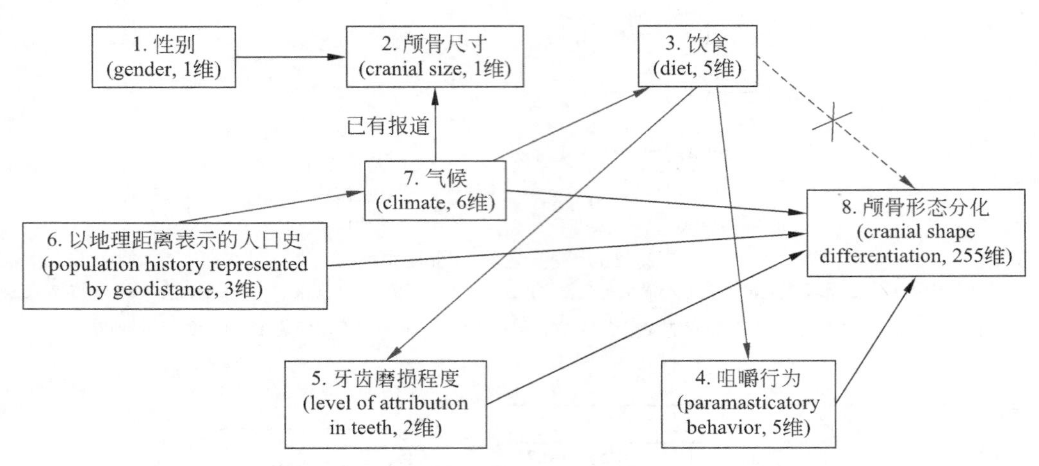

图 4 PC 算法在考古数据上学习到的因果网络图

FCI 算法是基于 PC 算法提出的解决有隐变量情况的方法，它可以从观察数据推断含隐混淆因子和选择偏差(selection bias)的模型中的因果关系。即使存在混淆因子，FCI 算法也能给出渐近正确的结果，但它的结果未必有足够信息，因为它往往包含较少的可确认的直接因果影响。若给定可靠的条件独立性检验方法，PC 和 FCI 类算法能处理各种类型的数据分布和因果关系，因而得到广泛使用。然而，PC 和 FCI 类算法不一定能提供完整的因果信息，即它们一般输出一组满足相同条件独立性的因果结构，这些结构都包含在对应的马尔可夫等价类中，如图 2(a)所示。在没有隐变量的情况下，还有通过优化合适的评分函数来找到因果结构的基于评分的方法(score-based methods)。贪婪等价类搜索(greedy equivalence search, GES)[3]算法是典型的基于评分的方法，它直接在等价类的空间上进行搜索。在 GES 的基础上，Ramsey 等人[4]提出了适用于高维数据的快速贪婪等价类搜索(fast greedy equivalence search, FGES)算法；Huang 等人[5]提出了一种不局限于特定模型约束的通用评分函数(generalized score function)的方法。

2.2 基于函数因果模型的因果推断方法

基于函数因果模型的方法则是从数据产生的因果机制出发，探索利用函数因果模型来识别因果关系，同时可以解决基于约束的方法的马尔可夫等价类问题。此类方法以结构方程模型(structure equation model, SEM)为基础。结构方程模型是一个可用于多元变量分析的框架，包括随机变量集与方程集：随机变量集包括观察变量集和误差变量集；一组结构方程对应节点为观察变量的有向图，表示模型的因果结构和结构方程的形式。当结构方程中结果变量 X_j 对应的原因变量 X_i 的系数不为零时，则 X_i 和 X_j 之间存在有向边

$X_i \to X_j$。结构方程模型可以形式化表示为

$$X_i = f\left(pa\left(X_i\right), n_i\right)$$

其中，$pa(X_i)$表示变量X_i的父亲变量的集合；n_i表示噪声变量。所有变量X_i及其父亲变量$pa(X_i)$构成的集合为独立变量集，所有噪声变量n_i构成的集合为误差变量集。结构方程和普通代数方程的差别在于，结构方程表示的并不只是方程左右两边的相等关系，而是在描述方程左边变量的值是如何被自然或者因果机制所决定的。

基于函数因果模型的代表性算法包括线性非高斯无环模型(linear non-Gaussian acyclic model, LiNGAM)[6]、后非线性(post-nonlinear, PNL)方法[7-8]、非线性条件下的加性噪声模型(additive noise model, ANM)[9-10]。

由 Shimizu 等人[11]提出的线性非高斯无环模型(LiNGAM)要求观察数据的产生机制必须满足以下 3 个条件。①有向图无环：观察变量$X_i, i \in \{1,2,\cdots,p\}$存在先后因果次序，即后面的变量不会影响前面的变量，这里定义此变量的因果次序为$k(i)$；②线性因果关系：变量X_i是其对应的原因变量的线性求和，外加噪声变量n_i和常数c_i，即$X_i = \sum_{k(j)<k(i)} b_{ij} X_j + n_i + c_i$；③噪声独立非高斯：噪声变量$n_i$服从非零方差的非高斯分布（或最多一个是高斯分布），且n_i彼此间相互独立，即$p\left(n_1, n_2, \cdots, n_p\right) = \prod_i p_i(n_i)$。在上述线性非高斯无环条件下，LiNGAM 可以形式化表示为

$$\boldsymbol{X} = \boldsymbol{B}\boldsymbol{X} + \boldsymbol{n} \tag{1}$$

其中，$\boldsymbol{X}$为p维的随机向量；$\boldsymbol{B}$为$p \times p$的连接矩阵；$\boldsymbol{n}$为p维的非高斯随机噪声变量。因为无环图假设，则存在置换矩阵$\boldsymbol{P} \in R^{m \times m}$使得$\boldsymbol{B}' = \boldsymbol{P}\boldsymbol{B}\boldsymbol{P}^{\mathrm{T}}$为严格的下三角矩阵且对角线的元素均为 0。

线性情况下变量间因果关系的不对称性的直观解释如图 5 所示。假设X_i是X_j的原因，同时满足 LiNGAM：$X_j = b_{i,j} X_i + n_j$。在左上角的图中，X_j对X_i进行线性回归后得到的残差n_j和X_i不相关。在右上角的图中，转换X_i和X_j，即X_i对X_j进行线性回归后的残差n_i与X_j不相关，即独立于X_j。因为在线性高斯情况下，不相关等于独立。在图的下半部分解释了非高斯的情况，X_i和噪声n_j符合均匀分布。我们可以看到，X_j对X_i进行线性回归后的残差n_j与X_i是互相独立的。如果用X对X_j做回归，其残差n_i和X_j是不独立的，可见这两个方向的判断存在因果不对称性，这个方法可以用来处理许多线性非高斯的情况。

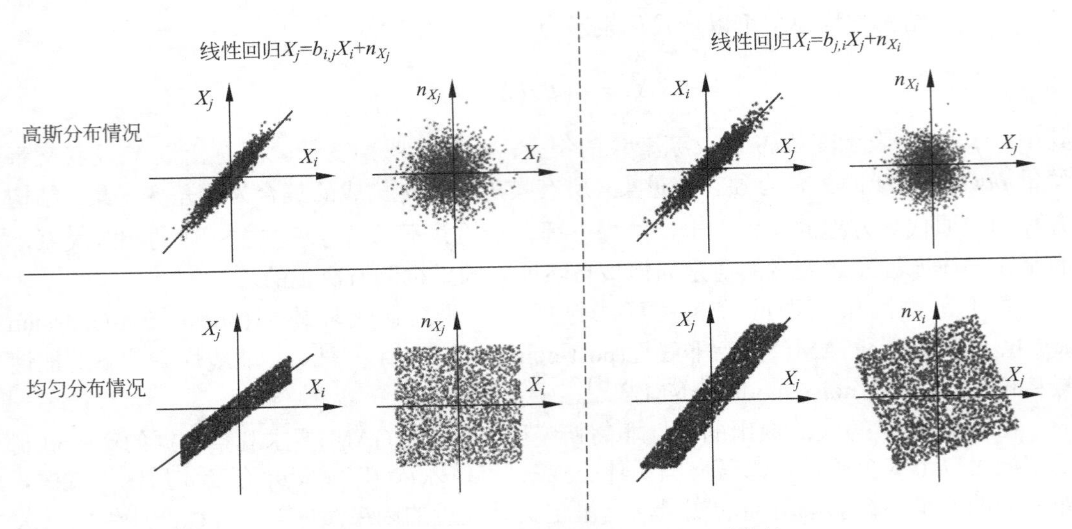

图 5 线性情况下变量间因果关系的不对称性

在非线性情况下，Zhang 等人[7-9]提出了后非线性（post nonlinear, PNL）模型，其示意图如图 6 所示。对于 PNL 模型，假设存在 $X_i \to X_j$，其因果机制可以用下式表示：

$$X_j = f_2(f_1(X_i) + n_j) \tag{2}$$

其中，原因变量 X_i 和噪声变量 n_j 互相独立；f_1 是不恒定的光滑函数；f_2 是可逆的光滑函数。该模型描述数据产生过程的能力较强——f_1 表示了因的非线性作用，而 f_2 可以刻画观察时的非线性变形。尽管该模型有广泛的适用性，可以看到如果 X_i 是 X_j 的因，也就是式(2)成立，那么相反方向一般不能满足噪声独立的假设。

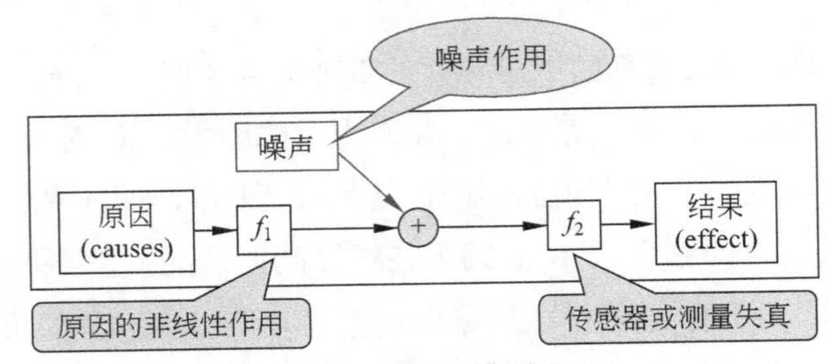

图 6 PNL 模型

为此，假设存在相反方向 $X_j \to X_i$，用 PNL 模型可以表示为

$$X_i = g_2(g_1(X_j) + n_i) \tag{3}$$

其中，原因变量 X_j 和噪声变量 n_i 互相独立； g_1 是不恒定的光滑函数； g_2 是可逆的光滑函数。文献[6]中已经证明，只有在特殊的函数和分布设置下，式(2)和式(3)产生的 X_i 和 X_j 分布才是相同的。在一般情况下，如果数据是按照式(2)的模型，则可以产生相同的变量分布。因此，因果关系方向就可以被识别。也就是说，根据 PNL 模型产生的数据，其变量之间的因果关系几乎在所有情况下都是可以被识别的。表 1 列出了按 PNL 模型产生的两个变量间因果关系的方向不可识别的所有情况[7]。

表 1　PNL 模型不可识别的所有情况

	p_{e2}	$p_{t1}(t_1 = g_2^{-1}(x_1))$	$h = f_1 \circ g_2$	备注
I	高斯(Gaussian)	高斯	线性(linear)	h_1 也是线性的
II	对数线性和指数混合(log-mix-lin-exp)	对数线性和指数混合(log-mix-lin-exp)	线性	h_1 严格单调，$h_1' \to 0$，随着 $z_2 \to +\infty$ 或 $z_2 \to -\infty$
III	对数线性和指数混合	单边渐近指数(one-sided asymptotically exponential)(但非对数线性和指数混合)	h 严格单调，$h' \to 0$，随着 $t_1 \to \infty$ 或 $t_1 \to -\infty$	—
IV	对数线性和指数混合	两个指数的广义混合(generalized mixture of two exponentials)	同上	—
V	两个指数的广义混合	双边渐近指数(two-sided asymptotically exponential)	同上	—

在识别变量之间的因果关系方向时，可以按照以下两个步骤进行：首先，使用条件独立性测试找到满足 d-分离的等价类；然后，利用 PNL 模型识别上一步中未确定的因果方向，对于每个包含等价类的因果结构，采用非线性独立成分分析的方法估计噪声变量 $\tilde{n}_i$，检查假设的原因变量 X_i 是否与假设的结果变量 X_j 对应的噪声变量 $\tilde{n}_i$ 互相独立，如果独立则说明 X_i 和 X_j 的因果关系方向判断为 $X_i \to X_j$，反之亦然。

3　隐变量场景下的因果推断

3.1　广义独立噪声约束

当观察变量是由隐变量产生的且隐变量间存在因果关系时，观察到的变量其实是隐变量的一种显式表示。此时，希望从观察变量中恢复出隐变量，甚至是隐变量之间的因果结构。现有方法常用的是利用独立噪声约束(independent noise condition, IN)来检测隐变量是否存在。

IN 约束：如果(Z,Y)满足独立噪声约束，当且仅当回归残差$Y-\omega^{\mathrm{T}}Z$独立于Z的时候成立。当Z是Y的原因变量集且不存在混淆因子时，独立噪声约束就会成立，否则不成立。

Shimizu 等人[12]利用独立噪声约束提出了一种直接估计 LiNGAM 的方法(direct method for learning a LiNGAM, DirectLiNGAM)。该算法提供了一套寻找外生变量的方法，以图 7 为例，如果先用$\{X_1\}$作为点集Z，X_2作为点集Y，则得到$(\{X_1\},X_2)$不满足独立噪声约束，即可以得出X_1不可能是外生变量。接着，用$\{X_2\}$作为点集Z，点X_1作为点集Y，则可以得到$(\{X_2\},X_1)$满足独立噪声约束。同时，如果剩下的点作为点集Y，皆满足独立噪声约束，那么可以得出X_2是外生变量，即X_2的因果次序大于其他变量。为了去除外生变量X_2对其他剩余变量的混淆影响，将刚刚算出来的残差作为变量的数据。最后依上述步骤迭代寻找外生变量，直到确定所有的因果次序为止。

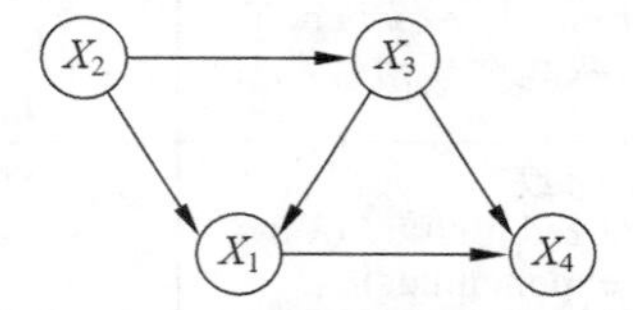

图 7　IN 约束的示例因果结构

在含有隐变量的问题上，无法利用上述寻找外生变量的方法学习因果结构。Cai 等人[13]提出了一种基于三分体(triad)约束的识别隐变量间因果结构的算法(learn the structure of latent variables based on triad constraints, LSTC)。基于线性非高斯性噪声的假设，提出了一个“伪残差”的概念，可以找到隐变量之间的不对称性，辅助推断因果关系。“伪残差”的定义为：观察变量$\{X_i,X_j,X_k\}$的数据遵循 LiNGAM，定义与X_i相关的$\{X_j,X_k\}$的“伪残差”为

$$E_{(i,j|k)}=X_k-\frac{\operatorname{Cov}\left(X_i,X_k\right)}{\operatorname{Cov}\left(X_i,X_j\right)}\cdot X_j$$

其中，$\operatorname{Cov}\left(X_i,X_k\right)$表示$X_i$和$X_k$的协方差。如图 8 所示，左示模型中$E_{(i,j|k)}$和$X_k$是不独立的，但是在右示模型中，两者互相独立，因此可以区分模型中两个隐变量的方向。

现实中很多变量是受多个隐变量影响的，基于上述 IN 约束和 triad 约束，Xie 等人[14]提出了基于广义独立噪声约束（generalize independent noise condition，GIN）的线性非高

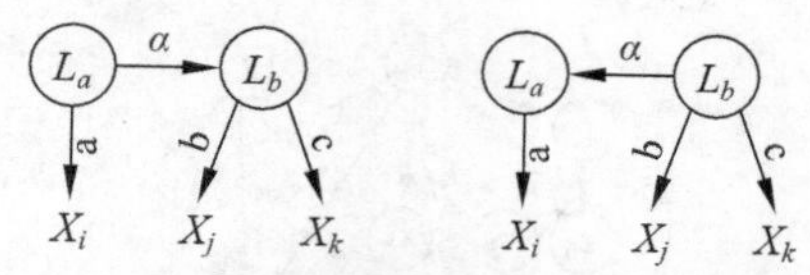

图 8 基于 triad 约束的隐变量间因果推断示意图

斯隐变量模型(linearnon-Gaussian Latent variable model, LiNGLaM)，解决了 N 个相同隐父亲变量影响多个观察变量的测量模型隐变量之间的因果结构学习问题。

GIN 约束：定义一个伪残差 $E_{Y\|Z}=\omega^{\mathrm{T}}\boldsymbol{Y}$，其中 $\omega^{\mathrm{T}}\mathbb{E}[\boldsymbol{YZ}^{\mathrm{T}}]=0$，当且仅当伪残差 $E_{Y\|Z}$ 与变量集 $\boldsymbol{Z}$ 独立时 GIN 约束成立。

LiNGLaM 模型是基于线性隐变量模型(linear latent variable model)提出的，需满足以下两个假设：①纯度假设，观察变量之间没有直接的因果边；②噪声是非高斯的。求解 LiNGLaM 算法流程分为识别纯类和识别隐变量间的因果方向两个阶段。如图 9 所示，利用 GIN 约束可以识别纯类 $\{X_1,X_2,X_3,X_4\}$，$\{X_5,X_6\}$ 和 $\{X_7,X_8\}$；接着识别隐变量间的方向为 $\{L_1,L_2\}\to\{L_3,L_4\},\{L_3\}\to\{L_4\}$。

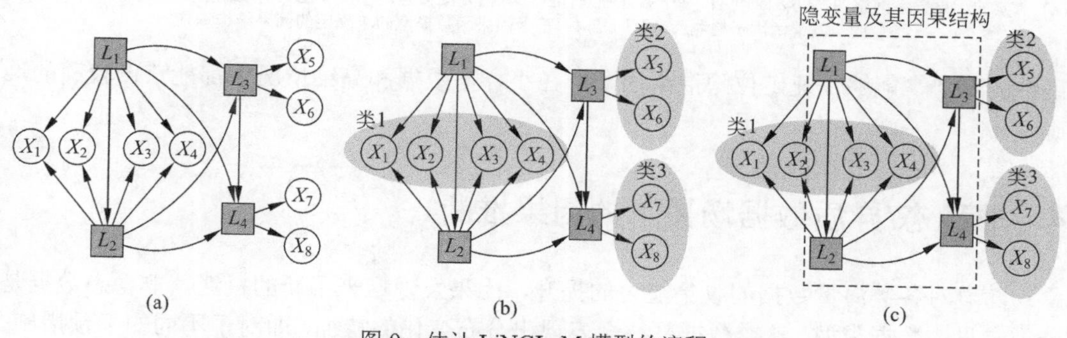

图 9 估计 LiNGLaM 模型的流程

(a) 真实的因果结构；(b) 识别纯类；(c) 识别隐变量间的因果方向

3.2 基于独立噪声条件的全局因果推断

3.1 节描述了隐变量场景下判别隐变量的约束条件，本节将基于这些条件并借助 FCI 算法的优势，阐述从含有隐变量的数据中推断包含隐变量和观察变量的全局因果结构的算法 FRITL[15]。FRITL 算法首先基于条件独立性检验，获得包含全局信息的部分祖先图；接着推断成对的观察变量之间的因果关系；然后利用 triad 约束，检测并合并同一个隐混淆因子；最后利用过完备的独立成分分析技术，估计未确定的局部因果结构。图 10 给出了 FRITL 算法的框架。该方法结合了条件独立性检测的可靠性和局部结构学习方法的准

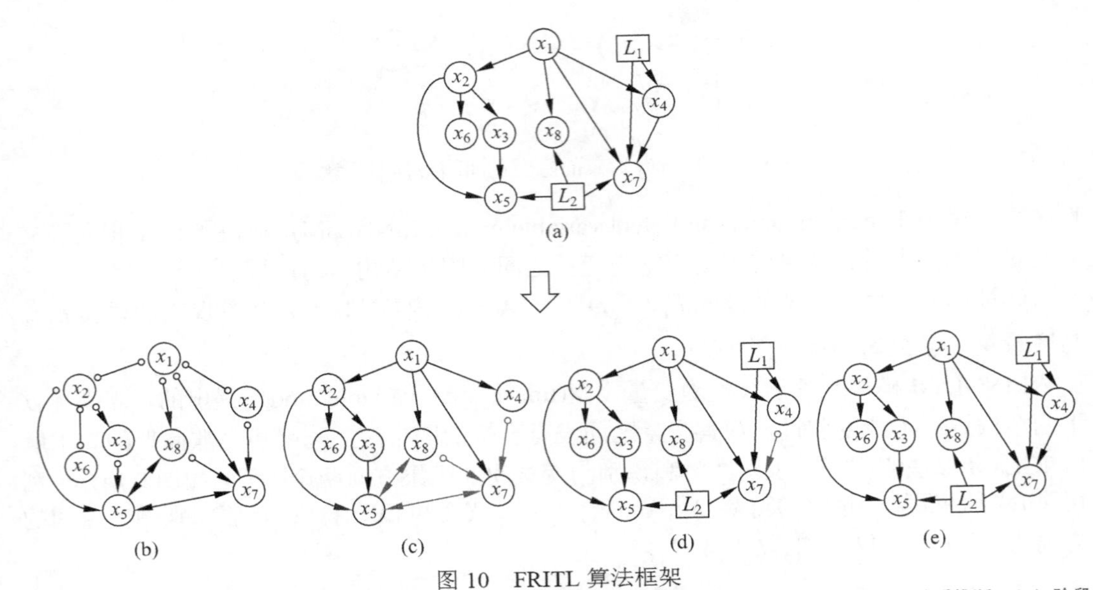

图 10 FRITL 算法框架

（a）真实的因果关系网络图；（b）阶段 1：基于条件独立性的部分祖先图构建；（c）阶段 2：局部因果关系推断；（d）阶段 3：同一隐混淆因子检测；（e）阶段 4：基于因果函数模型的未确定的因果结构识别

确性的优势，能够快速定位隐混淆因子，在小样本数据和高维情况下都能获得较高的准确度。

4 非稳态/异质数据场景下的因果推断

随着对各类海量数据的收集能力的提升，因果发现迎来了新的机遇。非稳态数据是一类常见的数据类型，这类数据往往会表现出分布变化的特征，即对于不同的干预措施、不同的数据收集条件，或者随着时间的推移，数据集之间的分布发生变化。以遥感图像数据为例，由于地面、植被、光照条件等物理因素的不同，在不同地区、不同时间采集的数据通常有不同的分布。其次，功能性磁共振成像记录通常是不稳定的：大脑中的信息流可能会随着刺激、任务、被试者的注意力等而改变。在这些情况下，许多现有的因果发现方法可能会失败，因为它们假设一个固定的因果模型，因此在观察到的数据基础上有一个固定的联合分布，从而导致这些因果发现方法无法取得良好的效果。

对于这类非稳态数据，Zhang 等人[16-17]提出了针对异构/非平稳数据的基于约束的因果发现方法(constraint-based causal discovery from heterogeneous/non-stationary data, CD-NOD)。这种方法在因果系统中引入代理变量 C，以表示跨领域或随时间变化的隐藏量，并表明与 C 相邻的变量具有非稳态的因果机制，具体来说，变量 X_i 的局部因果过程用结构方程模型表示：

$$X_i = f_i\left(PA^i, g^i(C), \theta_i(C), n_i\right)$$

其中，$g^i(C) \subseteq \left\{g_l(C)\right\}_{l=1}^{L}$ 表示影响 X_i 的一系列混淆因子（可为空）；$\theta_i(C)$ 表示不同的作用参数；n_i 表示独立同分布的噪声变量。

基于这种框架，Zhang 等人提出了一种检测伴随因果模块变化的变量并推断出因果图的骨架的方法，这种方法分为变化因果模块检测和因果骨架估计、利用数据的非稳态性确定因果方向两个阶段。首先利用 C 作为代理变量表示潜在的 $\left\{g_l(C)\right\}_{l=1}^{L} \cup \left\{\theta_m(C)\right\}_{m=1}^{n}$，再对系统内的因果变量与替代变量的并集使用基于约束的因果发现算法进行因果骨架学习。然后利用因果不变性和独立变化模块对第一阶段得到的因果骨架进行定向，最终输出定向因果图。

5 因果性学习

目前以深度学习(deep learning)为代表的机器学习受到学界越来越多的关注。然而，机器学习，尤其是深度学习的可解释性、泛化能力及对数据的过度依赖是目前公认的挑战。因此，学界越来越关注在机器学习中因果关系的运用，在半监督学习(semi-supervised learning，SSL)和领域自适应(domainadaption)等方面进行了尝试。相关的研究表明，因果推断理论给出了隐藏在观察数据背后的有用信息，为机器学习等领域的研究提供了新的思路和方向。

半监督学习是一种机器学习方法，它利用少量带标签数据和大量未带标签的数据进行有监督的学习或无监督的学习任务。在半监督学习中，马普所的 Schölkopf 等人[18]指出半监督学习在因果方向上的学习与反因果方向上的学习的区别，揭示了在没有混淆因子的情况下，无标签的数据只有在反因果方向下才是有效的，而在因果方向上是无效的。他们发现半监督学习方法的有效性和因果关系发现中原因变量的概率 P（原因）与给定原因变量的情况下结果变量的概率 P（结果|原因）的独立性有紧密的联系。

一般的机器学习问题认为训练集和测试集有相同的分布，然而在现实问题中可能存在不同数据集有不同分布的情况，由此可延伸为领域迁移问题。领域自适应是一个与机器学习和迁移学习相关的领域，其旨在为“从多个源域(source domain)数据预测分布不同但相关的目标域(target/test domain)数据”的问题提供解决方法，其示意图如图 11 所示。

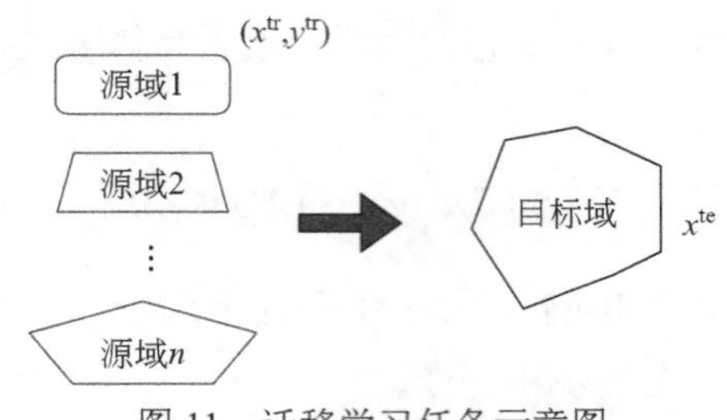

图 11 迁移学习任务示意图

现有的方法依赖于源域和目标域的数据具有相同的分布。但实际情况经常违背此假设，那么如何使用源域隐含的信息来预测目标域？解决领域自适应问题的关键在于找到一个具有高泛化力的方法来连接不同分布或场景。

在一个混杂环境中，各实体背后往往蕴含着海量的因果机制信息，以一个人影案例为例，如图 12 所示。如果看到了人物 A 的影子，便可根据这个影子推断出关于人物 A 的体型的一些信息；此外，还能推断得到光源方向的信息等。但是如果仅看到了人物 A，则难以通过人物 A 的形体得到关于人物 A 的影子的信息，如影子长度、影子映射方向等。影子的产生是因为有光照射到人的身上，影子蕴含了其产生的原因和产生的因果机制信息。从人影的例子可以得到结论：一般来说，在一个混杂的环境中，因果结构蕴含着丰富的信息，且从结果预测原因往往相对容易，反之则不然，这是因为结果中已经包含了一些关于原因的信息，这些信息仅需通过一些额外约束即可获得。

图 12 因果思维例子

在领域自适应研究中，确定从源域转移到目标领域及如何进行转移是一个至关重要的问题。在领域自适应的一些场景中，人们希望预测的变量 Y 往往是原因变量，如图 13 所示，因此引入因果结构信息可为领域自适应工作带来利好因素。Zhang 等人发现即使在这种条件不成立的情况下，仍然可以利用因果知识及一些技术条件来进行领域自适应。

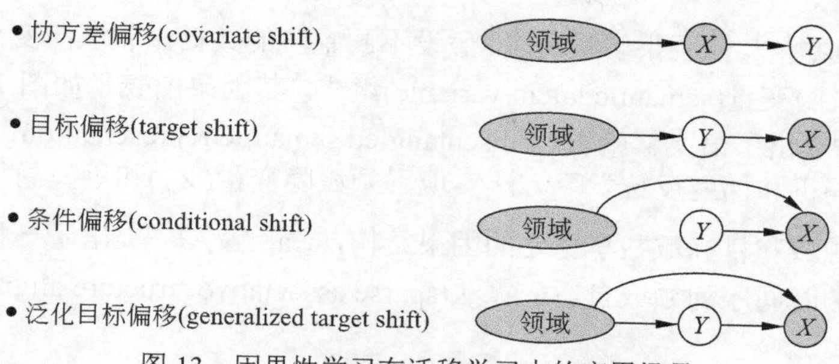

图 13　因果性学习在迁移学习中的应用场景

基本思想是如果它们之间没有混杂因子，则 P（原因）和 P（结果|原因）是真实因果过程的反映，并且变化是独立的，从而允许以简单的形式分别对变化进行参数化。在源域是多个的情况下，Zhang 等人[19-20]将已知的因果关系知识（数据背后产生的机制）融合到领域自适应中，提出了基于目标和条件转移的领域自适应方法。Gong 等人[21]提出了基于条件性的模块转移方法。

近期，Zhang 等人[22]提出了一种适用于多源域情况，将因果推断与领域自适应结合的基于图模型推理的方法 Infer（其流程如图 14 所示）。该方法中各领域的联合数据分布可分解为数个模块，而各模块的分布在跨域时是否会发生改变是事先未知的。该方法利用因果关系发现框架从数据中学习得到源域数据的联合分布中的变化模块，再使用图模型来对联合分布的变化性质进行编码，最终将领域自适应问题转换成为一个关于图模型的贝叶斯推断的问题，从而得到一种基于数据驱动的无监督的领域自适应的标准化框架。

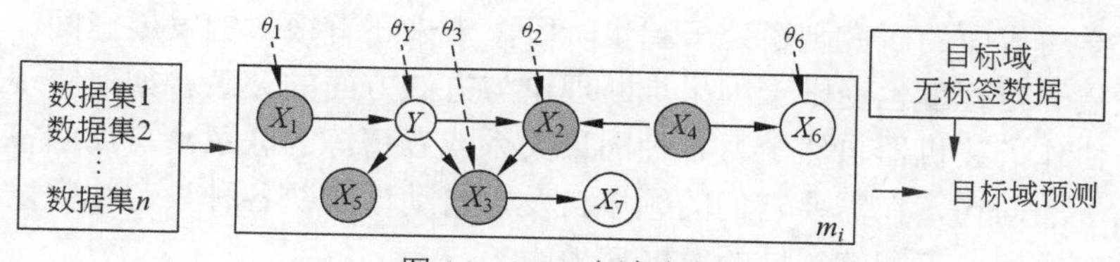

图 14　Infer 方法流程

在图 14 中，若变量 X_i 与变量 θ 相连，则意味着对于该变量，给定其父亲节点的条件分布可能会随着领域的变化而变化。相反地，如果变量 X_i 没有与任何变量 θ 相连，这意味着对于该变量，给定其父亲节点的条件分布不会随着领域的变化而变化，其中变量 θ 之间是彼此独立的。这种机制对源域数据的分布是如何跨域变化做出了解释，得到了具有高表现力的图模型，与领域自适应相结合，即可形成一种紧凑的领域自适应方法。

借助因果性学习的思想，Cai 等人则分别从因果解耦和因果机制不变两个角度探讨了领域自适应问题[23]。从因果解耦角度，假设不同领域的数据由领域隐变量(domain latent variables)和语义隐变量(semantic latent variable)组成，其因果图表示如图 15。基于这个假设，Cai 等人[23]提出了语义解耦表达(disentangled semantic representation，DSR)，模型采用变分自动编码机和梯度反转学习方法实现了领域隐变量(Z_d)和语义隐变量(Z_y)的重构和解耦。基于源域和目标域共享稳定的因果结构，Cai 等人[24]提出了一种新的基于稀疏关联结构对齐的时间序列领域自适应算法(sparse associative structure alignment, SASA)。

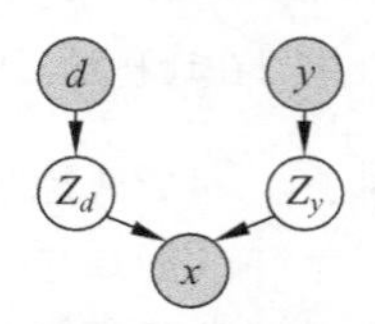

图 15　不同领域数据生成过程因果图

其中 Z_d 和 Z_y 分别表示领域隐变量和语义隐变量

6　小结及讨论

本文对因果推断及因果性学习的一些基本方法和最新进展进行了简要介绍。目前因果推断领域研究已经涌现出大量相关方法。除了提及的因果推断方法之外，还存在几类因果推断问题值得研究[25]，如非线性因果关系(Nonlinearities)[7,26-27]、分类变量或混合数据的情况(categorical variables or mixed cases)[5,28]、测量误差(measurement error)[29]、选择偏倚[30]、混淆因子[1,13,31]、缺失数据(Missing values)[32]、时序因果(causality in time series)[33-39]等。因果性学习方面的研究则还处于起步阶段，未来还有很大的发展空间。当然，因果性学习的发展一定程度也受限于因果推断理论与方法方面的突破。如何实现机器和人的无缝交流与协作，让机器自然学习各种因果关系或者精炼的联系，自如回答人类各种有关“为什么”的问句，在需要它们的领域自发、自动地解决各种设定或突发问题，是未来因果关系在通用人工智能领域的发展潜力。

参考文献

[1] SPIRTES P, GLYMOUR C N, SCHEINES R. Causation, prediction, and search[M]. 2nd edition. Cambridge, USA: MIT press, 2000.

[2] ZHANG K, PETERS J, JANZING D, et al. Kernel-based conditional independence test and application in causal discovery[C]//Proceedings of the 27th Conference on Uncertainty in Artificial Intelligence.

2011: 804-813.

[3] CHICKERING D M. Optimal structure identification with greedy search[J]. Journal of Machine Learning Research, 2002, 11(3): 507-554.

[4] RAMSEY J, GLYMOUR M, SANCHEZ-ROMERO R, et al. A million variables and more: the fast greedy equivalence search algorithm for learning high-dimensional graphical causal models, with an application to functional magnetic resonance images[J]. International Journal of Data Science and Analytics, 2017, 3(2): 121-129.

[5] HUANG B, ZHANG K, LIN Y, et al. Generalized score functions for causal discovery[C]//Proceedings of the 24th ACM SIGKDD International Conference on Knowledge Discovery & Data Mining. 2018: 1551-1560.

[6] SHIMIZU S, HOYER P O, HYVÄRINEN A, et al. A linear non-gaussian acyclic model for causal discovery[J]. Journal of Machine Learning Research, 2006, 10(7): 2003-2030.

[7] ZHANG K, HYVÄRINEN A. On the identifiability of the post-nonlinear causal model[C]//Proceedings of the 25th Conference on Uncertainty in Artificial Intelligence (UAI2009). 2009: 647-655.

[8] ZHANG K, CHAN L W. Extensions of ICA for causality discovery in the hongkong stock market[C]// Proceedings of the 13th international conference on Neural information processing-Volume Part III. 2006: 400-409.

[9] HOYER P O, JANZING D, MOOIJ J, et al. Nonlinear causal discovery with additive noise models[C]// Proceedings of the 21st International Conference on Neural Information Processing Systems. 2008: 689-696.

[10] PETERS J, JANZING D, SCHOLKOPF B. Causal inference on discrete data using additive noise models[J]. IEEE Transactions on Pattern Analysis and Machine Intelligence, 2011, 33(12): 2436-2450.

[11] HYVÄRINEN A, OJA E. Independent component analysis: algorithms and applications[J]. Neural Networks, 2000, 13(4-5): 411-430.

[12] SHIMIZU S, INAZUMI T, SOGAWA Y, et al. DirectLiNGAM: A direct method for learning a linear non-gaussian structural equation model[J]. Journal of Machine Learning Research, 2011, 12(2): 1225-1248.

[13] CAI R, XIE F, GLYMOUR C, et al. Triad Constraints for Learning Causal Structure of Latent Variables[C]//Proceedings of Advances in Neural Information Processing Systems. 2019: 12863-12872.

[14] XIE F, CAI R, HUANG B, et al. Generalized Independent Noise Condition for Estimating Linear Non-Gaussian Latent Variable Graphs[C]//Proceedings of the 33rd International Conference on Neural Information Processing Systems, 2020.

[15] CHEN W, ZHANG K, CAI R, et al. FRITL: A hybrid method for causal discovery in the presence of latent confounders[Z]. arXiv preprint arXiv: 2103.14238, 2021.

[16] ZHANG K, HUANG B, ZHANG J, et al. Discovery and visualization of nonstationary causal models[Z]. arXivpreprint arXiv:1509.08056, 2015.

[17] HUANG B, ZHANG K, ZHANG J, et al. Causal discovery from heterogeneous/nonstationary data[J]. Journal of Machine Learning Research, 2020, 21(89): 1-53.

[18] SCHÖLKOPF B, JANZING D, PETERS J, et al. On causal and anticausal learning[C]//Proceedings of the 29th International Coference on International Conference on Machine Learning. 2012: 459-466.

[19] ZHANG K, SCHÖLKOPF B, MUANDET K, et al. Domain adaptation under target and conditional shift[C]//Proceedings of the 30th International Conference on Machine Learning. 2013: 819-827.

[20] ZHANG K, GONG M, SCHÖLKOPF B. Multi-source domain adaptation: A causal view[C]// Proceedings of the 29th AAAI conference on artificial intelligence. 2015: 3150-3157.

[21] GONG M, ZHANG K, LIU T, et al. Domain adaptation with conditional transferable components[C]// Proceedings of the 33rd International Conference on Machine Learning. ICML, 2016: 2839-2848.

[22] ZHANG K, GONG M, STOJANOV P, et al. Domain adaptation as a problem of inference on graphical models[C]//Proceedings of the 33rd Advances in Neural Information Processing Systems. NeurIPS, 2020: 4965-4976.

[23] CAI R, LI Z, WEI P, et al. Learning disentangled semantic representation for domain adaptation[C]// Proceedings of the 29th International Joint Conference on Artificial Intelligence. IJCAI, 2019: 2060-2066.

[24] CAI R, CHEN J, LI Z, et al. Time series domain adaptation via sparse associative structure alignment[C]//Proceedings of the 35th AAAI Conference on Artificial Intelligence, 2021.

[25] ZHANG K, SCHÖLKOPF B, SPIRTES P, et al. Learning causality and causality-related learning: some recent progress[J]. National Science Review, 2018, 5(1): 26-29.

[26] ZhANG K, CHAN L W. Extensions of ICA for causality discovery in the hongkong stock market[C]// Proceedings of International Conference on Neural Information Processing. Springer, Berlin, Heidelberg, 2006: 400-409.

[27] HOYER P, JANZING D, MOOIJ J M, et al. Nonlinear causal discovery with additive noise models[C]// Proceedings of the 21st Advances in neural information processing systems. 2008, 21: 689-696.

[28] CAI R, QIAO J, ZHANG K, et al. Causal discovery from discrete data using hidden compact representation[C]//Proceedings of the 31st Advances in neural information processing systems. 2018: 2666-2674.

[29] ZHANG K, GONG M, RAMSEY J, et al. Causal discovery with linear non-gaussian models under measurement error: Structural identifiability results[C]//Proceedings of the 34th Conference on Uncertainty in Artificial Intelligence. 2018: 1063-1072.

[30] ZHANG K, ZHANG J, HUANG B, et al. On the identifiability and estimation of functional causal models in the presence of outcome-dependent selection[C]//Proceedings of the 32nd Conference on Uncertainty in Artificial Intelligence. 2016: 825-834.

[31] DING C, GONG M, ZHANG K, et al. Likelihood-free overcomplete ICA and applications in causal discovery[C]//Proceedings of the 32nd Advances in neural information processing systems, 2019.

[32] TU R, ZHANG C, ACKERMANN P, et al. Causal discovery in the presence of missing data[C]//Proceedings of the 22nd International Conference on Artificial Intelligence and Statistics. PMLR, 2019: 1762-1770.

[33] HYVÄRINEN A, SHIMIZU S, HOYER P O. Causal modelling combining instantaneous and lagged effects: an identifiable model based on non-Gaussianity[C]//Proceedings of the 25th international conference on Machine learning. 2008: 424-431.

[34] ZHANG K, HYVÄRINEN A. Causality discovery with additive disturbances: An information-theoretical perspective[C]//Proceedings of the Joint European Conference on Machine Learning and Knowledge Discovery in Databases. Springer, Berlin, Heidelberg, 2009: 570-585.

[35] HYVÄRINEN A, ZHANG K, SHIMIZU S, et al. Estimation of a structural vector autoregression model using non-gaussianity[J]. Journal of Machine Learning Research, 2010, 11(5): 1709-1731.

[36] DANKS D, PLIS S. Learning causal structure from undersampled time series[C]//Proceedings of the JMLR: Workshop and Conference Proceedings. 2014:1-10.

[37] GONG M, ZHANG K, SCHOELKOPF B, et al. Discovering temporal causal relations from subsampled data[C]//Proceedings of the 32nd International Conference on Machine Learning. 2015: 1898-1906.

[38] GONG M, ZHANG K, SCHÖLKOPF B, et al. Causal discovery from temporally aggregated time series[C]//Proceedings of the 33rd Conference on Uncertainty in Artificial Intelligence, 2017.

[39] GEIGER P, ZHANG K, SCHOELKOPF B, et al. Causal inference by identification of vector autoregressive processes with hidden components[C]//Proceedings of the 32nd International Conference on Machine Learning. 2015: 1917-1925.

机器学习中基于 Wasserstein 距离的分布鲁棒优化模型与算法

苏文藻

（香港中文大学系统工程与工程管理系，香港 999077）

1 引言

分布鲁棒优化（distributionally robust optimization，DRO）是一种利用数据来处理带不确定参数的决策问题的方法。简单来说，它首先从已有数据中构建一个参数的概率分布集合，然后求解能为这个集合提供最佳性能保证的决策[1]。近年来，分布鲁棒优化在机器学习领域渐渐受到关注，因为它为一些重要的学习问题提供了新的思路和方法，如结构风险最小化（structural risk minimization，SRM）[2]和对抗训练（adversarial training）[3]等。下面先以监督学习中结构风险最小化问题为例，介绍分布鲁棒优化模型的基本元素。

监督学习的目的是通过由输入与输出对组成的训练数据集（training dataset）来学习一个由输入到输出的映射（又称为模型）[4]。记 $\mathbb{X}$ 为输入空间（input space），$\mathbb{Y}$ 为输出空间（output space）。那么，训练数据集可以表示为

$$\mathcal{T}=\left\{\left(\hat{x}_1,\hat{y}_1\right),\left(\hat{x}_2,\hat{y}_2\right),\cdots,\left(\hat{x}_N,\hat{y}_N\right)\right\}$$

其中，$\hat{x}_1,\hat{x}_2,\cdots,\hat{x}_N\in\mathbb{X}$ 为输入的观测值；$\hat{y}_1,\hat{y}_2,\cdots,\hat{y}_N\in\mathbb{Y}$ 则为对应的输出观测值。在监督学习的场景下，输入和输出被看作随机变量并遵循一个定义在积空间 $\mathbb{X}\times\mathbb{Y}$、未知的联合分布 $\mathbb{P}$，训练数据则被看作根据 $\mathbb{P}$ 独立同分布产生的。学习的目的是要从训练数据集 $\mathcal{T}$ 中找出一个属于给定假设空间（hypothesis space）$\mathcal{H}$ 里最好的模型 $f:\mathbb{X}\longrightarrow\mathbb{Y}$。模型的好坏是由一个损失函数（loss function）$\ell:\mathbb{Y}\times\mathbb{Y}\longrightarrow\mathbb{R}_+$ 来衡量的。给定输入和输出对 $(x,y)\in\mathbb{X}\times\mathbb{Y}$ 和模型 $f\in\mathcal{H}$，根据损失函数 ℓ 来度量预测输出值 $f(x)\in\mathbb{Y}$ 和真实输出值

$y \in \mathbb{Y}$ 的误差，记作 $\ell(f(x), y)$。那么，通过求解下面的优化问题，可以得出假设空间里经验风险（empirical risk）最小的模型：

$$\inf_{f \in \mathcal{H}} \left\{ \mathbb{E}_{(x,y) \sim \hat{\mathbb{P}}_N} \left[\ell(f(x), y) \right] := \frac{1}{N} \sum_{i=1}^{N} \ell(f(\hat{x}_i), \hat{y}_i) \right\} \quad \text{问题(ERM)}$$

其中，$\hat{\mathbb{P}}_N := \frac{1}{N} \sum_{i=1}^{N} \delta_{(\hat{x}_i, \hat{y}_i)}$ 是训练数据的经验分布（empirical distribution）；$\mathbb{E}_{(x,y) \sim \hat{\mathbb{P}}_N} \left[\ell(f(x), y) \right]$ 是经验风险。

虽然问题（ERM）看起来非常直观，但因为它只考虑训练数据集，所以很容易出现过拟合（overfitting）现象。具体来说，由求解问题（ERM）得出的最优模型 f^* 一般只能较准确预测训练数据集里输入样本的输出值。对于不在训练数据集里的输入和输出对 $(x, y) \in \mathbb{X} \times \mathbb{Y}$，由模型 f^* 给出的预测输出值 $f^*(x)$ 和真实输出值 $y \in \mathbb{Y}$ 可以有很大的误差。这使模型 f^* 的泛化误差（generalization error）——样本内误差（in-sample error）$\mathbb{E}_{\hat{\mathbb{P}}_N} \left[\ell(f^*(x), y) \right]$ 与样本外误差（out-of-sample error）$\mathbb{E}_{\mathbb{P}} \left[\ell(f^*(x), y) \right]$ 之间的距离——变得很大。为了避免出现这种情况，一个惯常使用的策略是在经验风险后加上表示模型复杂度的正则项（regularizer 或 regularization term）$R: \mathcal{H} \longrightarrow \mathbb{R}_+$，然后求解下面的结构风险最小化问题：

$$\inf_{f \in \mathcal{H}} \left\{ \mathbb{E}_{(x,y) \sim \hat{\mathbb{P}}_N} \left[\ell(f(x), y) \right] + \varepsilon R(f) \right\} \quad \text{问题(SRM)}$$

其中，$\varepsilon \geqslant 0$，是正则系数（regularization parameter），用来权衡模型的拟合度和它的复杂度。对于不同的损失函数 ℓ 和正则项 R，文献里已有一些具有理论收敛保证的高效算法来求解问题（SRM）[5-8]。此外，实际经验告诉我们由求解问题（SRM）得出的最优模型通常都有比较好的泛化性能，但要严格解释这一现象并不容易，而且统计学习文献里对不同正则项的作用的解释都需要比较强的假设。关于这方面的研究可以参考文献[9]～文献[12]。

为了能从训练数据集中找出具有良好泛化性能的模型并更系统地解释不同正则项的作用，研究人员最近提出一种以分布鲁棒优化为基础的学习方法。此方法的起源可以追溯至一些早期从鲁棒优化角度去演绎正则项的工作，如文献[13]。直观来说，如果训练数据比较多，经验分布 $\hat{\mathbb{P}}_N$ 应该接近真实分布 $\mathbb{P}$。那么，为了降低样本内误差并控制样本外误差，其中一个自然的想法是求解下面的最坏期望损失（worst-case expected loss）最小化问题：

$$\inf_{f\in\mathcal{H}}\sup_{\mathbb{Q}\in\mathfrak{M}\left(\hat{\mathbb{P}}_N\right)}\mathbb{E}_{(x,y)\sim\mathbb{Q}}\left[\ell\left(f\left(x\right),y\right)\right] \qquad \text{问题(DR-RM)}$$

其中，$\mathfrak{M}\left(\hat{\mathbb{P}}_N\right)$为根据经验分布$\hat{\mathbb{P}}_N$构建的一个概率分布集合（又称为不确定性集合（ambiguity set））。问题（DR-RM）为分布鲁棒优化问题的一个实例。它的好处是能把从训练数据中得知的关于真实分布$\mathbb{P}$的信息整合到优化过程中。可是要实现这个想法，还需要解决以下问题：

（1）怎样刻画不确定性集合$\mathfrak{M}\left(\hat{\mathbb{P}}_N\right)$？粗略来说，希望$\mathfrak{M}\left(\hat{\mathbb{P}}_N\right)$是以经验分布$\hat{\mathbb{P}}_N$为中心、包含真实分布$\mathbb{P}$的一个邻域。这样，问题（DR-RM）的内层最大化部分就能有效地控制样本外误差。

（2）需要对假设空间$\mathcal{H}$、不确定性集合$\mathfrak{M}\left(\hat{\mathbb{P}}_N\right)$和损失函数$\ell$做什么样的假设才能为问题（DR-RM）设计出高效的求解算法。

（3）对于由上述分布鲁棒学习方法得出的模型，能否为它的泛化性能提供理论保证？

本文将介绍文献里提出的处理上述问题的一些思路，并集中介绍分布鲁棒优化理论如何为结构风险最小化和对抗训练提供原则性方法。最后，将介绍机器学习中的分布鲁棒优化这一研究方向的一些公开问题。

2 基于 Wasserstein 距离的分布鲁棒优化问题

在求解分布鲁棒优化问题（DR-RM）前，必须对不确定性集合$\mathfrak{M}\left(\hat{\mathbb{P}}_N\right)$进行刻画，其中主要的考虑是要权衡该集合对泛化性能的贡献和由它产生的问题（DR-RM）的计算复杂度。如前所述，希望$\mathfrak{M}\left(\hat{\mathbb{P}}_N\right)$包含跟$\hat{\mathbb{P}}_N$比较相似的分布，而这想法可以通过统计距离度量（statistical distance measure）来实现。文献里其中一个经常出现的统计距离度量是ϕ-散度（ϕ-divergence）。记$\Delta_m=\left\{p\in\mathbb{R}^m:\sum_{i=1}^m p_i=1,p\geqslant 0\right\}$为在$m$维欧氏空间里的单纯形（simplex）；$\phi:\mathbb{R}_+\longrightarrow\mathbb{R}$为一满足$\phi(1)=0$的凸函数，并定义

$$0\cdot\phi\left(\frac{a}{0}\right):=\begin{cases}a\cdot\lim_{t\to\infty}\dfrac{\phi(t)}{t}, & a>0\\ 0, & a=0\end{cases}$$

给定离散分布$\mathbb{P}=\left(p_1,p_2,\cdots,p_m\right),\mathbb{Q}=\left(q_1,q_2,\cdots,q_m\right)\in\Delta_m$，定义它们之间的$\phi$-散度为

$$I_{\phi}(\mathbb{P},\mathbb{Q}) := \sum_{i=1}^{m} q_i \cdot \phi\left(\frac{p_i}{q_i}\right)$$

需要注意的是，$I_{\phi}(\mathbb{P},\mathbb{Q})$ 不一定是对称的（即 $I_{\phi}(\mathbb{P},\mathbb{Q})$ 不一定等于 $I_{\phi}(\mathbb{Q},\mathbb{P})$）。因此，$\phi$-散度未必是一个度量（metric）。基于 ϕ-散度的不确定性集合可以表达为

$$\mathfrak{M}_{\phi}\left(\hat{\mathbb{P}}_N\right) := \left\{\mathbb{P} \in \Delta_N : I_{\phi}\left(\mathbb{P},\hat{\mathbb{P}}_N\right) \leqslant \varepsilon\right\} \tag{1}$$

其中，$\varepsilon \geqslant 0$ 是一个参数，用来控制 $\mathfrak{M}_{\phi}\left(\hat{\mathbb{P}}_N\right)$ 的大小。

利用 ϕ-散度来定义不确定性集合 $\mathfrak{M}_{\phi}\left(\hat{\mathbb{P}}_N\right)$ 有几个好处。①不少常见的散度——如 Kullback-Leibler（KL）散度（由 $\phi(t) = t\log t - t + 1$ 产生）和全变分（total variation）散度（由 $\phi(t) = |t-1|$ 产生）——都是 ϕ-散度的特例[14]。这些散度可以用来构建一个不确定性集合族。②基于 ϕ-散度的分布鲁棒优化问题可被等价变换为一个有限维优化问题，而且对一大类 ϕ-散度，该问题是可以被有效求解的[14-15]。③当不确定性集合取式（1）的形式时，问题（DR-RM）可以被理解为对经验风险加上一个模型方差（variance）的正则项，而且能为它给出一套统计理论[16]。可是，在学习的场景下，基于 ϕ-散度的不确定性集合也有它的局限。具体来说，如果 $\lim_{t\to\infty}\phi(t)/t = \infty$（如 KL 散度就满足这个条件），那么为了使 $I_{\phi}(\mathbb{P},\mathbb{Q}) < \infty$，当 q_i 等于 0 时，p_i 也必须等于 0。（在概率文献里，该条件叫做 $\mathbb{P}$ 相对于 $\mathbb{Q}$ 是绝对连续（absolutely continuous）的，并用 $\mathbb{P} \ll \mathbb{Q}$ 来表示）在这种情况下，对任意的 $\varepsilon \geqslant 0$，不确定性集合式（1）就不可能包含任何跟 $\hat{\mathbb{P}}_N$ 有不同支撑集（support set）的分布。换句话说，如果要求数据的真实分布 $\mathbb{P}$ 是在不确定性集合式（1）里面的话，就得假设 $\mathbb{P}$ 和 $\hat{\mathbb{P}}_N$ 有相同的支撑集，但是这一假设并不现实。

为了突破这一局限，研究人员最近提出利用 Wasserstein 距离来定义不确定性集合。Wasserstein 距离是一种统计距离度量，属于积分概率度量（integral probability metric）族。关于积分概率度量的定义和性质可以参考文献[17]。令 (Ξ, d) 为一度量空间，$\mathfrak{P}(\Xi)$ 为一支撑于 Ξ 的概率分布的集合。给定分布 $\mathbb{P},\mathbb{Q} \in \mathfrak{P}(\Xi)$，定义它们之间的 Wasserstein 距离为

$$W_d(\mathbb{P},\mathbb{Q}) := \inf_{\Pi \in \mathfrak{P}(\Xi\times\Xi)} \left\{ \int_{\Xi\times\Xi} d(\xi,\xi')\Pi(\mathrm{d}\xi,\mathrm{d}\xi') : \Pi(\mathrm{d}\xi,\Xi) = \mathbb{P}(\mathrm{d}\xi), \Pi(\Xi,\mathrm{d}\xi') = \mathbb{Q}(\mathrm{d}\xi') \right\} \tag{2}$$

换句话说，求解 $W_d(\mathbb{P},\mathbb{Q})$ 需要寻找一个定义在积空间 $\Xi\times\Xi$ 的联合分布 Π，使它的边缘分布分别为 $\mathbb{P}$ 和 $\mathbb{Q}$，而且在运输质量 $\xi \in \Xi$ 到 $\xi' \in \Xi$ 的成本为 $d(\xi,\xi')$ 的情况下，根据联

合分布 Π 计算的期望运输成本 $\mathbb{E}_{(\xi,\xi')\sim\Pi}\left[d\left(\xi,\xi'\right)\right]:=\int_{\Xi\times\Xi}d\left(\xi,\xi'\right)\Pi\left(\mathrm{d}\xi,\mathrm{d}\xi'\right)$ 最小。通过定义可以证明 Wasserstein 距离是一个度量。它在最优传输理论（optimal transport theory）里具有重要的角色，详情可参考文献[18]和文献[19]。基于 Wasserstein 距离的不确定性集合可以表达为

$$\mathfrak{M}_{W_d}\left(\hat{\mathbb{P}}_N\right):=\left\{\mathbb{P}\in\mathfrak{P}(\Xi):W_d\left(\mathbb{P},\hat{\mathbb{P}}_N\right)\leqslant\varepsilon\right\} \tag{3}$$

跟之前一样，$\varepsilon\geqslant 0$ 是一个用来控制 $\mathfrak{M}_{W_d}\left(\hat{\mathbb{P}}_N\right)$ 大小的参数。

一般来说，计算两个分布 $\mathbb{P},\mathbb{Q}$ 之间的 Wasserstein 距离比较困难，因为 $W_d(\mathbb{P},\mathbb{Q})$ 是一个无穷维优化问题的最优值。只有在一些特殊情况下，$W_d(\mathbb{P},\mathbb{Q})$ 才有显示公式。下面是其中一个例子。

例1 令 $\Xi=\mathbb{R}$，$d\left(\xi,\xi'\right)=\left|\xi-\xi'\right|$，那么，给定分布 $\mathbb{P},\mathbb{Q}\in\mathfrak{P}(\Xi)$，它们之间的 Wasserstein 距离可以写成[20]

$$W_{|\bullet|}(\mathbb{P},\mathbb{Q})=\int_{\mathbb{R}}\left|F_{\mathbb{P}}(u)-F_{\mathbb{Q}}(u)\right|\mathrm{d}u \tag{4}$$

其中，$F_{\mathbb{P}}$ 和 $F_{\mathbb{Q}}$ 分别是 $\mathbb{P}$ 和 $\mathbb{Q}$ 的分布函数（distribution function）。以下是式（4）一个简单的应用：令 $\mathbb{P}$ 为 $\{0,1\}$ 上的均匀分布，$\mathbb{Q}$ 为 $[0,1]$ 上的均匀分布。那么可以推出：

$$F_{\mathbb{P}}(u)=\begin{cases}0, & u<0\\ 1/2, & 0\leqslant u<1\\ 1, & u\geqslant 1\end{cases}\qquad F_{\mathbb{Q}}(u)=\begin{cases}0, & u<0\\ u, & 0\leqslant u\leqslant 1\\ 1, & u\geqslant 1\end{cases}$$

由此可得出：

$$W_{|\bullet|}(\mathbb{P},\mathbb{Q})=\int_0^1\left|\frac{1}{2}-u\right|\mathrm{d}u=\frac{1}{4}$$

关于其他场景下 $W_d(\mathbb{P},\mathbb{Q})$ 的计算可以参考文献[21]。

从例 1 可以看出，$\mathbb{P}$ 相对于 $\mathbb{Q}$ 可以不是绝对连续但 $W_d(\mathbb{P},\mathbb{Q})<\infty$。因此，利用式（3）定义的不确定性集合可以包含跟 $\hat{\mathbb{P}}_N$ 有不同支撑集的分布。这也克服了前述关于 ϕ-散度的一个缺点。可是，问题（DR-RM）里的内层最大化部分是一个无穷维优化问题，这使得求解问题（DR-RM）和分析其最优解的泛化性能变得很具挑战性。下面将介绍最近文献里关于克服这些难题的一些进展。

3　分布鲁棒监督学习

为了更好地了解不确定性集合式（3）的结构和问题（DR-RM）的难点，先考虑基于 Wasserstein 距离的分布鲁棒逻辑回归（logistic regression）问题。在传统的逻辑回归问题里，输入空间为 $\mathbb{X}=\mathbb{R}^n$，输出空间为 $\mathbb{Y}=\{-1,+1\}$，假设空间为 $\mathcal{H}=\{f:\mathbb{X}\longrightarrow\mathbb{R}|f(x)=\beta^T x,\beta\in\mathbb{R}^n\}$（即所有从 $\mathbb{R}^n$ 到 $\mathbb{R}$ 的线性函数），损失函数为 $\ell(u,v)=\ln\left[1+\exp(-uv)\right]$。给定输入和输出对 $(x,y)\in\mathbb{X}\times\mathbb{Y}$，$x$ 被称为特征向量（feature vector），y 则被称为标签（label）。因数据的真实分布 $\mathbb{P}$ 是定义在积空间 $\Xi=\mathbb{X}\times\mathbb{Y}=\mathbb{R}^n\times\{-1,+1\}$ 上的，为了定义在空间 $\mathfrak{P}(\Xi)$ 上的 Wasserstein 距离，需要在空间 Ξ 上选择一个合适的度量 d。其中一个选择是[22]

$$d(\xi,\xi')=\|x-x'\|+\frac{\kappa}{2}|y-y'|,\xi=(x,y)\in\Xi,\xi'=(x',y')\in\Xi \tag{5}$$

其中，$\|\cdot\|$ 是任意一个定义在 $\mathbb{R}^n$ 上的度量（如 p-范数 $\|\cdot\|_p$（$p\geqslant 1$））；$\kappa>0$ 控制在输出空间运输质量的成本。下面将讨论 κ 的选择对数据及结构风险优化的含义。

3.1　问题重构

在上述场景下，基于 Wasserstein 距离的分布鲁棒逻辑回归问题可以写成

$$\inf_{\beta\in\mathbb{R}^n}\sup_{\mathbb{Q}\in\mathfrak{M}_{W_d}(\hat{\mathbb{P}}_N)}\mathbb{E}_{(x,y)\sim\mathbb{Q}}\underbrace{\ln\left[1+\exp\left(-y\cdot\beta^{\mathrm{T}}x\right)\right]}_{\ell(\beta^{\mathrm{T}}x,y)} \qquad \text{问题(DR-LR)}$$

处理这类分布鲁棒优化问题的一个重要步骤是把内层的无穷维最大化问题变换为一个有限维的优化问题。利用式（2）和式（3），问题（DR-LR）内层的最大化问题可以写成

$$\sup_{\mathbb{Q}\in\mathfrak{M}_{W_d}(\hat{\mathbb{P}}_N)}\mathbb{E}_{\xi=(x,y)\sim\mathbb{Q}}\left[\ell\left(\beta^{\mathrm{T}}x,y\right)\right]=\begin{cases}\sup\limits_{\Pi\in\mathfrak{P}(\Xi\times\Xi)} & \int_{\xi\in\Xi}\ell\left(\beta^{\mathrm{T}}x,y\right)\cdot\underbrace{\Pi\left(\mathrm{d}\xi,\Xi\right)}_{\mathbb{Q}(\mathrm{d}\xi)}\\ \text{s.t.} & \int_{(\xi,\xi')\in\Xi\times\Xi}d\left(\xi,\xi'\right)\cdot\Pi\left(\mathrm{d}\xi,\mathrm{d}\xi'\right)\leqslant\varepsilon\\ & \Pi\left(\Xi,\mathrm{d}\xi'\right)=\hat{\mathbb{P}}_N\left(\mathrm{d}\xi'\right)\end{cases} \tag{6}$$

由于 $\hat{\mathbb{P}}_N=\frac{1}{N}\sum_{i=1}^{N}\delta_{(\hat{x}_i,\hat{y}_i)}$，得出：

$$\mathbb{Q}(\mathrm{d}\xi)=\Pi(\mathrm{d}\xi,\Xi)=\int_{\xi'\in\Xi}\Pi(\mathrm{d}\xi,\mathrm{d}\xi)=\sum_{i=1}^{N}\Pi\left(\mathrm{d}\xi|\xi'=(\hat{x}_i,\hat{y}_i)\right)\cdot\hat{\mathbb{P}}_N(\hat{x}_i,\hat{y}_i)=\frac{1}{N}\sum_{i=1}^{N}\mathbb{Q}^i(\mathrm{d}\xi)$$

其中，$\mathbb{Q}^i(\mathrm{d}\xi)=\Pi\left(\mathrm{d}\xi|\xi'=(\hat{x}_i,\hat{y}_i)\right)$是固定$\xi'=(\hat{x}_i,\hat{y}_i)$后$\xi$的条件分布。同样地，有

$$\Pi(\mathrm{d}\xi,\mathrm{d}\xi')=\Pi(\mathrm{d}\xi|\xi')\cdot\hat{\mathbb{P}}_N(\xi')=\frac{1}{N}\sum_{i=1}^{N}\delta_{(\hat{x}_i,\hat{y}_i)}(\xi')\cdot\mathbb{Q}^i(\mathrm{d}\xi)$$

因此，式（6）可以写成

$$\begin{aligned}\sup_{\mathbb{Q}^i\geqslant 0,i=1,2,\cdots,N}\quad&\frac{1}{N}\sum_{i=1}^{N}\int_{\xi\in\Xi}\ell(\beta^{\mathrm{T}}x,y)\cdot\mathbb{Q}^i(\mathrm{d}\xi)\\ \text{s.t.}\quad&\frac{1}{N}\sum_{i=1}^{N}\int_{\xi\in\Xi}d\left(\xi,(\hat{x}_i,\hat{y}_i)\right)\cdot\mathbb{Q}^i(\mathrm{d}\xi)\leqslant\varepsilon\\ &\int_{\xi\in\Xi}\mathbb{Q}^i(\mathrm{d}\xi)=1,\quad i=1,\cdots,N\end{aligned}\tag{7}$$

令$\mathbb{Q}_{+1}^i(\mathrm{d}x)=\mathbb{Q}^i(\mathrm{d}x,y=+1)$，$\mathbb{Q}_{-1}^i(\mathrm{d}x)=\mathbb{Q}^i(\mathrm{d}x,y=-1)$并利用$\xi=(x,y)\in\mathbb{R}^n\times\{-1,+1\}$，则有

$$\mathbb{Q}^i(\mathrm{d}\xi)=\mathbb{Q}_{+1}^i(\mathrm{d}x)+\mathbb{Q}_{-1}^i(\mathrm{d}x)\tag{8}$$

把式（8）代入式（7），得出：

$$\begin{aligned}\sup_{\mathbb{Q}_{\pm1}^i\geqslant 0,i=1,2,\cdots,N}\quad&\frac{1}{N}\sum_{i=1}^{N}\int_{x\in\mathbb{R}^n}\left(\ell(\beta^{\mathrm{T}}x,+1)\cdot\mathbb{Q}_{+1}^i(\mathrm{d}x)+\ell(\beta^{\mathrm{T}}x,-1)\cdot\mathbb{Q}_{-1}^i(\mathrm{d}x)\right)\\ \text{s.t.}\quad&\frac{1}{N}\sum_{i=1}^{N}\Big[\int_{x\in\mathbb{R}^n}d\left((x,+1),(\hat{x}_i,\hat{y}_i)\right)\cdot\mathbb{Q}_{+1}^i(\mathrm{d}x)+\\ &\int_{x\in\mathbb{R}^n}d\left((x,-1),(\hat{x}_i,\hat{y}_i)\right)\cdot\mathbb{Q}_{-1}^i(\mathrm{d}x)\Big]\leqslant\varepsilon\\ &\int_{x\in\mathbb{R}^n}\left(\mathbb{Q}_{+1}^i(\mathrm{d}x)+\mathbb{Q}_{-1}^i(\mathrm{d}x)\right)=1,\quad i=1,2,\cdots,N\end{aligned}\tag{9}$$

需要注意的是，式（9）里的决策变量$\mathbb{Q}_{+1}^i$和$\mathbb{Q}_{-1}^i$应被理解为定义在测度空间$(\mathbb{R}^n,\mathcal{B})$上（$\mathcal{B}$为在$\mathbb{R}^n$上的Borel σ-代数）的带号测度（signed measure）。由于$\hat{y}_1,\hat{y}_2,\cdots,\hat{y}_N\in\{-1,+1\}$，利用式（5），可以得出：

$$\begin{aligned}&\sum_{i=1}^{N}\int_{x\in\mathbb{R}^n}d\left((x,+1),(\hat{x}_i,\hat{y}_i)\right)\cdot\mathbb{Q}_{+1}^i(\mathrm{d}x)\\ =&\sum_{i:\hat{y}_i=+1}\int_{x\in\mathbb{R}^n}\|x-\hat{x}_i\|\cdot\mathbb{Q}_{+1}^i(\mathrm{d}x)+\sum_{i:\hat{y}_i=-1}\int_{x\in\mathbb{R}^n}(\|x-\hat{x}_i\|+\kappa)\cdot\mathbb{Q}_{+1}^i(\mathrm{d}x)\end{aligned}\tag{10}$$

同样地有

$$\begin{aligned}&\sum_{i=1}^{N}\int_{x\in\mathbb{R}^n} d\left(\left(x,-1\right),\left(\hat{x}_i,\hat{y}_i\right)\right)\cdot\mathbb{Q}_{-1}^i\left(\mathrm{d}x\right)\\&=\sum_{i:\hat{y}_i=+1}\int_{x\in\mathbb{R}^n}\left(\|x-\hat{x}_i\|+\kappa\right)\cdot\mathbb{Q}_{-1}^i\left(\mathrm{d}x\right)+\sum_{i:\hat{y}_i=-1}\int_{x\in\mathbb{R}^n}\|x-\hat{x}_i\|\cdot\mathbb{Q}_{-1}^i\left(\mathrm{d}x\right)\end{aligned}\tag{11}$$

把式（10）和式（11）代入式（9），得出式（6）等价于

$$\begin{aligned}\sup_{\mathbb{Q}_{\pm1}^i\geqslant 0,i=1,2,\cdots,N}\quad&\frac{1}{N}\sum_{i=1}^{N}\int_{x\in\mathbb{R}^n}\left(\ell\left(\beta^{\mathrm{T}}x,+1\right)\cdot\mathbb{Q}_{+1}^i\left(\mathrm{d}x\right)+\ell\left(\beta^{\mathrm{T}}x,-1\right)\cdot\mathbb{Q}_{-1}^i\left(\mathrm{d}x\right)\right)\\\text{s.t.}\quad&\frac{1}{N}\int_{x\in\mathbb{R}^n}\left[\kappa\left(\sum_{i:\hat{y}_i=+1}\mathbb{Q}_{-1}^i\left(\mathrm{d}x\right)+\sum_{i:\hat{y}_i=-1}\mathbb{Q}_{+1}^i\left(\mathrm{d}x\right)\right)+\right.\\&\left.\sum_{i=1}^{N}\|x-\hat{x}_i\|\left(\mathbb{Q}_{+1}^i\left(\mathrm{d}x\right)+\mathbb{Q}_{-1}^i\left(\mathrm{d}x\right)\right)\right]\leqslant\varepsilon\\&\int_{x\in\mathbb{R}^n}\left(\mathbb{Q}_{+1}^i\left(\mathrm{d}x\right)+\mathbb{Q}_{-1}^i\left(\mathrm{d}x\right)\right)=1,\quad i=1,2,\cdots,N\end{aligned}\tag{12}$$

现在一个重要的观察是，式（12）中目标函数和约束函数对决策变量$\left\{\mathbb{Q}_{\pm1}^i\right\}_{i=1}^N$是线性的。因此，式（12）是一个线性优化问题。可是，由于$\left\{\mathbb{Q}_{\pm1}^i\right\}_{i=1}^N$是测度而不是有限维欧氏空间里的向量，所以式（12）是一个半无限（semi-infinite）线性优化问题（即决策变量为无穷维空间里的向量，而约束条件的个数有限）。即使如此，还是可以通过式（12）的拉格朗日函数构造它的对偶问题①：

$$\begin{aligned}\inf_{\lambda\geqslant 0,s_1,s_2\cdots,s_N}\quad&\lambda\varepsilon+\frac{1}{N}\sum_{i=1}^{N}s_i\\\text{s.t.}\quad&\ell\left(\beta^{\mathrm{T}}x,+1\right)-\lambda\|x-\hat{x}_i\|-\frac{1}{2}\lambda\kappa\left(1-\hat{y}_i\right)\leqslant s_i,\forall x\in\mathbb{R}^n,i=1,2,\cdots,N,\\&\ell\left(\beta^{\mathrm{T}}x,-1\right)-\lambda\|x-\hat{x}_i\|-\frac{1}{2}\lambda\kappa\left(1+\hat{y}_i\right)\leqslant s_i,\forall x\in\mathbb{R}^n,i=1,2,\cdots,N.\end{aligned}\tag{13}$$

虽然一般的半无限线性优化问题与它的对偶问题不一定有强对偶（strong duality）关系，但是通过验证式（12）的 Slater 条件，可以证明它与式（13）之间的强对偶关系是

① 也可以通过以下启发式方法（heuristic method）来构造式（12）的对偶问题。先把式（12）里的积分离散化，得出一个有限维的线性优化问题。然后构造这一线性优化问题的对偶问题后再求极限，就能得出式（13）。必须强调的是，该方法并不严谨，但对理解式（13）的结构有一定帮助。

成立的。具体来说，记 $\mathcal{X}$ 为定义在 $(\mathbb{R}^n,\mathcal{B})$ 上的带号测度集合，$\mathcal{X}_+\subseteq\mathcal{X}$ 为所有在集合 $\mathcal{X}$ 里的非负测度。定义线性映射 $\mathcal{A}:\mathcal{X}^{2N}\longrightarrow\mathbb{R}^{N+1}$ 为

$$\begin{cases}\left[\mathcal{A}\left(\left\{\mathbb{Q}_{\pm1}^i\right\}_{i=1}^N\right)\right]_i := \int_{x\in\mathbb{R}^n}\left(\mathbb{Q}_{+1}^i(\mathrm{d}x)+\mathbb{Q}_{-1}^i(\mathrm{d}x)\right),\quad i=1,2,\cdots,N\\ \left[\mathcal{A}\left(\left\{\mathbb{Q}_{\pm1}^i\right\}_{i=1}^N\right)\right]_{N+1} := \dfrac{1}{N}\int_{x\in\mathbb{R}^n}\left[\kappa\left(\sum_{i:\hat{y}_i=+1}\mathbb{Q}_{-1}^i(\mathrm{d}x)+\sum_{i:\hat{y}_i=-1}\mathbb{Q}_{+1}^i(\mathrm{d}x)\right)+\right.\\ \qquad\qquad\left.\sum_{i=1}^N\|x-\hat{x}_i\|\left(\mathbb{Q}_{+1}^i(\mathrm{d}x)+\mathbb{Q}_{-1}^i(\mathrm{d}x)\right)\right]\end{cases}$$

那么，式（12）的约束条件可以表达为

$$b:=(1,2,\cdots,1,\varepsilon)\in\mathcal{A}\left(\left\{\mathbb{Q}_{\pm1}^i\right\}_{i=1}^N\right)+\underbrace{\left(\{0\}^n\times\mathbb{R}_+\right)}_{K},\quad \mathbb{Q}_{\pm1}^i\in\mathcal{X}_+,i=1,2,\cdots,N$$

注意 $\mathcal{A}(\mathcal{X}_+)=\mathbb{R}_+^n\times\mathbb{R}_+$（要证明这个关系，可以令 $\mathbb{Q}_{+1}^i=\omega_i^+\delta_{\hat{x}_i},\mathbb{Q}_{-1}^i=\omega_i^-\delta_{\hat{x}_i}$，然后选择合适的系数 $\omega_i^+,\omega_i^-\geqslant0$）。所以，对任意的 $\varepsilon>0$，式（12）满足下面的 Slater 条件：

$$b\in\operatorname{int}\left(\mathcal{A}(\mathcal{X}_+)+K\right)$$

这样就可以利用文献[23]中命题 3.4 的结果得出式（12）和式（13）的强对偶关系。这也说明了式（6）等价于式（13）。

需要注意的是，式（13）还是一个半无限优化问题，因为每一个 $x\in\mathbb{R}^n$ 对应一组约束条件。但是，观察到

$$\ell\left(\beta^{\mathrm{T}}x,\pm1\right)-\lambda\|x-\hat{x}_i\|-\frac{1}{2}\lambda\kappa(1\mp\hat{y}_i)\leqslant s_i,\quad\forall x\in\mathbb{R}^n$$
$$\Longleftrightarrow\sup_{x\in\mathbb{R}^n}\left\{\ell\left(\beta^{\mathrm{T}}x,\pm1\right)-\lambda\|x-\hat{x}_i\|\right\}-\frac{1}{2}\lambda\kappa(1\mp\hat{y}_i)\leqslant s_i$$

利用文献[22]中引理 1 的结果：

$$\forall\lambda>0,\quad\sup_{x\in\mathbb{R}^n}\left\{\ell\left(\beta^{\mathrm{T}}x,\pm1\right)-\lambda\|x-x'\|\right\}=\begin{cases}\ell\left(\beta^{\mathrm{T}}x',\pm1\right), & \|\beta\|_*\leqslant\lambda\\ -\infty, & \mathrm{o/w}\end{cases}$$

其中，$\|\cdot\|_*$ 是 $\|\cdot\|$ 的对偶范数，即 $\|\beta\|_*=\max\limits_{\pi:\|\pi\|\leqslant1}\pi^T\beta$，得出式（13）等价于

$$
\begin{aligned}
\inf_{\lambda \geqslant 0, s_1, s_2, \cdots, s_N} \quad & \lambda \varepsilon + \frac{1}{N} \sum_{i=1}^{N} s_i \\
\text{s.t.} \quad & \ell\left(\beta^{\mathrm{T}} \hat{x}_i, \hat{y}_i\right) \leqslant s_i, \quad i = 1, 2, \cdots, N \\
& \ell\left(\beta^{\mathrm{T}} \hat{x}_i, -\hat{y}_i\right) - \lambda \kappa \leqslant s_i, \quad i = 1, 2, \cdots, N \\
& \| \beta \|_* \leqslant \lambda
\end{aligned}
$$

总结上述推导，证明了下面的定理：

定理 1

$$
\inf_{\beta \in \mathbb{R}^n} \sup_{\mathbb{Q} \in \mathfrak{M}_{W_d}\left(\hat{\mathbb{P}}_N\right)} \mathbb{E}_{(x,y) \sim \mathbb{Q}} \left[\ell\left(\beta^{\mathrm{T}} x, y\right) \right] = \begin{cases}
\inf\limits_{\substack{\beta \in \mathbb{R}^n, \lambda \geqslant 0 \\ s_1, s_2, \cdots, s_N}} & \lambda \varepsilon + \dfrac{1}{N} \displaystyle\sum_{i=1}^{N} s_i \\
\text{s.t.} & \ell\left(\beta^{\mathrm{T}} \hat{x}_i, \hat{y}_i\right) \leqslant s_i, i = 1, 2, \cdots, N \\
& \ell\left(\beta^{\mathrm{T}} \hat{x}_i, -\hat{y}_i\right) - \lambda \kappa \leqslant s_i, i = 1, 2, \cdots, N \\
& \| \beta \|_* \leqslant \lambda
\end{cases}
$$

■

这里有两个有趣的观察。①定理 1 中的重构优化问题不仅是有限维而且是凸的。因此，只要对偶范数 $\|\cdot\|_*$ 可以被有效地表达，则问题（DR-LR）在计算上是可解的(computationally tractable)，而且可以用如 CVX[24]和 YALMIP[25]等现成的求解器对其进行求解。②当 $\kappa = +\infty$ 时，定理 1 说明问题（DR-LR）等价于下面的结构风险最小化问题：

$$
\inf_{\beta \in \mathbb{R}^n} \frac{1}{N} \sum_{i=1}^{N} \ell\left(\beta^{\mathrm{T}} \hat{x}_i, \hat{y}_i\right) + \varepsilon \| \beta \|_*
$$

这里的正则项恰巧就是输入空间运输成本度量 $\|\cdot\|$ 的对偶范数，正则系数则是用来控制不确定性集合 $\mathfrak{M}_{W_d}\left(\hat{\mathbb{P}}_N\right)$ 大小的参数 ε。由此可见，逻辑回归里的结构风险最小化问题可以理解为基于 Wasserstein 距离的分布鲁棒风险最小化问题。但需要注意的是，根据式(5)，当 $\kappa = +\infty$ 时，对于任意的 $\mathbb{Q} \in \mathfrak{M}_{W_d}\left(\hat{\mathbb{P}}_N\right)$，它与经验分布 $\hat{\mathbb{P}}_N$ 在输出空间 $\mathbb{Y}$ 上的边缘分布几乎必然（almost surely）是相等的。这意味着结构风险最小化方法做了一个隐含假设，就是给定特征向量 $x \in \mathbb{X}$，它的标签 $y \in \mathbb{Y}$ 是确定（deterministic）的。相比之下，分布鲁棒风险最小化方法并不需要做出这样的假设，因定理 1 里的等价结果对任意的 $\kappa > 0$ 都是成立的。

虽然我们一直集中讨论逻辑回归问题，但是上面介绍的重构和分析技巧可以推广到一大类分布鲁棒线性回归和分类问题，详情可参考文献[2]和文献[26]。关于重构一般基于 Wasserstein 距离的分布鲁棒优化问题的方法，可参考文献[27]。

3.2 泛化性能分析

为了保证由求解问题（DR-LR）得出的最优模型有良好的泛化性能，其中一个思路是选择合适的$\varepsilon > 0$，使不确定性集合$\mathfrak{M}_{W_d}\left(\hat{\mathbb{P}}_N\right)$包含数据的真实分布$\mathbb{P}$。这个思路可以通过刻画$\hat{\mathbb{P}}_N$收敛到$\mathbb{P}$的速度来实现。其中关键的一步是利用下面的测度集中（measure concentration）结果[28]。

定理 2 假设$\mathbb{P}$是一个轻尾分布（light-tailed distribution），即存在常数$a > 1, \varLambda < +\infty$使$\mathbb{E}_{\mathbb{P}}\left[\exp\left(\| 2x \|^a\right)\right] \leqslant \varLambda$。那么就存在与常数$a$、$\varLambda$、输入空间维数$n$和度量$d$相关的参数$c_1, c_2, c_3 > 0$，使对任意的样本数目$N \geqslant 1$和可信度$\eta \in (0,1]$，如果令

$$\varepsilon \geqslant \begin{cases} \left[\dfrac{\ln(c_1/\eta)}{c_2 N}\right]^{1/a}, & N < \ln\dfrac{(c_1/\eta)}{c_2 c_3} \\ \left[\dfrac{\ln(c_1/\eta)}{c_2 N}\right]^{1/n}, & N \geqslant \ln\dfrac{(c_1/\eta)}{c_2 c_3} \end{cases} \tag{14}$$

则有$\Pr\left(\mathbb{P} \in \mathfrak{M}_{W_d}\left(\hat{\mathbb{P}}_N\right)\right) \geqslant 1 - \eta$。 ■

由定理 2 可以推出，当$\mathbb{P}$是一个轻尾分布而ε满足式（14）时，对任意的样本数目$N \geqslant 1$和可信度$\eta \in (0,1]$，有如下的泛化性能界：

$$\Pr\left(\mathbb{E}_{\mathbb{P}}\left[\ell\left(\beta^{\mathrm{T}} x, y\right)\right] \leqslant \sup_{\mathbb{Q} \in \mathfrak{M}_{W_d}\left(\hat{\mathbb{P}}_N\right)} \mathbb{E}_{\mathbb{Q}}\left[\ell\left(\beta^{\mathrm{T}} x, y\right)\right] \ \ \forall \beta \in \mathbb{R}^n\right) \geqslant 1 - \eta$$

关于上述泛化性能分析的一些更详细讨论，以及其他基于 Wasserstein 距离的分布鲁棒学习方法的泛化性能分析，可以参考文献[2]、文献[26]～文献[27]和文献[29]。

3.3 求解问题（DR-LR）的高效一阶算法

定理 1 说明问题（DR-LR）基本上可以用现成的求解器来求解。可是，由于这些求解器里用的都是一般性的凸优化算法，所以在求解如问题（DR-LR）这一类具有结构的凸优化问题时效率并不高，特别是当问题的规模比较大的时候。此外，由于问题（DR-LR）的约束条件比较复杂，大部分现有的一阶算法均不能有效地求解每次迭代产生的子问题。

因此，一个很自然的问题是如何设计高效的算法来求解问题（DR-LR）。

为了方便讨论，考虑一个比较简单的场景，就是输入空间运输成本度量为 1-范数 $\|\cdot\|_1$。注意它的对偶范数为 ∞-范数 $\|\cdot\|_\infty$。令 $\mathcal{G}=\left\{(\beta,\lambda)\in\mathbb{R}^n\times\mathbb{R}:\|\beta\|_\infty\leqslant\lambda\right\}$ 为 ∞-范数的上镜图（epigraph）且

$$\mathbb{I}_{\mathcal{G}}(\beta,\lambda)=\begin{cases}0, & (\beta,\lambda)\in\mathcal{G}\\ +\infty, & (\beta,\lambda)\notin\mathcal{G}\end{cases}$$

为 $\mathcal{G}$ 的指示函数（indicator function）。利用逻辑损失函数（logistic loss）的定义 $\ell(u,v)=\ln\left[1+\exp(-uv)\right]$，可以把问题（DR-LR）的凸重构写成

$$\inf_{\substack{\beta\in\mathbb{R}^n,\\ \lambda\in R}} \quad F(\beta,\lambda):=\lambda\varepsilon+\frac{1}{N}\sum_{i=1}^{N}\left[\ell\left(\beta^{\mathrm{T}}\hat{x}_i,\hat{y}_i\right)+\max\left\{\hat{y}_i\beta^{\mathrm{T}}\hat{x}_i-\lambda\kappa,0\right\}\right]+\mathbb{I}_{\mathcal{G}}(\beta,\lambda) \tag{15}$$

如果对任意给定的 $\lambda\geqslant0$，都可以有效地求解 $q(\lambda):=\inf\limits_{\beta\in\mathbb{R}^n}F(\beta,\lambda)$，那么自然地，可以使用合适的一维搜索算法找出函数 q 的最小解 $\lambda^*\geqslant0$，然后再找出 $\beta^*\in\mathbb{R}^n$ 使 $q\left(\lambda^*\right)=F\left(\beta^*,\lambda^*\right)$。这样，$\left(\beta^*,\lambda^*\right)$ 就是式（15）的最优解。

为了实现这个想法，首先需要证明函数 q 的最小解是可以通过一维搜索算法求出的。由于 F 对 (β,λ) 是一个凸函数，所以 q 是一个单峰（unimodal）凸函数。另外，通过考虑式（15）的 KKT 条件，可以得出 $0\leqslant\lambda^*\leqslant\lambda^U:=0.2785/\varepsilon$，详情可参见文献[30]中的命题 3.1。因此，如果对任意的 $\lambda\geqslant0$ 都能算出 $q(\lambda)$ 的值的话，那么就可以使用黄金分割搜索法（golden-section search）求出函数 q 的最小解。

根据定义，观察到 $q(\lambda)$ 是一个优化问题的最优值，所以接下来的挑战是如何设计能有效求解这一优化问题的算法。为此，对任意的 $\lambda\geqslant0$，考虑下面的优化问题：

$$\begin{aligned}\inf_{x\in\mathbb{R}^d,\,y\in\mathbb{R}^n} \quad & H(x,y):=h(y)+P(y)+\mathbb{I}_{\mathcal{G}_\lambda}(x)\\ \text{s.t.} \quad & \boldsymbol{A}x=y\end{aligned} \tag{16}$$

其中，$h:\mathbb{R}^n\longrightarrow\mathbb{R}$ 是一个连续可微的凸函数，而其梯度 ∇h 为 L_h-Lipschitz 连续的；$P:\mathbb{R}^n\longrightarrow\mathbb{R}\cup\{+\infty\}$ 是一个正常（proper）闭凸函数；$\boldsymbol{A}\in\mathbb{R}^{n\times d}$ 是一个给定的矩阵；$\lambda\geqslant0$ 是一个给定的常数；$\mathcal{G}_\lambda:=\left\{x\in\mathbb{R}^d:\|x\|_\infty\leqslant\lambda\right\}$ 是半径为 λ 的 ∞-范数球。注意如果令 $\boldsymbol{A}\in\mathbb{R}^{N\times n}$ 的第 i 行为 $\hat{y}_i\hat{x}_i^{\mathrm{T}}$（$i=1,2,\cdots,N$），那么就有

$$\inf_{\beta\in\mathbb{R}^n} F(\beta,\lambda)=\begin{cases}\inf_{\beta\in\mathbb{R}^n,\mu\in\mathbb{R}^N} & \underbrace{\lambda\varepsilon+\frac{1}{N}\sum_{i=1}^{N}\ln\left[1+\exp(\mu_i)\right]}_{h(\mu)}+\underbrace{\frac{1}{N}\sum_{i=1}^{N}\max\{\mu_i-\lambda\kappa,0\}}_{P(\mu)}+\mathbb{I}_{\mathcal{G}_\lambda}(\beta)\\ \text{s.t.} & \boldsymbol{A}\beta=\mu\end{cases}$$

换句话说，问题 $\inf_{\beta\in\mathbb{R}^n} F(\beta,\lambda)$ 是式（16）的特例。后者可以通过提出的一种交替方向乘子法（alternating direction method of multipliers, ADMM）来求解。首先，记式（16）的增广拉格朗日（augmented Lagrangian）函数为

$$\mathcal{L}_\rho(x,y;w):=h(y)+P(y)+\mathbb{I}_{\mathcal{G}_\lambda}(x)-w^{\mathrm{T}}(Ax-y)+\frac{\rho}{2}\|Ax-y\|_2^2$$

其中，$w\in\mathbb{R}^n$ 是式（16）里线性约束条件的对偶变量；$\rho>0$ 是惩罚系数。然后，令

$$\hat{h}(y;y^k):=h(y^k)+\nabla h(y^k)^{\mathrm{T}}(y-y^k)$$

为 h 在 y^k 的线性近似。那么，可以给出如下迭代步骤：

$$\begin{cases}x^{k+1}=\arg\min_{x\in\mathbb{R}^d}\mathcal{L}_\rho(x,y^k;w^k)\\ \quad=\arg\min_{x\in\mathbb{R}^d}\left\{\frac{\rho}{2}\left\|Ax-y^k-\frac{w^k}{\rho}\right\|_2^2+\mathbb{I}_{\mathcal{G}_\lambda}(x)\right\}\\ y^{k+1}=\arg\min_{y\in\mathbb{R}^n}\left\{\hat{h}(y;y^k)-(w^k)^{\mathrm{T}}(\boldsymbol{A}x^{k+1}-y)+\frac{\rho}{2}\|\boldsymbol{A}x^{k+1}-y\|_2^2+P(y)\right\}\\ \quad=\arg\min_{y\in\mathbb{R}^n}\left\{\frac{\rho}{2}\left\|y-\left(\boldsymbol{A}x^{k+1}-\frac{w^k+\nabla h(y^k)}{\rho}\right)\right\|_2^2+P(y)\right\}\\ w^{k+1}=w^k-\rho(\boldsymbol{A}x^{k+1}-y^{k+1})\end{cases}\tag{17}$$

称式（17）为线性化近端交替方向乘子法（linearized proximal ADMM，LP-ADMM）。它对 x 的更新和一般的 ADMM 算法是一样的。注意求解 x^{k+1} 等价于求解一个带边界约束（box-constrained）的二次优化问题。这类问题有不同的高效算法，如加速投影梯度法（accelerated projected gradient method）[31]、坐标下降法（coordinate descent）[32]、共轭梯度积极集算法（active set conjugate gradient method）[33]等。至于对 y 的更新，由于 $\mathcal{L}_\rho$ 对 y 是强凸的，采用 h 在 y^k 的线性近似，而不是一般 ADMM 所用的二次近似。这样就不用在更新 y 时选择步长，而且在实际计算中可以令算法收敛得更快。用 UCI Adult 数据集①

① 可从 LIBSVM（https://www.csie.ntu.edu.tw/~cjlin/libsvmtools/datasets/binary.html）下载。

测试了 LP-ADMM 和 YALMIP 在求解分布鲁棒逻辑回归式（15）时所需的 CPU 时间，结果见表 1。值得注意的是，在提出 LP-ADMM 之前，式（15）基本上只能通过现成的求解器如 YALMIP 来求解。从表 1 可见，LP-ADMM 在求解大规模问题时有相当大的优势。

表 1　LP-ADMM 与 YALMIP 在 UCI Adult 数据集测试中的 CPU 时间对比

数据集	样本数（N）	特征数（n）	CPU 时间/s		提速比例
			LP-ADMM	YALMIP	
a1a	1605	123	**2.93**	25.63	**9**
a2a	2265	123	**3.53**	39.20	**11**
a3a	3185	123	**4.26**	57.79	**14**
a4a	4781	123	**4.56**	105.32	**23**
a5a	6414	123	**4.39**	155.42	**35**
a6a	11220	123	**4.68**	413.65	**88**
a7a	16100	123	**5.41**	738.12	**137**
a8a	22696	123	**5.81**	1396.45	**240**
a9a	32561	123	**7.08**	2993.30	**423**

现在给出 LP-ADMM 算法式（17）的收敛结果。

定理 3　假设式（16）的最优解集合是非空的，而 LP-ADMM 算法（式（17））里的惩罚系数 ρ 满足 $\rho>\left(\sqrt{3}+1\right)L_h$。对任意的初始点 $\left(x^0,y^0,w^0\right)\in\mathbb{R}^d\times\mathbb{R}^n\times\mathbb{R}^n$，令 $\left\{\left(x^k,y^k,w^k\right)\right\}_{k\geqslant 0}$ 为 LP-ADMM 算法(式(17))产生的无穷序列。那么，序列 $\left\{\left(x^k,y^k,w^k\right)\right\}_{k\geqslant 0}$ 收敛到式（17）的 KKT 点 $\left(x^*,y^*,w^*\right)$，即

$$\boldsymbol{A}^{\mathrm{T}}w^*\in\partial\mathbb{I}_{\mathcal{G}_\lambda}\left(x^*\right),\quad -w^*\in\nabla h\left(y^*\right)+\partial P\left(y^*\right),\quad \boldsymbol{A}x^*-y^*=0$$

令 $\bar{x}^k=\dfrac{1}{K}\sum_{k=1}^{K}x^k$，$\bar{y}^k=\dfrac{1}{K}\sum_{k=1}^{K}y^k$，那么有

$$H\left(\bar{x}^k,\bar{y}^k\right)-H\left(x^*,y^*\right)\leqslant\frac{(1/2\rho)\|w^0\|_2^2+(\rho/2)\|y^*-y^0\|_2^2+\left[(\rho-2L_h)/4\right]\|y^0-y^1\|_2^2}{K}$$

即序列 $\left\{H\left(\bar{x}^k,\bar{y}^k\right)\right\}_{k\geqslant 0}$ 以次线性的速度收敛到式（16）的最优值。■

关于 LP-ADMM 算法的细节和更多的数值结果，以及定理 3 的证明，可以参考文献[30]。值得注意的是，上述 LP-ADMM 算法框架可以处理 1-范数以外的输入空间运输成本度量

$\|\cdot\|$，而算法的效率取决于对 x 的更新（见式（17））是否有高效算法。

4 对抗训练

众所周知，神经网络已被广泛应用在不同领域里的学习任务上[34-35]。可是，基于神经网络的学习系统一般容易受对抗样本（adversarial example）影响，导致学习出来的模型表现出现偏差[36-37]。为了减轻对抗样本对模型的影响，研究人员最近提出利用分布鲁棒优化方法来处理训练数据。其中一个思路是考虑下面的分布鲁棒优化问题：

$$\inf_{\beta\in\Theta}\sup_{\mathbb{Q}\in\mathfrak{M}_{W_d}(\hat{\mathbb{P}}_N)}\mathbb{E}_{z\sim\mathbb{Q}}\left[\mathcal{L}(\beta,z)\right] \qquad \text{问题(DR-AT)}$$

注意问题（DR-AT）跟问题（DR-RM）相似：这里的假设空间为 $\Theta\subseteq\mathbb{R}^n$；数据遵循一个定义在空间 Ξ 上但未知的分布 $\mathbb{P}$，训练数据则被看为根据 $\mathbb{P}$ 独立同分布产生的；$\mathcal{L}:\Theta\times\Xi\longrightarrow\mathbb{R}_+$ 为损失函数。不同的地方是，Wasserstein 距离里的运输成本 $d:\Xi\times\Xi\longrightarrow\mathbb{R}_+\cup\{+\infty\}$ 现在被理解为由一个样本生成一个对抗样本的成本。由于在神经网络学习的场景下损失函数 $\mathcal{L}$ 一般都是非凸的，故第 3 节介绍的凸重构技巧一般不适用于问题（DR-AT）。为了克服这一难题，文献[3]提出利用拉格朗日松弛（Lagrangian relaxation）法来构建一个具有对抗鲁棒性的模型。首先，从文献[38]里的定理 1 或文献[39]里的定理 1 的结果可以得知，当 $\mathcal{L}$ 和 d 都是连续时，下面的强对偶关系就会成立：

$$\sup_{\mathbb{Q}\in\mathfrak{M}_{W_d}(\hat{\mathbb{P}}_N)}\mathbb{E}_{\mathbb{Q}}\left[\mathcal{L}(\beta,z)\right]=\inf_{\gamma\geqslant 0}\left\{\gamma\varepsilon+\sup_{\mathbb{Q}\in\mathfrak{P}(\Xi)}\left\{\mathbb{E}_{\mathbb{Q}}\left[\mathcal{L}(\beta,z)\right]-\gamma W_d\left(\mathbb{Q},\hat{\mathbb{P}}_N\right)\right\}\right\} \tag{18}$$

此外，对任意的 $\gamma\geqslant 0$，记 $\phi_\gamma(\beta,z)=\sup_{w\in\Xi}\left\{\mathcal{L}(\beta,w)-\gamma d(w,z)\right\}$，则有

$$\mathbb{E}_{\hat{\mathbb{P}}_N}\left[\phi_\gamma(\beta,z)\right]=\sup_{\mathbb{Q}\in\mathfrak{P}(\Xi)}\left\{\mathbb{E}_{\mathbb{Q}}\left[\mathcal{L}(\beta,z)\right]-\gamma W_d\left(\mathbb{Q},\hat{\mathbb{P}}_N\right)\right\} \tag{19}$$

这启发文献[3]的作者利用下面的经验风险最小化问题（对任意给定的 $\gamma\geqslant 0$）作为对问题（DR-AT）的近似：

$$\inf_{\beta\in\Theta}\left\{\mathbb{E}_{\hat{\mathbb{P}}_N}\left[\phi_\gamma(\beta,z)\right]:=\frac{1}{N}\sum_{i=1}^{N}\phi_\gamma(\beta,\hat{z}_i)\right\} \qquad \text{问题(DR-AT-R)}$$

现在一个重要的观察是：①如果对任意的 $\beta\in\Theta$，函数 $z\mapsto\mathcal{L}(\beta,z)$ 是可微的且其梯度 $\nabla_z\mathcal{L}(\beta,\cdot)$ 是 Lipschitz 连续的；②对任意的 $z\in\Xi$，函数 $w\mapsto d(w,z)$ 是 1-强凸的，那

么当γ足够大的时候，对任意的$\beta \in \Theta$和$z \in \Xi$，函数$w \mapsto \mathcal{L}(\beta, w) - \gamma d(w, z)$就是强凹的。那么，计算$\phi_\gamma(\beta, z)$就等价于求解一个强凹最大化问题。直观来说，这问题是可以被有效求解的。因此，文献[3]的作者提出使用随机梯度下降法（stochastic gradient descent, SGD）来求解问题（DR-AT-R）。该算法的迭代步骤如下：

$$\begin{cases} \text{随机采样} i \in \{1, \cdots, N\}, \ \text{求} \phi_\gamma\left(\beta^k, \hat{z}_i\right) = \sup\limits_{w \in \Xi} \left\{ \mathcal{L}\left(\beta^k, w\right) - \gamma d\left(w, \hat{z}_i\right) \right\} \text{的（近似）最优解} \hat{w}^k \\ \beta^{k+1} = \Pi_\Theta\left(\beta^k - \alpha_k \nabla_\beta \mathcal{L}\left(\beta^k, \hat{w}^k\right)\right) \end{cases}$$

其中，Π_Θ为对Θ的投影算子；$\alpha_k > 0$为步长。文献[3]的作者还证明了在适当的条件下，上述的 SGD 算法会以$O\left(1/\sqrt{K}\right)$的速度收敛。该证明的基本思路跟文献里关于求解非凸优化问题的随机一阶算法收敛性分析相似，详情可参考文献[3]。

虽然上述的 SGD 算法可以用来求解问题（DR-AT）的近似解，但是问题（DR-AT）内层的最大化问题只是刻画了对经验分布$\hat{\mathbb{P}}_N$的对抗鲁棒性，而不是对数据真实分布$\mathbb{P}$的对抗鲁棒性。因此，一个自然的问题是能否为后者提供理论保证。从式（18）可以看出，对任意的$\beta \in \Theta$和$\gamma \geqslant 0$，有

$$\sup_{\mathbb{Q} \in \mathfrak{M}_{W_d}\left(\hat{\mathbb{P}}_N\right)} \mathbb{E}_{\mathbb{Q}}\left[\mathcal{L}(\beta, z)\right] \leqslant \gamma \varepsilon + \mathbb{E}_{\hat{\mathbb{P}}_N}\left[\phi_\gamma(\beta, z)\right]$$

有趣的是，这个上界基本上也可以用来控制对数据真实分布$\mathbb{P}$的对抗鲁棒性。具体来说，以下不等式高概率成立：

$$\sup_{\mathbb{Q} \in \mathfrak{M}_{W_d}(\mathbb{P})} \mathbb{E}_{\mathbb{Q}}\left[\mathcal{L}(\beta, z)\right] \leqslant \gamma \varepsilon + \mathbb{E}_{\hat{\mathbb{P}}_N}\left[\phi_\gamma(\beta, z)\right] + O\left(\frac{1}{\sqrt{N}}\right), \quad \forall \beta \in \Theta$$

关于上述结果的证明，以及通过求解问题（DR-AT-R）得出的模型的泛化性能分析可以参考文献[3]。

5 总结和展望

本文介绍了基于 Wasserstein 距离的分布鲁棒优化模型及其在机器学习中结构风险最小化问题和对抗训练的应用。值得注意的是，文献里关于求解上述分布鲁棒优化模型的算法是非常有限的。因此，一个重要的研究方向是为机器学习中的分布鲁棒优化模型设计高效算法，并分析它们的优化与泛化性能。最近相关的文献可参考文献[30]和文献[40]。另外，神经网络学习中经常会用到非凸非光滑的损失函数，但是文献里的结果（第 4 节）

基本上都要求损失函数是光滑的。如何处理带非凸非光滑损失函数的分布鲁棒优化模型就成了一个很自然的研究问题。关于非光滑优化问题的一些基本分析和算法设计技巧可以参考文献[41]。

参考文献

[1] RAHIMIAN H, MEHROTRA S. Distributionally Robust optimization: A review[Z]. arXiv: 1908.05659v1.

[2] SHAFIEEZADEH-ABADEH S, KUHN D, MOHAJERIN-ESFAHANI P. Regularization via mass transportation[J]. J. Mach. Learn. Res., 2019, 20(103): 1-68.

[3] SINHA A, NAMKOONG H, VOLPI R, et al. Certifying some distributional robustness with principled adversarial training[C]//Proceedings of ICLR, 2018.

[4] 李航. 统计学习方法[M]. 北京：清华大学出版社，2012.

[5] SRA S, NOWOZIN S, WRIGHT S J. Optimization for machine learning[M]//Neural Information Processing Series. Cambridge, Massachusetts: MIT Press, 2012.

[6] XIAO L, ZHANG T. A proximal stochastic gradient method with progressive variance reduction[J]. SIAM J. Opt., 2014, 24(4): 2057-2075.

[7] LOH P L, WAINWRIGHT M J. Regularized M-estimators with nonconvexity: Statistics and algorithmic theory for local optima[J]. J. Mach. Learn. Res., 2015, 16: 559-616.

[8] YUE M C, ZHOU Z, SO A M C. A family of inexact SQA methods for non-smooth convex minimization with provable convergence guarantees based on the Luo-Tseng error bound property[J]. Math.Prog., 2019, 174(1-2): 327-358.

[9] BOUSQUET O, ELISSEEFF A. Stability and generalization[J]. J. Mach. Learn. Res., 2002, 2: 499-526.

[10] KAKADE S M, SRIDHARAN K, TEWARI A. On the complexity of linear prediction: Risk bounds, margin bounds, and regularization[C]//Proceedings of NIPS. 2008: 793-800.

[11] JAKUBOVITZ D, GIRYES R, RODRIGUES M R D. Generalization error in deep learning[M]//Boche et al. Compressed sensing and its applications: Applied and numerical harmonic analysis. Switzerland: Springer Nature Switzerland AG, 2019: 153-193.

[12] WEI C, LEE J D, LIU Q, et al. Regularization matters: Generalization and optimization of neural nets v.s. their induced kernel[C]//Proceedings of NIPS, 2019.

[13] BERTSIMAS D, COPENHAVER M S. Characterization of the equivalence of robustification and regularization in linear and matrix regression[J]. European J. Oper. Res., 2018, 270(3): 931-942.

[14] BEN-TAL A, DEN HERTOG D, DE WAEGENAERE A, et al. Robust solutions of optimization problems affected by uncertain probabilities[J]. Manag. Sci., 2013, 59(2): 341-357.

[15] NAMKOONG N, DUCHI J C. Stochastic gradient methods for distributionally robust optimization with f-divergences[C]//Proceedings of NIPS, 2016.

[16] DUCHI J C, GLYNN P W, NAMKOONG H. Statistics of robust optimization: A generalized empirical likelihood approach[J]. Math.Oper.Res., 2021.

[17] SRIPERUMBUDUR B K, FUKUMIZU K, GRETTON A, et al. On integral probability metrics, ϕ -divergences and binary classification[Z]. arXiv: 0901.2698.

[18] VILLANI C. Topics in optimal transportation[M]//Graduate studies in Mathematics. Providence, Rhode Island: American Mathematical Society, 2003.

[19] KOLOURI S, PARK S R, THORPE M, et al. Optimal mass transport: Signal processing and machine-learning applications[J]. IEEE Signal Process. Mag., 2017, 34(4): 43-59.

[20] VALLENDER S S. Calculation of the Wasserstein distance between probability distributions on the line[J]. Theory Probab. Appl., 1974, 18(4): 784-786.

[21] RÜSCHENDORF L. The Wasserstein distance and approximation theorems[J]. Z. Wahrsch. verw. Gebiete, 1985, 70(1): 117-129.

[22] SHAFIEEZADEH-ABADEH S, MOHAJERIN-ESFAHANI P, KUHN D. Distributionally robust logistic regression[C]//Proceedings of NIPS, 2015.

[23] SHAPIRO A. On duality theory of conic linear problems[M]//Goberna M Á, López M A. Semi-Infinite Programming: Recent advances: Nonconvex optimization and its applications. Springer Science+Business Media, 2001: 135-165.

[24] GRANT M, BOYD S. CVX: Matlab software for disciplined convex programming Beta[Z]. http://cvxr.com/cvx.

[25] LÖFBERG J. YALMIP: A toolbox for modeling and optimization in MATLAB[C]//Proceedings of IEEE ICRA. 2004: 284-289.

[26] GAO R, CHEN X, KLEYWEGT A J. Wasserstein distributionally robust optimization and variation regularization[Z]. arXiv: 1712.06050v3.

[27] MOHAJERIN-ESFAHANI P, KUHN D. Data-driven distributionally robust optimization using the Wasserstein metric: Performance guarantees and tractable reformulations[J]. Math. Prog., 2018, 171(1-2): 115-166.

[28] FOURNIER N, GUILLIN A. On the rate of convergence in Wasserstein distance of the empirical measure[J]. Probab. Theory Relat Fields, 2015, 162(3-4): 707-738.

[29] GAO R. Finite-sample guarantees for Wasserstein distributionally robust optimization: Breaking the curse of dimensionality[Z]. arXiv: 2009.04382.

[30] LI J, HUANG S, SO A M C. A first-order algorithmic framework for Wasserstein distributionally robust logistic regression[C]//Proceedings of NIPS, 2019.

[31] TSENG P. On accelerated proximal gradient methods for convex-concave optimization[R]. 2008.

[32] HSIEH C J, CHANG KW, LIN C J, et al. A dual coordinate descent method for large-scale linear SVM[C]//Proceedings of ICML. 2008: 408-415.

[33] CHENG W, LIU Q, LI D. An accurate active set conjugate gradient algorithm with project search for bound constrained optimization[J]. Opt. Lett., 2014, 8(2): 763-776.

[34] GOODFELLOW I, BENGIO Y, COURVILLE A. Deep learning[M]. Massachusetts: MIT Press, 2016.

[35] 周志华. 机器学习[M]. 北京：清华大学出版社，2016.

[36] KURAKIN A, GOODFELLOW I J, BENGIO S. Adversarial machine learning at scale[C]//Proceedings of ICLR, 2017.

[37] FINLAYSON S G, BOWERS J D, ITO J, et al. Adversarial attacks on medical machine learning[J]. Science, 2019, 363(6433): 1287-1289.

[38] GAO R, KLEYWEGT A J. Distributionally robust stochastic optimization with Wasserstein distance [Z]. arXiv: 1604.02199, 2016.

[39] BLANCHET J, MURTHY K. Quantifying distributional model risk via optimal transport[J]. Math.Oper. Res., 2019, 44(2): 565-600.

[40] LI J, CHEN C, SO A M C. Fast epigraphical projection-based incremental algorithms for Wasserstein distributionally robust support vector machine[C]//Proceedings of NIPS, 2020.

[41] LI J, SO A M C, MA W K. Understanding notions of stationarity in non-smooth optimization[J]. IEEE Signal Process. Mag., 2021, 37(5): 18-31.

基于环境模型的强化学习研究进展*

俞 扬

（计算机软件新技术国家重点实验室（南京大学），南京 210023）

1 引言

强化学习主要研究如何让智能体在与环境交互的过程中自主学习完成特定任务的最优策略，即从状态到动作的映射，从而最大化长期累积回报奖赏[1]，是机器学习的主要研究领域之一[2]。强化学习由于其学习框架的普适性，在许多序列决策任务中得到了广泛应用，并取得了前所未有的巨大成功。例如，在经典的 Atari 游戏中，利用强化学习训练得到的智能体策略已经达到甚至超越了绝大多数人类玩家的水平[3-4]。在人类传承数千年的围棋领域，2016 年由 DeepMind 公司研发的人工智能围棋程序 AlphaGo[5]在与人类顶尖棋手的对弈中取得了突破性的胜利，宣告了在围棋领域人工智能水平已经超越人类，AlphaGo 项目的成功也突显出强化学习所具备的强大决策能力。

然而，强化学习的成功案例也多限于游戏场景，在现实场景中取得应用效果的任务则很少。究其原因，强化学习算法需要通过大量试错来总结学习出完成特定目标任务的最优行动策略。在虚拟数字世界中，智能体与环境的交互极其方便且成本很低，因此，利用在线式的强化学习方法，通过大规模采样直接与环境进行交互学习即可得出优化策略。强化学习在游戏棋牌类应用中率先取得了有目共睹的进展[3-6]。当试图将强化学习应用到现实世界任务场景中时，通常不再有可供其高效采样的虚拟环境，而且受现实因素制约，策略也不能直接通过与真实物理环境交互进行学习。一方面，强化学习自身的试错机制会导致在学习过程中出现不可避免的错误行为和代价风险，而真实世界是无法重

* 本文得到科技创新 2030——“新一代人工智能”重大项目(2020AAA0107200)和国家自然科学基金(61876077)资助。

置的，所有的代价和后果不会像数字游戏那样通过重开一局游戏而消除；另一方面，强化学习算法需要大规模的策略采样才能学出最优策略，而真实世界是无法加速的，所有采样只能按照真实时间线依次获取，如此低效率的采样显然无法匹及数字虚拟环境在计算机中通过高并发和模拟加速实现的高效采样，最终无法满足强化学习算法所需的大规模采样需求，比如在经典的 Atari 游戏中使用 DQN 算法训练策略需要亿级的采样量。

面对需要大量试错数据的制约，强化学习研究者试图借助现实应用场景中累积的历史交互数据对策略进行训练优化，其中，环境重构是解决此类问题的有效方案。基于环境重构的强化学习可以简单总结为三个步骤：首先从历史数据中学习环境模型，然后在环境模型中训练智能体策略，最后将得到的策略应用于现实任务。如此一来，强化学习只在环境模型中进行训练，策略的学习只用到了历史积累的数据，避免了在实际环境中进行试错的代价。在这一技术途径中，如何从历史交互数据有效学习环境模型，并且使环境模型中训练的策略模型可迁移到实际环境中，是面临的关键问题。

本文的主旨是将作者在基于环境模型的强化学习方面的研究做一个概括性的介绍，第 2 节介绍强化学习和环境模型学习的相关背景，第 3 节介绍基于对抗生成的环境模型的学习，第 4 节和第 5 节将分别介绍在两种推荐任务上环境模型的学习与策略学习的结果，第 6 节总结本文并讨论值得进一步研究的问题。

2 相关背景

2.1 强化学习

如图 1 所示，在一个标准的强化学习框架中，智能体通过感知信息、执行动作、获得反馈和环境发生交互。具体而言，每一步中，智能体通过传感器感知得到当前自己在环境中所处的状态，然后通过自己的决策程序选择一个动作进行执行，动作执行之后，智能体在环境中所处的状态发生变化，而且还会感知到环境提供的一个反馈信号，即瞬时奖赏。值得一提的是，智能体所处的新状态将会在下一时刻被智能体感知到。

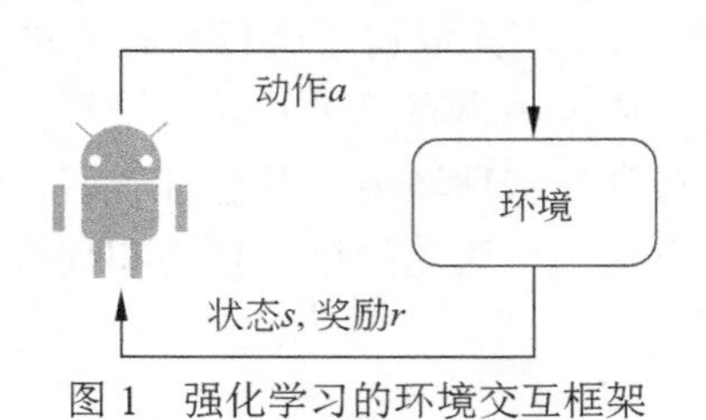

图 1　强化学习的环境交互框架

强化学习面临的决策过程通常可以抽象为马尔可夫决策（Markov decision process，MDP）过程，可以用一个五元组（S, A, T, R, γ）来表示。其中，S 是整个任务场景下所有可能状态的集合，即状态空间；A 是策略可选择动作行为的集合，即动作空间；T: $S\times A\to S$ 是状态转移模型，它描述了在当前状态 s 下执行动作 a 转移到下一个状态 s'的规律；$R: S\times A\to R$ 是奖赏函数，负责给出在当前状态 s 下执行动作 a 得到的奖赏值 r；γ 是长期累积回报奖赏的折扣系数，通常取接近并小于 1 的值，以保证累积奖赏能够收敛，通常取 0.99，0.999 等。

在与环境交互的过程中，智能体观测到环境的状态 s_t，并由策略 π 给出动作 a_t 在环境中执行，然后观测到环境的下一个状态 s_{t+1} 及该步的反馈奖赏 r_t，如此循环，直到环境进入终止状态为止。强化学习旨在让智能体通过与环境的交互来优化策略 $\pi: S\to A$，即从状态到动作的映射，从而最大化长期累积 γ 折扣回报奖赏 $r_1+\gamma r_2+\gamma^2 r_3+\cdots$。

强化学习算法通常遵循“评估-改进”迭代的框架：在评估步，当下的策略模型与环境交互，通过交互的结果对策略模型的好坏进行评分；在改进步，找到策略模型改进的方向，并由此更新策略模型。两个步骤反复迭代，不断改进策略模型的性能。由于在评估步需要实际与环境进行交互，即用当下的策略进行试错，可见在反复迭代的过程中，强化学习需要不断试错，这使强化学习在实际环境中的学习面临极高的试错代价。

强化学习算法按照是否对环境模型进行还原可以分为两大类别，即基于环境模型（model-based）的方法和免模型（model-free）的方法。基于模型的强化学习算法根据在环境中的交互数据，收集转移概率和奖赏函数的观测值，尝试对 T 和 R 进行还原，之后可以使用传统的求解 MDP 过程的方法（如动态规划算法）或免模型的强化学习算法在 MDP 模型(环境模型)中进行最优策略的求解,这一类的方法包括PEGASUS[7]和RMAX[8]等。在模型无关的强化学习算法中，算法并不需要显式地对 T 和 R 进行建模，而是将其隐式地表示在其他模型构件中，通过对这些模型构件进行求解以找到最优策略。

2.2 策略模仿学习

除了从试错数据中学习策略的强化学习方法外，获得策略模型的另一个途径是模仿专家教师的行为，这一分支领域称为模仿学习[9]，也称从示范中学习[10]、伴随学习[11]等。模仿学习的设定和传统的强化学习有所区别，强化学习通常从经历中学习，是一种基于策略采集数据和更新策略循环迭代的过程，而模仿学习主要关注如何充分利用教师提供的示例。

在模仿学习的设定中，教师提供某个具体的强化学习任务执行成功的示例，智能体则从这些示例中学习导师使用的策略。教师数据形如 $s_1, a_1, s_2, a_2, \cdots, s_m$ 给出了每个状态下的教师行为。下面通过跟随教师行为学习策略模型，将教师数据整理为标记数据集，形如

$$(x_1=s_1, y_1=a_1), (x_2=s_2, y_2=a_2), \cdots, (x_{m-1}=s_{m-1}, y_{m-1}=a_{m-1})$$

接下来，可以通过监督学习技术，从标记数据中学习到从状态到动作的映射，即通过最小化 loss($\pi(x)$, y)，得到近似教师行为的策略模型 π。使用监督学习对状态上的动作进行逼近的这一方法称为“行为克隆”，是最简单直接的模仿学习方法。

2.3 环境模型学习

对于环境模型，其中最为关键的是转移函数 T（在很多实际应用中，奖励函数 R 是可以直接获得的），也就是说，特别关注状态转移模型 $T: S\times A\to S$ 的学习。策略与真实环境交互的数据通常呈现为状态与动作交替的序列，形如 $s_1, a_1, s_2, a_2, \cdots, s_m$。在这样的序列中，可以很直接地看到一个时刻的状态 s_t 在执行指定动作 a_t 后达到的下一个状态 s_{t+1}。也就是说，在历史数据上，可以直接观测到 T 的输入和输出。接下来，最为直接的想法则是将交互数据序列组织为标记数据集，形如

$$(x_1=(s_1,a_1),y_1=s_2),(x_2=(s_2,a_2),y_2=s_3),\cdots,(x_m=(s_{m-1},a_{m-1}),y_m=s_m)$$

这样的标记数据集可以直接交给监督学习，学到从 x 到 y 的映射函数，即作为转移函数。监督学习的目标可以简单表达为寻找模型 T，使损失函数 loss($T(x)$, y)在所有样本上的加和值最小。这里的损失函数可以有多种选择，但都是关于预测的状态与数据上的下一时刻状态的差异度量。

对于以往的环境模型学习方法，在学习的不同环节进行改进，但都没有逃脱以上的框架。例如，世界模型[12]中使用了 CNN 进行状态特征的提取并使用 LSTM 进行时序关系的学习，然而这也是在相同的学习框架下进行的改进。

3 基于对抗生成的环境模型学习

3.1 环境模型的多智能体视角

在经典强化学习中，智能体是通过与环境交互进行学习的。从多智能体角度来看，强化学习中的交互亦可视为两个智能体进行的交互，其中环境即为一个智能体，负责对外界交互行为做出反馈，两个智能体之间存在着“互为环境”的关系，如图 2 所示。例如，在商品推荐系统中，用户和推荐系统是两个互相交互的智能体，用户将推荐系统视为环境，而推荐系统也将用户视为环境。

强化学习中的环境重构即可将环境视为一个智能体策略，这样一来，前面背景中介绍的环境的学习与策略模仿学习就有了直接的对应关系。基于监督学习的环境模型学习方法，即对应行为克隆模仿学习方法。

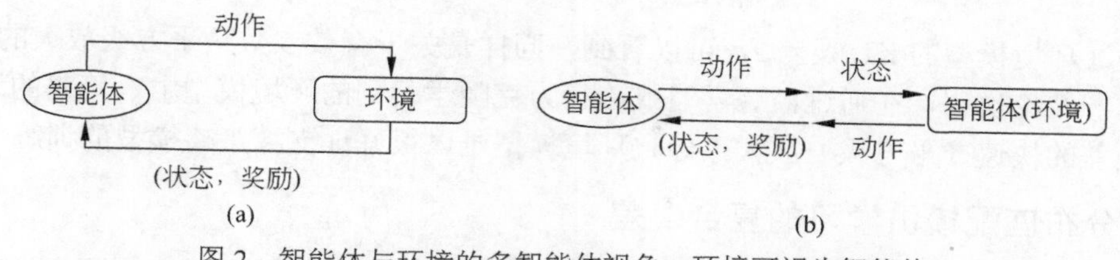

图 2 智能体与环境的多智能体视角：环境可视为智能体
(a) 强化学习经典视角；(b) 强化学习多智能体视角

3.2 行为克隆模仿学习的复合误差

读者到这里可能有一个疑问，既然行为克隆可以直接用于学习环境模型，那么环境模型的学习是否就已完成？下面在环境模型中用强化学习训练策略是否可以？注意到我们的最终目标是要把环境模型中训练的策略应用到真实环境中，那么我们自然会期待环境模型中能够取得最大奖励回报的策略模型在真实环境中也能取得高回报。然而，无论如何，我们学到的环境模型总是存在误差，不会与真实环境完全一致。那么，当环境不一致时，在环境模型中训练的策略回报会有多少损失是需要关注的问题。

Xu 等人[13]对此问题进行了详细的理论分析。下面简要概括分析的结论。在模仿学习中，如果以最小化每个状态上教师动作与策略模型输出动作的 KL 距离取得了 KL 距离为 ε 的策略模型，那么，策略模型的回报值相比教师策略的回报值损失上界为

$$O\left(\frac{1}{(1-\gamma)^2}\sqrt{\varepsilon}\right)$$

其中，γ 是折扣因子，$R_{\max}$ 是奖励值上界。这里需要说明的是，如果把回报计算中所有的折扣系数加起来，$1+\gamma+\gamma^2+\cdots$，那么正好等于 $1/(1-\gamma)$，称为有效步数。这里的损失上界表明，行为克隆学习策略模型的误差在执行策略时会随着步数的增加而放大，且呈平方级放大。当 γ 取值为 0.999 时，误差被放大了 1 000 000 倍。更加不幸的是，平方级的误差放大性质同样存在于回报损失的下界，这就意味着此方法无法改善误差被平方级放大的缺陷。

根据强化学习的多智能体视角，Xu 等人[13]将上述行为克隆策略模型学习的结果拓展到环境模型的学习上，得到同一策略模型在环境模型中的回报和在真实环境中的回报误差存在的上界为

$$O\left(\frac{\gamma}{(1-\gamma)^2}\sqrt{\varepsilon}+\frac{2}{(1-\gamma)^2}\sqrt{\varepsilon_\pi}\right)$$

其中，ε_π 是给定的策略与环境模型数据采样策略的差异，可以认为是与 ε 无关的常数。

我们关注环境模型的 KL 误差 ε，可以看到，同样是关于有效步数呈平方级放大的关系。

这一结论说明，在通过监督学习或称行为克隆学习到的环境模型中，策略的运行会导致误差的快速（平方级）放大，导致环境模型难以很好地支撑策略模型的训练。

3.3 分布匹配模仿学习的复合误差

幸运的是，虽然以往环境模型学习方法较单一，但在策略模仿学习方面，还有更加有效的方法。对抗生成模仿学习[14]是一类不同的模仿学习方法，其不同之处在于不使用通常监督学习的损失函数，而是考虑整个数据的分布是否与真实数据的整体分布匹配。假设模仿学习策略在环境中交互产生的数据分布与教师数据的分布 JS 距离为 ε，Xu 等人[13]得到策略模型的回报值相比教师策略的回报值损失上界为

$$O\left(\frac{1}{1-\gamma}\sqrt{\varepsilon}\right)$$

可见通过分布匹配，这里策略误差的系数仅仅是随着步数增加线性放大了，当 γ 取值为 0.999 时，误差被放大了 1000 倍，相比行为克隆方法，误差为其 1/1000。

通过强化学习的多智能体视角，Xu 等人[13]得到了基于分布匹配环境模型学习的结果，同一策略模型在环境模型中的回报和在真实环境中的回报误差存在的上界为

$$O\left(\frac{1}{1-\gamma}\sqrt{\varepsilon}+\frac{1}{(1-\gamma)^2}\sqrt{\varepsilon_\pi}\right)$$

可见，损失关于环境模型学习误差的部分得到了改善，误差仅被线性放大，环境模型的可用性得到了提高。Xu 等人[13]的以上理论结果首次揭示了降低误差放大系数的可行途径，因此，使用分布匹配的方式进行环境模型的学习是潜在的有效途径。

3.4 基于对抗生成的环境学习方法

生成对抗网络（GANs）[15]及其一系列变体[16-17]快速推动了无监督机器学习的发展。最早的生成对抗网络通过求解一个二元零和博弈的过程来学习一个生成式模型：

$$\arg\min_{G}\arg\max_{D\in(0,1)}\mathbb{E}_{x\sim p_E}[\log D(x)]+\mathbb{E}_{z\sim p_z}[\log(1-D(G(z)))]$$

其中，p_z 是随机噪声分布。在这个博弈训练过程中，判别器 D 通过监督学习训练，其中真实样本标记为 1，生成样本标记为 0，以此来尽可能区分开生成样本和真实样本，从而提高分类正确率；而生成器 G 学习生成服从真实样本分布 p_E 的样本（用 x 表示），也就是尽可能生成比较接近真实的样本，从而降低判别器 D 的分类正确率。当判别器 D 无法再对两类样本正确区分时，这个博弈过程结束，此时生成器 G 已经学会如何生成服从真

实分布的样本。生成对抗网络的训练过程可以看作在一个高维参数空间中寻找纳什均衡，其本质是优化生成样本分布和真实样本分布之间的 JS 散度，匹配分布。

生成对抗模仿学习（GAIL）[14]将 GAN 的框架拓展到模仿学习任务上。在模仿任务中，生成器对应为强化学习策略模型 π，生成的样本对应为策略在环境中进行的采样，同时判别器的实现与 GAN 中的相同，并对如下目标进行 D 和 π 的交替优化，从而获得基于分布匹配的模仿策略。

$$\underset{\pi}{\arg\min}\underset{D\in(0,1)}{\arg\max}\,\mathbb{E}_{\pi}[\log D(s,a)]+\mathbb{E}_{\pi_E}[\log(1-D(s,a))]-\lambda H(\pi)$$

进一步地，透过强化学习的多智能体视角，可以将生成对抗模仿学习方法用于环境模型的学习。这里需要注意的是，如果仅给定历史数据，则表示我们也无法直接取得生成这些历史数据使用的策略模型。因此与模仿学习的一个关键不同点在于进行环境模型学习时，策略也同样是未知的，因此，需要同时对环境模型和策略进行模仿学习。

图 3 是基于生成对抗的环境模型学习框架，从整体来看，该学习框架依照生成对抗网络的设计机制，主要包括两个模块：生成器（generator）和判别器（discriminator）。其中生成器模块是根据实际特定的业务场景定义的多智能体交互环境，负责模拟生成轨迹数据集合；判别器模块作为二分类判别器，其输入是形如（s, a）的状态动作对数据，输出是该数据来自于真实状态动作联合分布的概率。通过借助这种对抗训练的机制，判别器模型即定义了两个分布的差异度量函数 L，从而可以实现环境重构目标优化问题的求解。

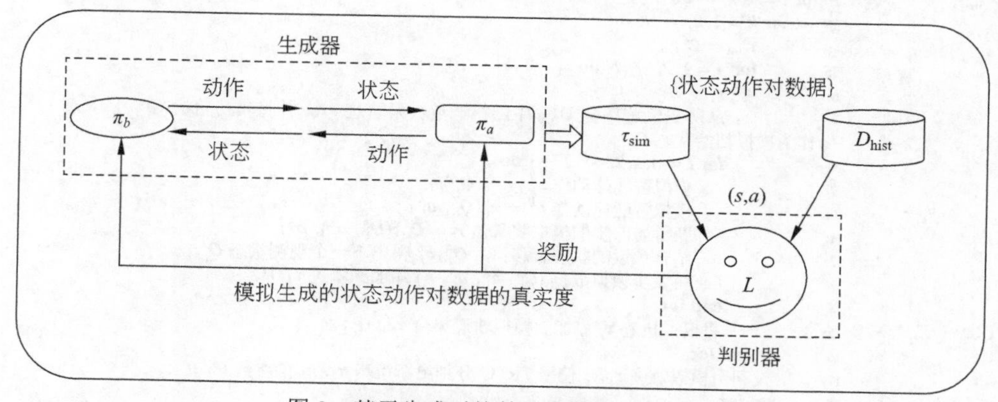

图 3　基于生成对抗的环境模型学习框架

在迭代训练过程中，每轮迭代主要包括三步：模拟、判别和策略更新。

模拟就是利用生成器模块中的多智能体环境，按照预定的推荐业务交互逻辑执行策略，模拟产生多条交互轨迹数据。特别地，所有模拟轨迹的初始状态均直接从历史轨迹集合的初始状态中采样而得，从采样出的真实状态开始触发模拟，由策略 π_a 和 π_b 依次产生交互状态行为序列，同时由判别器网络负责给出每步模拟的状态行为数据的仿真奖赏值，如此循环，直至达到预设的终止状态或者最大轨迹长度 T。

判别器主要对模拟生成的状态动作对数据进行真实度的判别。通过将历史真实状态动作对数据标记为 1，模拟状态动作对数据标记为 0 的方式对其进行监督式学习训练。在进行模仿学习更新策略的过程中，判别器模型作为生成器中策略模拟生成轨迹数据时的奖赏函数，其输出值的实际意义就是生成数据的真实度，该值可以作为策略模拟时的回报奖赏，以此来指导生成器中每个智能体策略通过强化学习的方法最大化生成轨迹的期望 T 步累计奖赏之和，即模拟产生的整条轨迹的仿真度，这正与环境重构的目标契合。

策略更新主要是利用上述模拟产生的轨迹数据，使用强化学习算法对生成器中智能体策略进行更新优化，使其能在后续迭代过程中产生更接近真实分布的轨迹数据，直到最终训练收敛。此时，判别器无法对生成的模拟数据与真实数据进行正确分类，即生成器中的智能体环境可以模拟产生具有极高真实度的轨迹，从而实现模拟环境重构。

图 4 给出了算法的伪代码，基于强化学习算法 TRPO 和生成对抗框架，进行环境模型的学习。

输入: $D_{\text{real}} = \{\tau_1, \tau_2, \cdots, \tau_n\}$：历史观测 T 步轨迹集合；
N：每轮迭代中模拟生成轨迹数量；
K：每轮迭代中策略更新的次数；
1: 随机初始化策略 π_a, π_b 的参数 θ^a, θ^b，判别器 **D** 的参数 σ；
2: **for** $i = 1, 2, \cdots$ **do**
3:   **for** $k = 1, 2, \cdots, K$ **do**
4:     $\tau_{\text{sim}} = \varnothing$；
5:     **for** $j = 1, 2, \cdots, N$ **do**
6:       $\tau_j = \varnothing$；
7:       从历史轨迹集合 D_{real} 中随机选取一条轨迹 τ_r，并将其初始状态作为模拟初始状态 O_0^A；
8:       **for** $t = 0, 1, 2, \cdots, T-1$ **do**
9:         模拟智能体动作 $a_t^A = \pi_a(O_t^A)$；
10:         模拟智能体动作 $a_t^B = \pi_b(O_t^A, a_t^A)$；
11:         获得当前步的模拟奖赏值 $r_t = D_\sigma((O_t^A, a_t^A), a_t^B)$；
12:         结合预设的状态转移，由 O_t^A, a_t^B 得出下一个观测状态 O_{t+1}^A；
13:         将该步模拟数据 $\{O_t^A, a_t^A, a_t^B, r_t\}$ 添加到轨迹 τ_j 中；
14:       **end for**
15:       将模拟轨迹 τ_j 添加到模拟轨迹集合 τ_{sim} 中；
16:     **end for**
17:     利用模拟轨迹 τ_{sim}，使用 TRPO 分别更新策略 π_a, π_b 的参数 θ^a, θ^b；
18:   **end for**
19:   通过最小化如下损失函数来优化判别器 **D** 的参数 σ：

$$\mathbb{E}_{\tau_{\text{sim}}}[\log(D_\sigma((O^A, a^A), a^B))] + \mathbb{E}_{x\sim\tau_{\text{real}}}[\log(1 - D_\sigma((O^A, a^A), a^B))]；$$

20: **end for**
输出: 训练稳定的策略 π_b 即是重构之后的模拟环境。

图 4 基于多智能体视角的环境学习算法 MAIL

4 在淘宝推荐任务中的环境模型学习

本节介绍基于环境模型的强化学习方法在淘宝推荐任务中的使用，详细内容见参考文献[18]。

4.1 问题建模

淘宝搜索是建立在数十种排序因子之上的，对于用户的每一次搜索请求 $u = (\text{user, query})$，淘宝的排序引擎会依次计算每个文档 d 的 n 个排序因子$(x_1(\text{d}), x_2(\text{d}), \cdots, x_n(\text{d}))$，然后将这些因子进行加权求和得出最后的总分 $s(u, \text{d})$，用 $s(u, \text{d})$进行排序，最后展示排名靠前的商品。排序因子的质量和排序权重 $w(u)$的选择都决定着排序结果的好坏。这里，我们重点关注排序权重的选择问题。

淘宝推荐任务可建模为马尔可夫决策过程（MDP）$\text{M} =< S, A, P, R, \gamma>$。

- 状态空间 S：将用户的一些特征和查询信息 u 作为状态 s，用户特征选择了用户的性别、年龄、购买力，查询信息只提取到查询行业的粒度。
- 动作空间 A：将排序的权重 $w \in \text{R}^n$ 作为动作 a。
- 奖励函数 R：如果用户购买了商品，会返回一个正的奖励，否则返回 0。
- 策略 π 定义参数化策略 $\pi_\theta : S \to A$。

智能体在状态 s_t 下做了动作 a_t 之后，应该转移到哪一个状态？即时的奖赏应该是多少？在很多强化学习问题（例如，Atari 游戏，围棋）中，都可以很容易得到答案，因为在这些环境中，智能体与环境进行交互非常方便，只需要做出一个动作，然后观察环境反馈的结果就可以了，并不会带来更多的开销。而在电商场景下，与环境的交互是昂贵且耗时的，大量探索式的交互是不切实际的。

我们希望用模仿学习的方法学习用户的意图，也就是模拟线上环境。将环境，也就是在线的用户，视为专家教师。专家历史数据（每天交易产生的大量日志）可以从中学得用户的策略作为我们的环境。注意到，我们日志的量虽然很大，但它是高度有偏的，因为只有发生购买行为的 p_v 才会被记录，所以基于行为克隆的模仿学习是不适用的。

为了区别于训练引擎策略的 MDP 过程 M，我们用 $\text{Me} =< \text{Se, Ae, Pe, Re}, \gamma\text{e}>$表示模拟环境时的 MDP 过程。

- 状态空间 Se：将提取的用户特征和引擎权重作为状态。
- 动作空间 Ae：将用户购买行为作为环境的动作。
- 奖励函数 Re。
- 判别器 Dw，用来判断$< u, w >$是否来自真实数据，用 Dw 的输出作为奖励。
- 策略 π_e 定义参数化策略 $\pi_{\text{e}\theta} : s \to a$。

4.2 算法设置

图 5 给出了整个算法的结构，该结构是 MAIL 算法的直接应用。

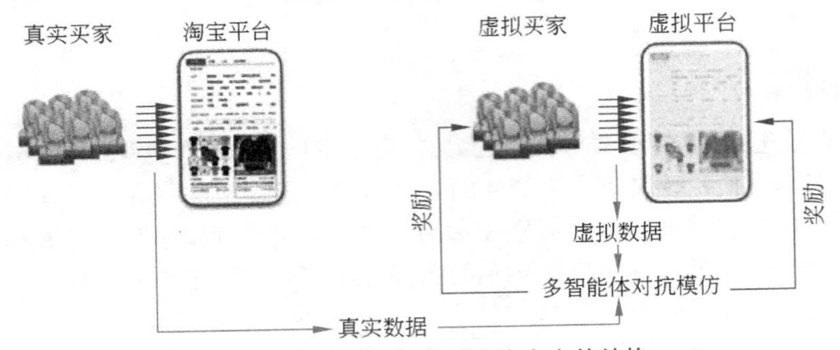

图 5 对抗生成环境学习在淘宝上的结构

用户模型和推荐平台分别由两个网络实现，称为模拟器网络和引擎网络。首先，经过引擎网络，输出引擎的动作，然后，用户特征及引擎动作经过模拟器网络产生用户行为，即是否购买。另外，判别器网络会根据用户特征及引擎权重给出奖励。

在环境模型中直接训练强化学习算法，最终得到的策略可能发现并利用环境模型存在的误差，从而降低实际使用回报。在最初构建的淘宝环境模型中训练策略时，获得了 30%的转换率提升，从经验上来说这个值几乎是不可能达到的。因此使用约束动作（ANC）的修正奖励函数，即当策略输出动作向量的长度超过数据上动作向量长度时，降低奖励：

$$R'(s,a) = R(s,a) / (1 + \max\{\|a\| - \text{数据均值}, 0\})$$

4.3 实验效果

在使用了约束动作奖励函数后，转换率从 30%下降到 11%，经验判断进入合理范围，图 6（a）给出了约束动作奖励函数在线测试的结果，相比不使用的情况可提高接近 2%的营收和 0.5%的销售商品数。图 6（b）显示出相对于基于监督学习的方法，强化学习获得的策略可取得 2%左右的提升。

另外，我们特别关注环境模型学习的有效性，并对比了使用行为克隆学习的环境模型在当天、1 天后、1 周后、1 月后数据上的转换率，结果见表 1。在当天数据上，行为克隆的环境模型有很好的性能，但在之后的日期上，数据分布发生变化后，性能快速衰退。相对而言，基于对抗生成的环境模型则具有稳定的性能。

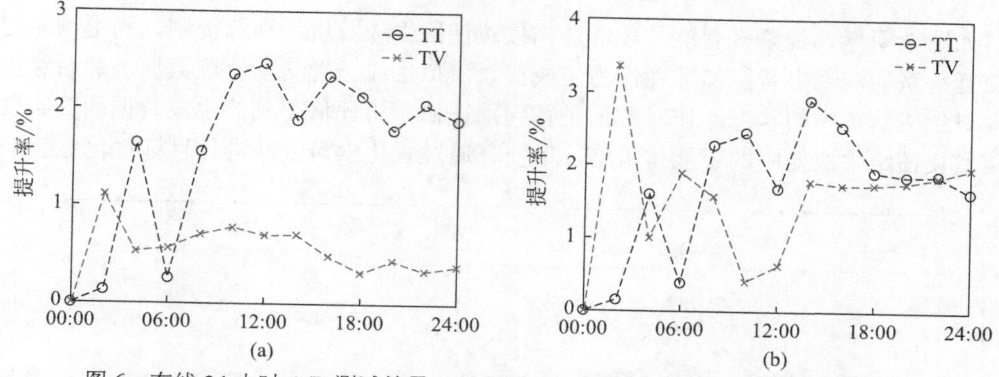

图6 在线24小时A/B测试结果（TT和TV分别是销售总价格和销售总单数）

（a）使用约束动作奖励函数TRPO-ANC相比不使用TRPO的提升率；（b）强化学习策略相对于监督学习策略的提升率

表1 基于行为克隆的环境模型和基于对抗生成的环境模型泛化性能比较

	0天	1天	1周	1个月
基于行为克隆的环境模型	21.65%	8.33%	1.04%	–0.54%
基于对抗生成的环境模型	18.56%	16.67%	8.34%	8.49%

5 在滴滴出行推荐任务中的环境模型学习

本节概括介绍基于环境模型的强化学习方法在滴滴出行司机规划推荐任务中的使用，详细内容见参考文献[19]。

5.1 算法设置

在滴滴出行的业务中，我们进一步关注环境模型的构建。在淘宝推荐的应用中，假设在整个推荐场景中只包含用户和推荐系统两个智能体的假设。从现实条件下的用户角度来看，用户往往接收到许多无法观测到的信息，这些信息在用户做出行为反馈的过程中也起到一定的作用。此类现实任务场景是十分复杂的，在整个序列决策过程中，由于只有部分可观测信息记录在历史交互数据中，也就意味着潜在影响用户行为的混杂因子无法直观地体现在历史交互数据中。如果依然按照前述假设“整个任务场景的所有信息都是可观测的”，那么由于忽略了数据中潜在的混杂效应，环境重构过程将会受到数据中表现出的伪相关性的影响而产生带有严重偏差的用户策略模型，从而误导后续强化学习策略的优化。因此，在环境重构过程中考虑隐藏的混杂因子是非常重要的。

如图 7 所示，混杂多智能体环境重构依然使用生成对抗的训练机制，利用预先定义好交互关系的混杂多智能体环境作为生成器，模拟生成一批交互轨迹数据，然后经由判别器对生成轨迹中的状态动作对数据进行判别评估，将评估得到的生成数据的真实度作为奖赏反馈给生成器中的智能体策略，以便其通过强化学习方法进行目标优化更新。

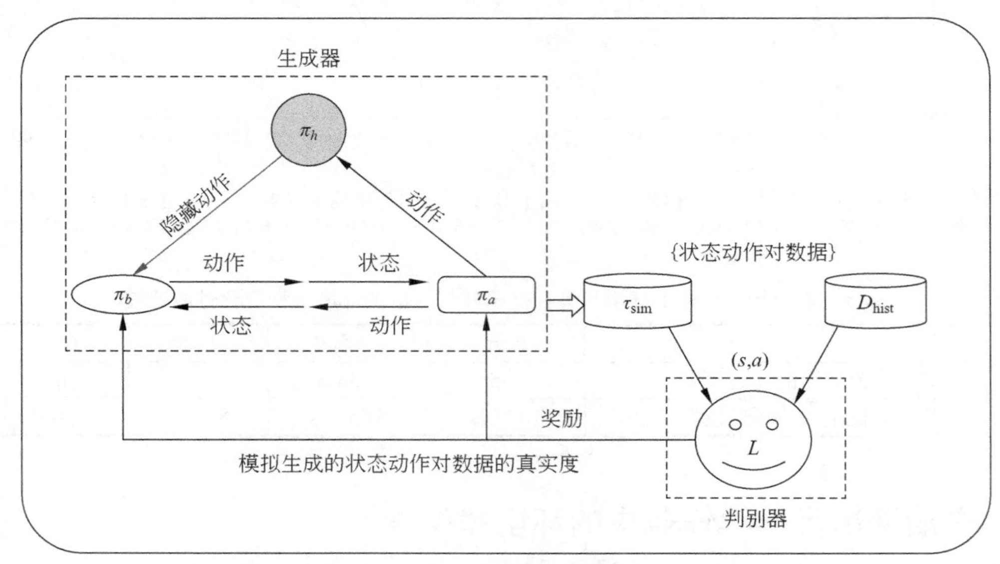

图 7　混杂因子多智能体环境重构框架（DEMER）

由于混杂因子自身的不可观测属性，在定义的混杂多智能体环境中也存在一个无法利用观测数据直接指导更新的混杂智能体策略。该智能体策略在模拟生成轨迹的过程中还原了混杂因子的真实作用过程，但是在历史观测数据中，并没有关于该混杂因子的记录信息。因此，在判别器进行判别更新的过程中，其输入依然是可观测智能体策略的状态行为数据，而混杂智能体策略在模拟轨迹生成过程中的仿真效果无法直接评估。换言之，由于真实轨迹数据中缺少混杂变量的数据，导致判别器在判别过程中没有直接评估混杂策略行为好坏的依据标准，因而无法直接对生成器中的混杂智能体策略进行更新优化。在 MAIL 框架的基础上，可以直接增加不可观测因素的模型，使用相似的训练方法即可完成训练。

5.2　实验效果

首先我们关心的是不可观测的因素是否能够有效还原。为了可以直接观测到环境模型学习的好坏，我们构建了一个低维人造环境，其中司机模型为 π_{d}，推荐系统模型为 π_{p}，

隐藏因素模型为 π_{h}。图 8 中 real π_{p}，real π_{d} 和 real π_{h} 为构造的人造函数，DEMER 和 MAIL 则为两个算法分别还原出的函数。可以观测到，DEMER 还原的 π_{p} 和 π_{d} 均明显优于 MAIL 的结果，并且 DEMER 学到的 π_{h} 在趋势上与真实 π_{h} 较为一致，值得注意的是，DEMER 的学习过程并没有看到 π_{h} 的输入输出数据。因此可以认为，DEMER 对隐藏环境因素的还原是可行的。

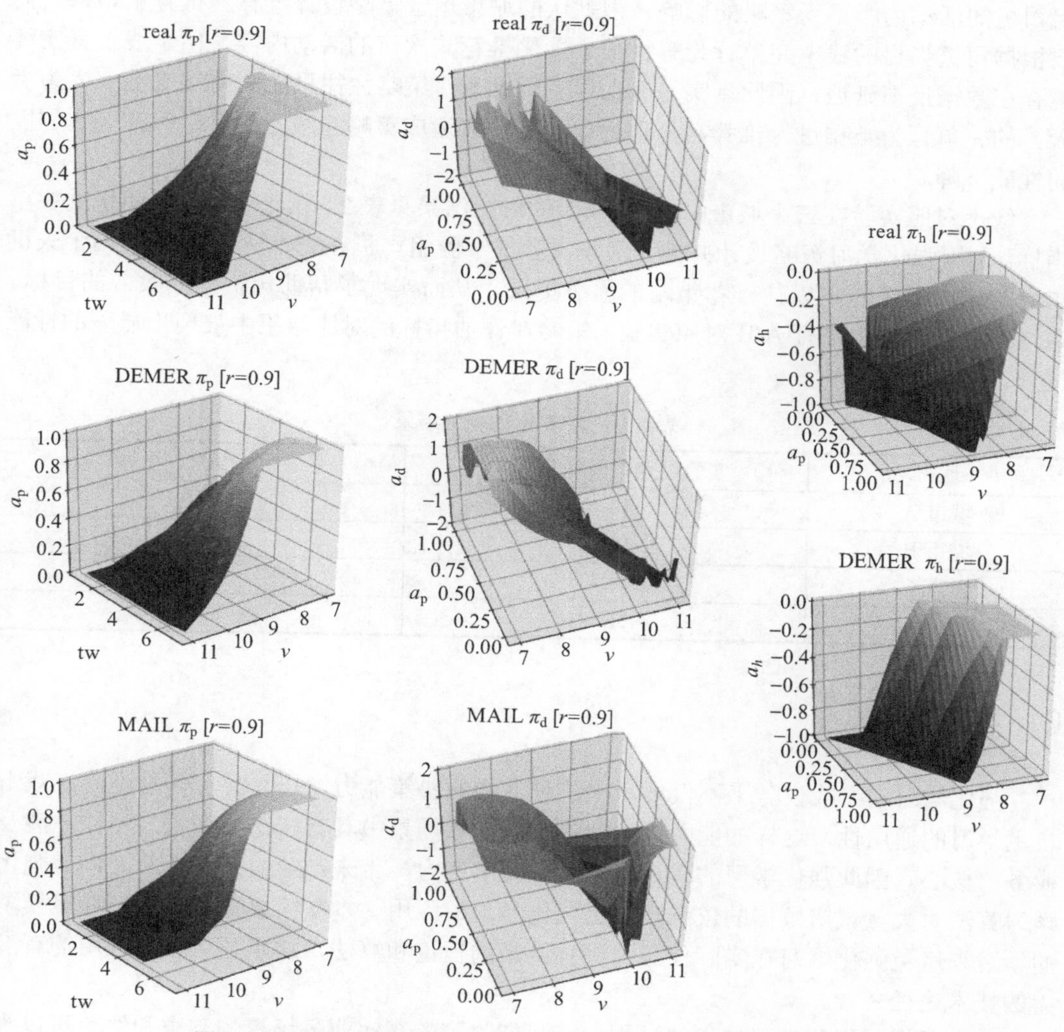

图 8　人造环境上 DEMER 与 MAIL 方法环境模型学习结果对比

最终策略有效性的验证是在真实环境中进行 A/B 测试。将强化学习推荐策略部署到大规模网约车出行平台滴滴出行的司机活动推荐系统中，并在 3 个不同规模的城市 A，B，C（大、中、小规模）中进行了小流量 A/B 对照实验。在每个城市中，将实验司机群体随机划分为人数相等的两组，分别称为对照组 G_c（control group）和实验组 G_t（treat group）。在实验进行过程中，实验组司机的活动内容由强化学习策略设计并推荐，对照组司机的活动内容不受策略的影响，由默认的城市运营策略设计推荐。推荐策略每天需要根据司机当日的线上完单行为等特征个性化推荐其次日的活动内容。由于线上真实环境存在数据反馈延迟，因此真实的线上环境运行时，策略往往只能利用 $T-2$ 日（T 为当天）的完单行为数据进行推荐，从理论上讲，这会造成策略在线上执行时存在一定程度的性能下降。

线上对照实验在三个城市中进行 14 天的对比结果见表 2。相比于对照组司机的完单指标，利用强化学习策略设计并推荐活动内容的实验组司机在完单量和 GMV 两个关键指标上均取得了超越以往推荐策略的显著提升，从而实现了帮助司机提高收入的目标。通过在真实环境中进行 A/B 对照实验，策略在评估指标的对比结果中表现出显著的性能提升。

表 2　三城市 A/B 测试结果

城市	实验组完单量提升率/%	实验组 GMV 提升率/%
城市 A	10.73	6.16
城市 B	10.16	9.38
城市 C	18.47	17.84
总计	11.74	8.71

6　结束语

强化学习在游戏任务上不仅展示出超越人类的决策能力，还具备在多种不同游戏上自主学习的通用性，这样的能力在各行各业的业务场景中具有广泛的需求，还往往涉及业务的核心。因此强化学习的落地可望转化为生产力，带来产业的升级革新。但是强化学习算法对大量试错数据的依赖阻碍了强化学习的应用。本文简要介绍了通过环境模型的学习来避免强化学习在实际任务中进行试错的理论和算法，这是强化学习应用的有潜力的技术途径之一。

在尝试应用强化学习技术的过程中，我们收到的真实业务场景的需求和约束可以为未来的研究提供方向。这里的一些关键需求包括部署前验证、首次即成功、决策可解释等。部署前验证在监督学习模型的部署过程中十分常见，在训练数据上学到的模型在使用前必须在验证数据上检验，以减少使用时出错的风险。由于策略模型目标在于做出更好的决策，因此无法直接使用历史上的决策数据对新的策略决策进行评估。强化学习领域对模型的验证研究很少，最近在异策略评估（off-policy evaluation）方面有一些研究进展，然而从实验效果上看，尚不具备精确的验证能力。另一种对策略进行评估的方法是从验证数据学习验证环境模型，在训练环境模型中学到的策略放到验证环境模型中评估性能，这一方法对环境模型的学习又提出了要求。实际业务环境通常都期待策略首次上线即成功，难以容忍策略在线探索，这就要求我们从历史数据中稳定地学到优于历史策略的结果。另外，对策略模型给出做决策的合理解释，对关键业务决策上取得责任人的信任十分重要。以往对预测模型的解释往往通过规则抽取或可视化技术对预测模型产生输出的机制进行解释，然而在决策任务中，策略模型做出决策的原因是应对环境的转移变化，对决策的解释不仅要从策略模型中寻找，还要从环境的转移中寻找，因此环境模型的学习也不可避免。

基于环境模型的强化学习算法一度因为在游戏环境中的表现不如免模型强化学习算法，因而受到的关注不多。从实际业务的需求中可以看到，环境模型对强化学习的应用有多个维度上的支撑，具有难以替代的作用。可以预见，更加高效的环境模型学习是未来强化学习大规模应用的关键技术之一。

参考文献

[1] SUTTON R S, BARTO A G. Reinforcement learning: An introduction[M]. 2nd Edition. MIT Press, 2018.

[2] 周志华. 机器学习[M]. 北京：清华大学出版社, 2016.

[3] MNIH V, KAVUKCUOGLU K, SILVER D, et al. Human-level control through deep reinforcement learning[J]. Nature, 2015, 518: 529-533.

[4] MNIH V, KAVUKCUOGLU K, SILVER D, et al. Playing Atari with deep reinforcement learning[Z]. arXiv, vol. abs/1312.5602, 2013.

[5] SILVER D, HUANG A, MADDISON C J, et al. Mastering the game of go with deep neural networks and tree search[J]. Nature, 2016, 529: 484-489.

[6] BROWN N, SANDHOLM T. Safe and nested subgame solving for imperfect-information games[C]//Proceedings of Advances in Neural Information Processing Systems. 2017, 30: 689-699.

[7] NG A Y, JORDAN M P. A policy search method for large MDPs and POMDPs[C]//Proceedings of the 16th Conference on Uncertainty in Artificial Intelligence. 2000: 406-415.

[8] BRAFMAN R I, Tennenholtz M. R-MAX——A general polynomial time algorithm for near-optimal reinforcement learning[J]. Journal of Machine Learning Research, 2003, 3: 213-231.

[9] SCHAAL S. Is imitation learning the route to humanoid robots[J]. Trends in Cognitive Sciences. 1999, 3(6): 233-242.

[10] ATKESON C, SCHAAL S. Robot learning from demonstration[C]//Proceedings of the 14th International Conference on Machine Learning. 1997: 12-20.

[11] ABBEEL P, NG A Y. Apprenticeship learning via inverse reinforcement learning[C]//Proceedings of the 21st International Conference on Machine Learning. 2004: 1-8.

[12] HA D, SCHMIDHUBER J. Recurrent world models facilitate policy evolution[C]//Proceedings of Advances in Neural Information Processing Systems. 2018, 31: 2455-2467.

[13] XU T, LI Z, YU Y. Error bounds of imitating policies and environments[C]//Proceedings of Advances in Neural Information Processing Systems. 2020: 33.

[14] HO J, ERMON S. Generative adversarial imitation learning[C]//Proceedings of Advances in Neural Information Processing Systems. 2016, 29: 4565-4573.

[15] GOODFELLOW I J, POUGET-ABADIE J, MIRZA M, et al. Generative adversarial nets[C]//Proceedings of Advances in Neural Information Processing Systems. 2014, 27: 2672-2680, 2014.

[16] ARJOVSKY M, CHINTALA S, BOTTOU L. Wasserstein GAN[Z]. arXiv 1701.07875, 2017.

[17] GULRAJANI I, AHMED F, ARJOVSKY M, et al. Improved training of Wasserstein GANs[C]//Proceedings of Advances in Neural Information Processing Systems. 2017, 30: 5769-5779.

[18] SHI J C, YU Y, DA Q, et al. Virtual-Taobao: Virtualizing real-world online retail environment for reinforcement learning[C]//Proceedings of the 33rd AAAI Conference on Artificial Intelligence, 2019.

[19] SHANG W, YU Y, LI Q, et al. Environment reconstruction with hidden confounders for reinforcement learning based recommendation[C]//Proceedings of the 25th ACM SIGKDD Conference on Knowledge Discovery and Data Mining, 2019.

自适应迭代与采样的黑盒对抗攻击方法

韩亚洪　石育澄

（天津大学智能与计算学部，天津 300350）

1　引言

对抗样本[1-2]揭示了深度神经网络内在的脆弱性。基于攻击方对于目标模型[3]的知识，对抗攻击可以分为白盒攻击与黑盒攻击。在黑盒攻击中，攻击方只能对目标模型进行查询并得到返回的硬标签，而无法得知目标模型的完整知识。基于迁移的攻击[4-7]、基于决策的攻击[8-10]及基于零阶优化的攻击[11-14]是三类主流的黑盒攻击方法。其中，基于迁移的攻击方法使用在本地替代模型上生成的对抗样本，并利用其迁移能力在目标模型上实现攻击[5-6,15]。基于决策的攻击方法通过在原始图像的输入空间随机搜索以减小噪声幅度。多篇近期的研究[10,16-18]表明，构建对抗样本不是简单地欺骗深度神经网络，而是定量地评估目标模型的鲁棒性。通过不断压缩对抗扰动，可以逐步实现准确估计错误分类的最小噪声幅度。对于一个图像分类器，错分每张图像所需的最小噪声、在一次攻击过程的每个阶段的合理查询方向，甚至一张图像中每个像素的敏感度都是不同的[19]。因此，对一个目标模型鲁棒性的准确计算需要根据每张图像及其攻击过程进行适应性评估。然而，目前主流的基于迁移和基于决策的攻击都存在严重的缺陷。

现有的基于迁移的攻击的问题具有两重性。首先，黑盒场景下模型之间的决策边界相距较远。基于迁移的攻击的迭代轨迹难以在带有少量噪声的情况下穿越目标模型的决策边界，这是因为它们是基于沿着替代模型梯度上升方向的单调搜索，这损害了对抗样本的迁移性[20]。其次，尽管噪声幅度决定了攻击方法的性能，基于迁移的方法产生的对抗样本包含一定量的冗余噪声，并且这些噪声不能通过简单地增加迭代次数来完全消除。我们采用一种后迭代压缩机制来排出冗余的对抗噪声。

基于决策的攻击的问题主要存在于噪声压缩过程中的常采样中。一方面，白盒攻击通过反向传播直接建模像素和类别之间的相关性[4,21]。对于黑盒攻击，攻击者可用的唯一线索是历史查询，它是对每个像素噪声敏感性的无偏特征描述。然而，现有的大多数基于决策的攻击[10, 22]都使用与历史查询或当前噪声无关的常数采样设置，这严重阻碍了噪声压缩的效率。另一方面，基于决策的攻击中失败的采样包含决策边界的位置信息[23]。虽然失败的采样（例如，样本落在了真实类别）不能直接用于压缩噪声，但它们以更大的概率描述了穿过决策边界的方向。既然我们想要尽可能多的样本落在决策边界的另一侧，这些信息就能被用于调整采样分布，并使新样本远离失败概率高的方向。但是现有的基于决策的攻击总是在一个恒定的分布上采样，并且在攻击过程中从不改变采样。此外，现有基于决策的攻击中单步修改的步长也是一个常数。随着噪声幅度的减小，查询的成功率会逐渐降低，在步长不变的情况下，噪声压缩的效率会进一步受到影响。

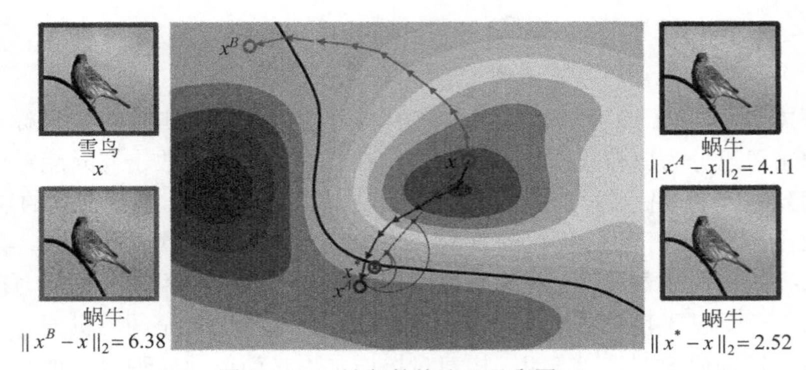

图 1　Curls 迭代的轨迹（见彩图 7）

背景是交叉熵损失的等高线。颜色越红，损失越小。连续的黑色曲线表示“雪鸟”和“蜗牛”两类之间的决策边界，绿色和紫色的两条多段线分别表示简单梯度上升和二分搜索的 Curls 迭代的轨迹，蓝色和红色的环表示在二分搜索后发现的原始图像 x 和对抗样本。表示原始图像和两侧的三个对抗样本的环对应于与边框颜色相同的图像

本文提出了 Curls &Whey 黑盒攻击。在 Curls 迭代过程中，沿着替代模型损失函数的梯度上升和下降方向迭代，如图 1 中的绿色和紫色多段线所示。双向设置“Curls”的迭代轨迹，更可能在较近的距离跨越目标模型的决策边界，这有效地增强了对抗样本的多样性和迁移性。沿梯度上升方向单调增加噪声所引起的边际递减效应也被削弱。Curls 迭代的末尾和开头分别包含了压缩对抗噪声（图 1 中的红色弧线）和引导初始方向的机制。通过对抗性扰动的鲁棒性，Whey 优化被用来进一步压缩噪声幅度。首先根据像素值将对抗性扰动分为若干组，并尝试滤除每组的噪声。然后对每个像素点进行随机压缩，从而一点一点地剔除冗余噪声。

此外，还验证了为了在一步边界攻击后最小化噪声幅度，多元正态分布的方差应与当前噪声的绝对值线性相关，如图 2（a）所示，而不是对每个维度使用单位方差。此外，通过噪声压缩的单调性，分析了基于迁移的攻击初始化对抗噪声相对于随机初始化的优势。基于噪声压缩的这一特性，采取适应性步长调整策略。在当前噪声和历史查询的指导下，提出了适应性对抗边界攻击（CAB）——一种基于决策的攻击，它根据每个像素的噪声敏感度适应性调整采样分布。CAB 通过历史失败样本调整样本分布的平均值，如图 2（b）中的黄色十字所示。通过这种方式，新样本被导向远离失败率高的方向。

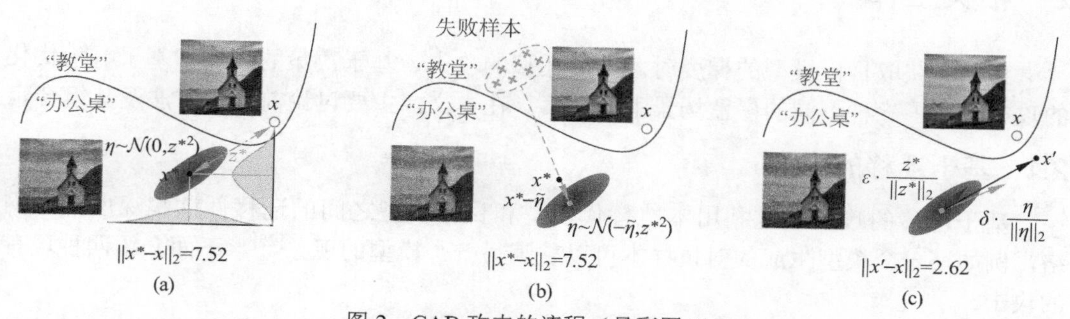

图 2　CAB 攻击的流程（见彩图 8）

（a）方差调整；（b）均值调整；（c）采样与噪声压缩

蓝色曲线表示“教堂”和“办公桌”类别之间的决策边界。在每一个图中，绿圈和黑点分别表示原始图像和当前对抗样本。图（b）中的黄色十字是一组历史失败样本 $\tilde{x}$。CAB 通过图（a）中的当前噪声 z^* 和图（b）中的与历史失败样本方向相反的平均值 $-\tilde{\eta}$ 来适应性调整正态分布的方差，然后对调整后的分布进行随机采样，形成球形方向。最后，将球形（图（c）中的黄色箭头）和源方向（红色箭头）组合在一起，形成图（c）中的完美的对抗样本 x'

在 Imagenet[24]和 Tiny-Imagenet[22]上的实验证明，在相同的查询限制下，我们的方法在 ℓ_2 范数上生成的对抗样本具有较高的迁移性和较小的扰动。我们还系统地研究了各个参数对该方法性能的影响。此外，我们的方法对集成模型和对抗训练模型展现出很强的迁移性[25]。

长期以来，黑盒场景中的针对性错分一直被认为是难解的问题[20]，因为替代模型和目标模型在决策边界和分类空间上的差异阻碍了对抗样本从源类到目标类的迁移。现有的大多数迭代攻击都试图通过简单地将非针对性错分中的梯度下降替换为向目标类的梯度上升来解决这个问题[5-6]。本文将插值集成进迭代过程，使原始图像向目标方向增强，显著降低了针对性错分的难度。

本文的贡献总结如下：

（1）为了最大化降噪期望，边界攻击中的采样方差应该与当前噪声成正比。

（2）提出了 Curls 迭代，这是一种旨在通过结合梯度上升和梯度下降方向来提高迭代轨迹的多样性和对抗样本的迁移性的黑盒攻击方法。

（3）提出 Whey 优化，一种利用扰动的鲁棒性的新颖的噪声压缩方法。

（4）开发了 CAB，这是一种基于决策的攻击，它利用当前的噪声和失败样本来调整采样过程中的正态分布。在多个数据集和模型上的大量实验表明，CAB 的性能优于其他基于决策的攻击。

（5）将所提出的方法扩展到针对性攻击，显著提高了黑盒场景下迭代方法的攻击效果。

2 相关工作

当无法获取目标模型的梯度时，基于迁移的攻击、基于决策的攻击和基于零阶优化的攻击给出了三种不同的黑盒场景下的方案。在本文，主要讨论前两种方法及其组合。

2.1 基于迁移的攻击

基于迁移的攻击通过利用本地替代模型和目标模型之间的迁移性欺骗深度神经网络，例如，一个模型生成的对抗样本可以欺骗另一个模型的现象[26]。下面介绍四种现有的攻击。

（1）Fast gradient sign method (FGSM)。作为一种经典的一步攻击，FGSM[4]通过计算交叉熵损失的梯度 $J(x, y_T)$ 来确定噪声的方向：

$$x' = x + \tau \cdot \text{sign}\left(\nabla J\left(x, y_T\right)\right) \tag{1}$$

（2）Iterative FGSM (I-FGSM)。I-FGSM[5]将噪声上界 τ 分解为若干小步长 μ，并逐步添加噪声：

$$\text{target } x'_{t+1} = \text{Clip}_{x,\tau}\left\{x'_t + \mu \cdot \text{sign}\left(\nabla J\left(x'_t, y_T\right)\right)\right\} \tag{2}$$

I-FGSM 在白盒场景下的迭代攻击中具有最高的攻击效果。它的主要缺点是迭代步的边际效应递减。换而言之，随着迭代次数 t 的增加和步长 μ 的减小，不断增加迭代步数对攻击效果的改善不大。

（3）Momentum iterative FGSM (MI-FGSM)。MI-FGSM[6]引入动量项，使噪声添加方向的调整更加平滑，但迭代次数的边际效应递减现象仍然存在：

$$m_{t+1} = \theta \cdot m_t + \frac{\nabla J\left(x'_t, y_T\right)}{\|\nabla J\left(x'_t, y_T\right)\|} \tag{3}$$

$$x'_{t+1} = \text{Clip}_{x,\tau}\left\{x'_t + \mu \cdot \text{sign}\left(g_{t+1}\right)\right\} \tag{4}$$

（4）Variance-reduced iterative FGSM (vr-IGSM)。Vr-IGSM[15]将原始图像的梯度替换为带高斯噪声的原始图像的平均梯度。

$$G_{t+1}=\frac{1}{m}\sum_{i=1}^{m}\nabla J\left(x_t+\xi_i\right),\xi_i\sim\mathcal{N}\left(0,\sigma^2 I\right) \tag{5}$$

$$x'_{t+1}=\mathrm{Clip}_{x,\tau}\left\{x'_t+\mu\cdot\mathrm{sign}\left(G_{t+1}\right)\right\} \tag{6}$$

高斯噪声消除了替代模型中的局部波动，因此提高了模型的迁移性。

基于迁移的攻击效果可能会受到集成对抗训练[25]或其他提高目标模型鲁棒性的防御方法[27-29]的影响。一种更合理的策略是将黑盒攻击分为两个阶段：首先通过基于迁移的攻击产生对抗样本作为起点，然后通过基于决策的攻击[10]进一步压缩其冗余噪声。

2.2 基于决策的攻击

基于决策的攻击在原始图像的邻域内采样，在不跨越决策边界的情况下寻找较小的噪声幅度。基于决策的攻击不依赖于替代模型，而是使用各种策略来寻找对抗样本。大多数基于决策的攻击都需要一个已经被错分的初始对抗样本作为起点。下面介绍几种最新的基于决策的攻击。

（1）边界攻击。边界攻击[8]从一个对抗样本开始，同时沿两个方向搜索，即源方向和球方向：

$$x_{t+1}=x_t+\delta\cdot\frac{\eta}{\|\eta\|_2}+\varepsilon\cdot\frac{x-x_t}{\|x-x_t\|_2},\eta\sim\mathcal{N}(0,I) \tag{7}$$

其中，x_t是在t步边界攻击之后具有最小噪声的对抗样本；η和$\left(x-x_t\right)$分别指球面方向和源方向；δ是球面方向的步长；ε是源方向的步长。由于对每个维度不加区分地使用标准正态分布，边界攻击无法评估和利用像素之间的噪声敏感性差异。

（2）有偏边界攻击。有偏边界攻击[10]将边界中的正态分布替换为 Perlin 分布，集中在输入空间的低频域，使对抗示例更“自然”。

（3）进化攻击。进化攻击[9]通过双线性插值和将噪声限制在图像的中心部分来降低采样空间的维数。进化攻击在涉及强先验知识的任务（如人脸识别）中表现更好。

还有一些其他攻击包括零阶优化的攻击[11-14,30-31]。它们主要针对黑盒场景，在黑盒场景中可以获得每个类别的分数，或者查询预算相对充足。在本文中，只讨论有限查询的黑盒场景，目标模型只输出硬标签。

3 CURLS&WHEY 攻击

本节提出了一种新的迁移攻击，包括 Curls 迭代和 Whey 优化两个阶段。首先，明确对抗攻击的形式化表达；然后，展示对 Curls 迭代和 Whey 优化的见解，并对这两个阶段进行详细的描述；最后，将基于迁移的方法扩展到黑盒场景下的针对性攻击。

3.1 符号系统

考虑一个黑盒攻击下基于深度神经网络的目标模型：$F: X^N \to Y^C$，其中 X 表示输入空间，N 表示维度（对于图像数据，$N=$宽度×高度×通道），Y 表示具有 C 个类别的分类空间。成功的对抗攻击会在向原始图像添加尽可能少的噪声后改变目标模型的原始分类结果[32]：

$$\min \| x' - x \|_v, \quad \text{s.t.} F(x) \neq F\left(x'\right) \tag{8}$$

其中，v 指用于测量噪声幅度的范数，包括 ℓ_1，ℓ_2 和 ℓ_∞ 范数。本文讨论了噪声的 ℓ_2 范数。一些现有的文献[6, 33-34]将错分率与固定的 ℓ_∞ 范数进行了比较，但我们主要关注不同攻击对同一幅图像产生的对抗噪声的质量。使用替代模型[26]的黑盒攻击被用于解决目标模型不能反向传播的问题。第 t 步的梯度信息是指替代模型的损失函数 J_{sub} 相对于对抗样本 x_t' 的梯度值，本文中使用交叉熵损失。

3.2 迭代步数的边际效应递减

迭代攻击在白盒场景中表现良好，白盒场景中几乎可以保证 100%的迁移性[35]。然而，当攻击黑盒目标模型时，迭代攻击的缺点逐渐暴露出来。首先，替代模型和目标模型在决策边界上的差异削弱了迁移性[36]。迭代攻击总是朝着替代模型损失函数增加的方向走。但是不同的模型在分类空间上存在巨大的差距，它们的梯度方向甚至可能是相互正交的[20]。因此，简单地沿着替代模型的梯度上升方向搜索对抗样本可能不再适用于黑盒攻击。而且，迭代次数的边际效应是递减的。假设为了最小化噪声幅度，每一步的步长 μ 与总迭代次数成反比。在 I-FGSM 中，当迭代次数 T 增加 1 时，噪声幅度减小的边际增益为

$$\sum_{t=1}^{T+1} \frac{1}{T+1} \cdot \nabla J_{\text{sub}}\left(x_t\right) - \sum_{t=1}^{T} \frac{1}{T} \cdot \nabla J_{\text{sub}}\left(x_t\right) \tag{9}$$

一般来说，随着 T 的增大和单步步长的缩短，迭代轨迹趋于一致和平滑，并逐渐收敛，如图 3 所示。考虑到黑盒攻击中对目标模型的查询次数也是有限的，如果迭代次数已经很高，增加迭代次数对对抗性噪声压缩的效率很低。

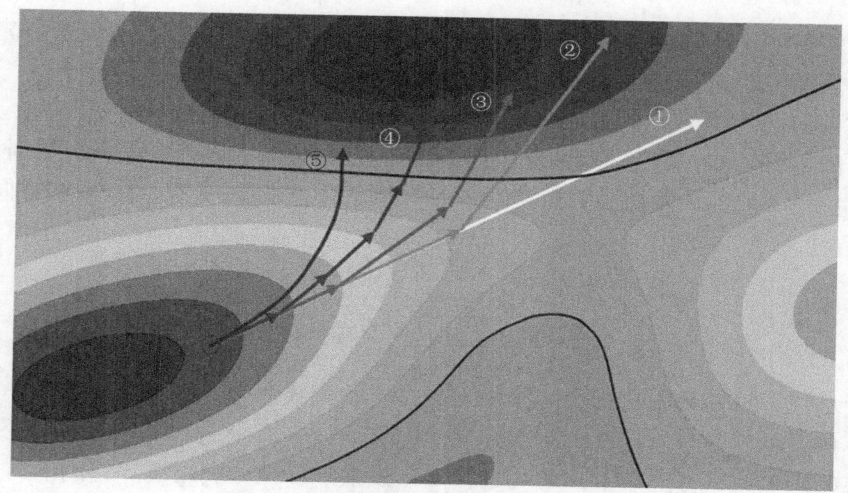

图 3 迭代次数为 T 的边际效应递减

左下角的小圆环表示原始图像，标记①~⑤的 5 条多段线是 $T=1,2,3,5,\infty$ 的跨越决策边界的迭代轨迹

3.3 Curls 迭代

当前黑盒场景下基于迁移的攻击的迭代轨迹是单调的。首先，单调地沿替代模型的损失函数的梯度上升方向更容易使迭代轨迹进入替代模型的局部最优点，而不是穿过目标模型的决策边界。其次，单纯依赖替代模型和目标模型之间的迁移性，而忽略每次查询目标模型的反馈，使迭代轨迹缺乏适应性。

“Curl”并使迭代轨迹多样化可能是一种更具成本效益的解决方案[37]。图 1 展示了目标模型损失函数的一种可能分布。在损失函数沿梯度上升方向缓慢上升的情况下，如绿色轨迹所示，可能可以从附近的起点找到跨越判定边界的捷径，如图 1 中的紫色多段线所示。放弃基于梯度上升的单调搜索策略来增加迭代轨迹的多样性：

$$x_0'=x,x_1'=\mathrm{Clip}_{x,\tau}\left\{x_0'-\mu\cdot\nabla J_{\mathrm{sub}}\left(x_0'\right)\right\} \tag{10}$$

$$g_{t+1}=\begin{cases}-\nabla J_{\mathrm{sub}}\left(x_t'\right), & J\left(x_t'\right)<J\left(x_{t-1}'\right)\\ \nabla J_{\mathrm{sub}}\left(x_t'\right), & J\left(x_t'\right)\geqslant J\left(x_{t-1}'\right)\end{cases} \tag{11}$$

$$x_{t+1}'=\mathrm{Clip}_{x,\tau}\left\{x_t'+\mu\cdot g_{t+1}\right\} \tag{12}$$

其中，$J_{\mathrm{sub}}\left(x_t'\right)$和$J\left(x_t'\right)$分别表示替代模型和目标模型相对于对抗样本$x_t'$的交叉熵损失。首先，沿梯度下降方向更新原始图像一步。当目标模型上当前对抗样本的交叉熵损失低于前一步时，通常是“谷底”，例如，尚未达到损失函数的局部最小值。因此，当目标模型上的损失仍在下降时，继续沿梯度下降方向更新，反之亦然。我们把这种“先降后升”的迭代方法称为 Curls 迭代。

在 Curls 的基础上，在每轮迭代前后引入两种启发式策略。对图像而言，最接近的对抗性示例更有可能在特征空间中以大致相同的方向分布。因此，记录并更新一张图像的所有对抗样本的平均方向 $\overline{R}$，并在计算每轮梯度时，在第一步添加指向该方向的向量：

$$\overline{R}=\frac{1}{K}\sum_{i=1}^{K}x',\text{s.t.}F(x)\neq F\left(x'\right) \tag{13}$$

$$x_1'=\text{Clip}_{x,\tau}\left\{x_0'+\mu\cdot\nabla J\left(x_0'+\mu\cdot\overline{R}\right)\right\} \tag{14}$$

由于迭代轨迹在高维特征空间中不可能是一条直线，图 1 中红色弧线所示的情况是存在的：在已找到的对抗样本和原始图像之间存在具有更小 ℓ_2 范数的对抗样本。在每一轮后在原始图像 x 和对抗样本 x' 之间进行二分搜索，以充分利用这一轮的潜力：

$$\text{left}=x,\text{right}=x' \tag{15}$$

$$\text{BS(left, right)}=\begin{cases}\text{BS(left, (left + right) / 2)}, & F(x)\neq((\text{left}+\text{right})/2)\\ \text{BS((left + right) / 2, right)}, & F(x)=F((\text{left}+\text{right})/2)\end{cases} \tag{16}$$

在 Curls 迭代的实际实现中，为了防止对抗噪声更新的振荡，不直接根据目标模型的损失函数确定梯度符号，而是将每轮迭代分为两个阶段。在第一阶段，对原始图像进行梯度下降。一旦目标模型的交叉熵低于前一步，第二步开始进行梯度上升，直到最后一步。同时进行直接梯度上升的正态迭代轨迹。另外，受 vr-IGSM[15]的启发，在梯度计算过程中加入高斯噪声，从而提高迁移性。算法 1 描述了 Curls 迭代的细节。

算法 1 Curls 迭代

输入： 目标 DNN $F(x)$，替代模型 Sub(x)

原始图像 x 以及标签 y

初始噪声幅度限制 T

迭代步 T 以及高斯噪声的方差 var

步长 μ 和二分搜索步 bs

---

1. 初始化 $\overline{R}$ 以及两个起点
2. $\overline{R}=0, x_0^A=x, x_0^B=x$
3. downhill = True // 设置梯度下降标志为 True
4. for $t=0$ to T do

续表

5. $\xi_t^A, \xi_t^B \sim \mathcal{N}\left(0, \mathrm{var}^2 I\right)$
6. 在替代模型上计算梯度
7. $g_t^A = \nabla J_{\mathrm{sub}}\left(x_t^A + \xi_t^A + \mu \cdot \overline{R}\right)$
8. $g_t^B = \nabla J_{\mathrm{sub}}\left(x_t^B + \xi_t^B + \mu \cdot \overline{R}\right)$
9. $x_{t+1}^A = \begin{cases} \mathrm{Clip}_{x,\tau}\left\{x_t^A - \mu \cdot g_t^A\right\} & \mathrm{downhill} = \mathrm{True} \\ \mathrm{Clip}_{x,\tau}\left\{x_t^A + \mu \cdot g_t^A\right\} & \mathrm{downhill} \neq \mathrm{True} \end{cases}$
$x_{t+1}^B = \mathrm{Clip}_{x,\tau}\left\{x_t^B + \mu \cdot g_t^B\right\}$
10. if downhill = True and $J\left(x_{t+1}^A\right) > J\left(x_t^A\right)$ then
11. downhill = False
12. end if
13. if $F\left(x_{t+1}^A\right) \neq F(x)$ or $F\left(x_{t+1}^B\right) \neq F(x)$ then
14. 使用式(13)更新 $\overline{R}$
15. end if
16. end for
17. if $F\left(x_T^A\right) \neq F(x)$ or $F\left(x_T^B\right) \neq F(x)$ then
$x' = \begin{cases} x_T^A & \lVert x_T^A - x \rVert_2 < \lVert x_T^B - x \rVert_2 \\ x_T^B & \mathrm{else} \end{cases}$
18. 使用式(16)压缩 x'
19. end if
20. return x'
输出：对抗样本 x'

3.4 Whey 优化

通常，一个基于迁移的攻击一旦发现对抗样本或迭代次数用完就结束。然而，生成的对抗样本在迭代之后仍然可能包含冗余的“Whey”噪声。或者在确保对抗样本仍能欺骗目标模型的同时，最大程度地降低噪声[38]：

$$\max\left(\lVert x' - x \rVert_2 - \lVert x^{\circ} - x \rVert_2\right), \text{s.t. } F\left(x'\right) = F\left(x^{\circ}\right)$$

其中，x，x' 和 x° 分别指原始图像、目前发现的对抗样本和与原始图像最接近的对抗样本。

由于已经在 x 和 x' 之间进行了二分搜索，冗余噪声较少的对抗样本更可能存在于与

$x'-x$ 线性无关的方向上。我们提出了 Whey 优化，以在黑盒攻击中挤出冗余噪声剩余的“Whey”。Whey 优化在噪声压缩幅度和压缩次数之间保持平衡。一次压缩过多的噪声可能会使对抗样本回到原来的分类。然而，增量压缩使优化不可能在有限的查询数内完成。一个折中的解决方案是先将对抗噪声分为若干组，然后尝试一组一组地降低噪声：

$$z_0 = x' - x \tag{17}$$

$$z_{t+1}^{\text{whc}} = z_t^{\text{whc}} / 2, \text{s.t.}\ z_t^{\text{whc}} = L\left(V\left(z_0\right), t\right) \tag{18}$$

其中，z 是噪声；$L(V,t)$ 表示在像素值集合 V 中具有第 t 个最大绝对值的数字：

$$V(z) = \left\{v \mid v = z^{\text{whc}}, w \in [0,W], h \in [0,H], c \in [0,C]\right\}$$

其中，W，H，C 分别代表原始图像 x 的宽度、高度和通道。Whey 优化将噪声 z 按像素值分成若干组，每次按降序选取一组，将 z 中所有等于 $L(V,t)$ 的像素值减半，并检查是否仍能欺骗目标模型。

分组压缩后，执行更细粒度的压缩。Whey 优化的最后一步将每个像素的值以概率 ω 设置为 0:

$$z_{t+1} = z_t \cdot \text{mask}_t \tag{19}$$

$$\text{mask}^{\text{whc}} = \begin{cases} 0, & \text{random}\,() \leqslant \omega \\ 1, & \text{否则} \end{cases} \tag{20}$$

其中，mask 的形状与 z 相同。算法 2 给出了 Whey 优化的细节。

算法 2 Whey 优化

输入： 目标 DNN $F(x)$ 以及对抗样本 x'

原始图像 x 以及标签 y

两个挤压步骤的最大尝试次数，T_1，T_2

$x'-x, V$ 的像素值集合

范围 $[0,1]$ 的随机数生成器，random( )

置零概率，ω

1. $z = x' - x$
2. $t_1 = 0, t_2 = 0$
3. for p in V and $t_1 < T_1$ do // 步骤 1：在分组中挤压
4. 将像素值减半

续表

5. $z[z = p] /= 2$
6. if $F(z) = y$ then
7. 取消该步的更新
8. end if
9. $t_1 = t_1 + 1$
10. end for
11. while $t_2 < T_2$ do // 步骤 2：随机挤压
12. 产生与图像形状相同的随机掩码

$$\text{mask}^{\text{whc}} = \begin{cases} 0, & \text{random}() \leqslant \omega \\ 1, & 否则 \end{cases}$$

13. $z = z \cdot \text{mask}$[// 逐元素乘积
14. if $F(z) = y$ then
15. 取消该步的更新
16. end if
17. $t_2 = t_2 + 1$
18. end while
19. $x' = z + x$
20. return x'

输出：压缩的对抗样本 x'

3.5 针对性攻击

与非针对性攻击不同，针对性攻击不仅要求目标模型对对抗样本错误分类，而且还需要将其错分到指定的类别中。这在黑盒攻击中尤其困难，因为不同模型之间的决策边界差别很大，梯度方向甚至相互正交[20]。即使每个迭代步的更新从相对于原始类别的梯度上升方向 $\nabla J_{\text{sub}}\left(x', y_{\text{ori}}\right)$ 更改为相对于目标类别的梯度下降方向 $-\nabla J_{\text{sub}}\left(x', y_{\text{target}}\right)$[6]，原始图像的迭代轨迹几乎不可能到达目标类别空间，这是由于目标模型和替代模型的梯度值不同。

放弃从原始图像开始攻击的策略，将插值与基于迁移的攻击相结合，以获得更好的初始更新方向。首先，收集一个合理的图像 x_{target}，它可以被目标模型分类为目标类别。其次，使用二分搜索在原始图像 x 和 x_{target} 之间找到一个图像 x_0'，确保 x_0' 也能被分类到目标类别中。在此之后，使用x_0'引导从x开始的第一个梯度上升步：

$$x_0' = (1-s)\cdot x + s\cdot x_{\text{target}} \tag{21}$$

$$x_1' = \text{Clip}_{x,\tau}\left\{x - \mu\cdot\nabla\cdot J\left(x_0'\right)\right\} \tag{22}$$

$$x_{t+1}' = \text{Clip}_{x,\tau}\left\{x_t' - \mu\cdot\nabla J\left(x_t'\right)\right\}, t\geqslant 1 \tag{23}$$

其中，$0<s<1$表示通过二分搜索确定的插值系数。通过这种方式，将原始样本增强到目标类别的方向。在第一个增强步骤之后，就像在非针对性攻击中一样继续应用 Curls 和 Whey 攻击。

4　适应性对抗边界攻击

在本节中，提出了 CAB——一种新的基于决策的攻击，它利用历史采样和对抗噪声来适应性调整采样分布。首先通过分析边界攻击中多元正态分布的方差与降噪期望之间的关系，给出对 CAB 的一些见解。此外，本文还分析了基于决策的攻击噪声压缩过程的单调性，并用于边界攻击的步长。最后，详细描述了所提出的攻击。

4.1　方差与噪声抑制

假设x^*是找到的具有最小噪声幅度的对抗样本。基于决策的攻击的目标函数可以描述为

$$\max_{x'} \| x^* - x \|_2 - \| x' - x \|_2, \text{s.t.}\ F(x) \neq F\left(x'\right) \tag{24}$$

其中，x和x'分别表示原始图像和该步骤后生成的新对抗样本。用原始图像x、对抗噪声z^*和z^*+z之和替换对抗样本x^*和x'，其中z^*和z分别是当前具有最小幅度的对抗性噪声和该步之后添加的噪声。由于x和x^*是固定的，所以在式(24)中的目标函数可等价地重新形式化为

$$\min_{z} \| z^* + z \|_2, \text{s.t.}\ F(x) \neq F\left(x + z^* + z\right) \tag{25}$$

注意，ℓ_2距离是在目标模型错分对抗样本的前提下计算的。选择ℓ_2范数作为距离度量，是因为它比ℓ_∞范数更能准确地描述一个模型的鲁棒性[19]。

考虑在式(7)中边界攻击的一步噪声更新，重写z为$\delta\cdot\frac{\eta}{\|\eta\|_2}+\varepsilon\cdot\frac{x-x^*}{\|x-x^*\|_2}$，且$\left(1-\frac{\varepsilon}{\|z^*\|_2}\right)$为$\alpha$。因为$x-x^*=-z^*$，有

$$
\begin{aligned}
&\| z^{*}+z\|_{2}\\
&=\left\| z^{*}+\delta\cdot\frac{\eta}{\|\eta\|_{2}}+\varepsilon\cdot\frac{-z^{*}}{\|-z^{*}\|_{2}}\right\|_{2}\\
&=\left\|\alpha\cdot z^{*}+\delta\cdot\frac{\eta}{\|\eta\|_{2}}\right\|_{2}\\
&=\sqrt{\|\alpha\cdot z^{*}\|_{2}^{2}+2\cdot\delta\cdot\alpha\cdot\left(z^{*}\cdot\frac{\eta}{\|\eta\|_{2}}\right)+\left\|\delta\cdot\frac{\eta}{\|\eta\|_{2}}\right\|_{2}^{2}}
\end{aligned}
$$

其中，$\cdot$代表标准内积。因为z^{*}，ε和δ都是固定的，$\left\|\delta\cdot\frac{\eta}{\|\eta\|_{2}}\right\|_{2}^{2}\equiv\delta^{2}$，式(25)的目标实际上就是最小化$z^{*}\cdot\frac{\eta^{\|\eta\|}}{\|\eta\|_{2}}$。根据 Cauchy-Schwarz 不等式，有

$$
-\| z^{*}\|_{2}\cdot\|\eta\|_{2}\leqslant z^{*}\cdot\eta\leqslant\| z^{*}\|_{2}\cdot\|\eta\|_{2} \tag{26}
$$

其中，$\beta=\frac{\eta}{\|\eta\|_{2}}$，这使得

$$
\begin{aligned}
&\| z^{*}+z\|_{2}^{2}\\
&\geqslant\|\alpha\cdot z^{*}\|_{2}^{2}-2\cdot\alpha\cdot\delta\cdot\left(\| z^{*}\|_{2}\cdot\|\beta\|_{2}\right)+\|\delta\cdot\beta\|_{2}^{2}\\
&=\left(\|\alpha\cdot z^{*}\|_{2}-\|\delta\cdot\beta\|_{2}\right)^{2}\\
&\Rightarrow\| z^{*}+z\|_{2}\geqslant\|\alpha\cdot z^{*}\|_{2}-\|\delta\cdot\beta\|_{2}
\end{aligned}
$$

当$z^{*}=-k\beta,k\in\mathbb{R}^{+}$时，该等式成立。换而言之，当$\frac{\eta}{\|\eta\|_{2}}$和当前噪声$z^{*}$正好相反时，边界攻击之后的总噪声量$\| z^{*}\|_{2}$被最小化。假设$\eta\sim\mathcal{N}(0,\Sigma)$是一个$N$维随机向量，它服从均值为零、协方差矩阵$\boldsymbol{\Sigma}=\operatorname{diag}\left(\sigma_{1}^{2},\sigma_{2}^{2},\cdots,\sigma_{N}^{2}\right)$（在边界攻击中，$\Sigma=I_{N}$）的正态分布。$\eta$的每个元素是一个单变量正态分布，它们的平均值设置为零，以便在采样空间中更好地搜索[9]。因为$\beta=\frac{\eta}{\|\eta\|_{2}}$且$\beta_{i}=\frac{\eta_{i}}{\sqrt{\eta_{i}^{2}+\cdots+\eta_{N}^{2}}}$，$\beta_{i}^{2}$的期望值的比率满足

$$\begin{aligned}&E\left(\beta_1^2\right):E\left(\beta_2^2\right):\cdots:E\left(\beta_N^2\right)\\=&E\left(\eta_1^2\right):E\left(\eta_2^2\right):\cdots:E\left(\eta_N^2\right)\\=&\mathrm{Var}\left(\eta_1\right):\mathrm{Var}\left(\eta_2\right):\cdots:\mathrm{Var}\left(\eta_N\right)\\=&\sigma_1^2:\sigma_2^2:\cdots:\sigma_N^2\end{aligned}$$

作为一种拒绝采样[23]，边界攻击只会在一个采样的降噪量大于零时查询目标模型，例如，$z^*\cdot\eta\leqslant 0$。因此，当$\sigma_i\propto\left|z_i^*\right|,1\leqslant i\leqslant N$时，一步边界攻击后新噪声$x'$的期望被最小化。

为了更直观地展示σ对降噪的影响，将x'的分布在二维空间中可视化，如图 4 所示。蓝色向量表示$x^*=(3,1)$。红色标记表示在正态分布下，以$\sigma_1:\sigma_2=3:1$（图 4（a））和$\sigma_1:\sigma_2=1:1$（图 4（b））、$(3,1)$为中心的 1000 次抽样后x'的分布。红色越深，附近的样本就越密集。$x_1=0$和$x_2=4$处的黑线图分别是x_2和x_1的独立概率分布$\boldsymbol{P}$。当二维方差比$\sigma_1:\sigma_2=x_1^*:x_2^*$时，在图 4（a）中，$x'$集中在$x^*$的相反方向。然而，图 4（b）中每个维度的方差相等，x^*在所有方向上均匀分布，这阻碍了噪声压缩的效率。这种关系表明，与所有维度的标准正态分布相比，在通过当前噪声调整的正态分布上进行采样增加了降噪的期望。

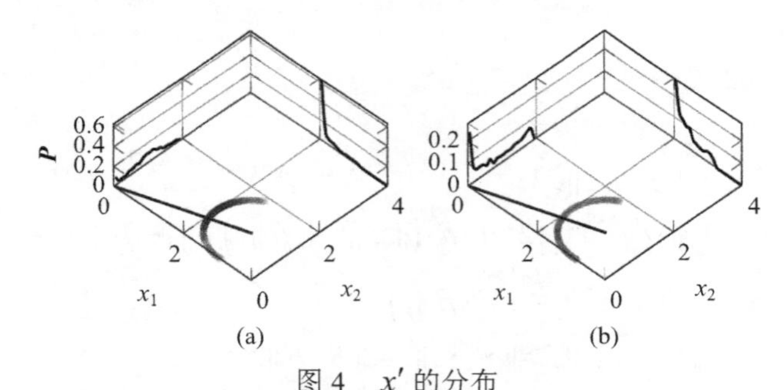

图 4　x'的分布

（a）$N=2$且$\sigma_1:\sigma_2=3:1$条件下z的分布；（b）$N=2$且$\sigma_1:\sigma_2=1:1$条件下z的分布

4.2　初始噪声和步长的自适应调整

本节通过分析基于决策的攻击中噪声压缩的单调性来调整初始噪声和步长。在错分概率随与原始图像的距离单调增加的假设下，最终噪声幅度与初始噪声幅度呈正相关。一方面，这解释了通过基于迁移的攻击来初始化对抗噪声的有效性。另一方面，基于这一特征，在边界攻击中采用自定义步长的策略。

根据文献[19]中的设置，$\rho_{F,x}(\lambda)$表示目标模型F对与原始图像x相距λ的随机点的错分概率：

$$\rho_{F,x}(\lambda) = \mathbb{P}_{z \sim \lambda \mathbb{S}}\{F(x) \neq F(x+z)\} \tag{27}$$

其中，$\lambda\mathbb{S}$ 表示圆心为 0 且半径为 λ 的球面上的统一度量，即满足 $\|z\|_2 = \lambda$ 的点集。由于错分的风险通常随着与原始图像的距离的增加而增加，假设在特定范围 $\Delta_{\text{adv}}(x;F) \leqslant \lambda \leqslant \Delta_{\text{unif},\kappa}(x;F)$ 内，$\rho_{F,x}(\lambda)$（缩写为 $\rho(\lambda)$）随 λ 单调增加：

$$\forall \lambda_1, \lambda_2 \in \left[\Delta_{\text{adv}}(x;F), \Delta_{\text{unif},\kappa}(x;F)\right]$$
$$\lambda_1 \geqslant \lambda_2 \to \rho(\lambda_1) \geqslant \rho(\lambda_2)$$

如文献[19]所定义，$\Delta_{\text{adv}}(x;F)$ 表示导致错误分类的全局最小对抗扰动的 ℓ_2 范数。$\Delta_{\text{unif},\kappa}(x;F)$ 表示 F 对随机均匀噪声的 k-鲁棒性。为简单起见，假设在基于决策的攻击中经过 t 步查询之后，要查询的对抗样本 x' 的噪声幅度服从 $\Delta_{\text{adv}}(x;F)$ 到当前最小噪声幅度 $\|z^*\|_2$ 的均匀分布。

命题 1 假设当 λ 在范围 $\Delta_{\text{adv}}(x;F) \leqslant \lambda \leqslant \Delta_{\text{unif},\kappa}(x;F)$ 内时，错分概率 $\rho(\lambda)$ 单调增加。对于原始图像 x 任意两个对抗样本 x_1' 和 x_2'，如果相应的噪声幅度满足 $\Delta_{\text{adv}}(x;F) \leqslant \lambda_2 < \lambda_1 \leqslant \Delta_{\text{unif},\kappa}(x;F)$，那么在一步决策攻击之后的噪声幅度期望满足 $\mathbb{E}(\lambda_2) < \mathbb{E}(\lambda_1)$。

证明 考虑噪声幅度 $\lambda = \|z\|_2$ 与错分概率 $\rho(\lambda)$ 之间的关系，在 $t+1$ 步之后，期望的噪声幅度是

$$\mathbb{E}(\lambda) = \int_{\Delta_{\text{adv}}(x;F)}^{\lambda} a\rho(a) \frac{1}{\lambda - \Delta_{\text{adv}}(x;F)} da \tag{28}$$

由于不可能在基于决策的攻击完全收敛前获得 $\Delta_{\text{adv}}(x;F)$ 的确切值，且当 $\lambda \leqslant \Delta_{\text{adv}}(x;F)$ 时 $\rho(\lambda) = 0$，式(28) 重写为

$$\mathbb{E}(\lambda) = \int_0^{\lambda} a\rho(a) \frac{1}{\lambda} da \tag{29}$$

对于满足 $\Delta_{\text{adv}}(x;F) \leqslant \lambda_2 < \lambda_1 \leqslant \Delta_{\text{unif},\kappa}(x;F)$ 的 x_1' 和 x_2'，在基于决策的攻击的一步迭代之后的噪声幅度的差是

$$\begin{aligned}
&\mathbb{E}(\lambda_1) - \mathbb{E}(\lambda_2) \\
&= \int_0^{\lambda_1} a\rho(a) \frac{1}{\lambda_1} da - \int_0^{\lambda_2} a\rho(a) \frac{1}{\lambda_2} da \\
&= \frac{1}{\lambda_1} \int_{\lambda_2}^{\lambda_1} a\rho(a) da - \left(\frac{1}{\lambda_2} - \frac{1}{\lambda_1}\right) \int_0^{\lambda_2} a\rho(a) da
\end{aligned}$$

给定 $\lambda_2 < \lambda_1$ 及 $\rho(\lambda)$ 的单调性，有

$$\begin{aligned}
&\mathbb{E}(\lambda_1) - \mathbb{E}(\lambda_2) \\
&\geqslant \frac{1}{\lambda_1}\int_{\lambda_2}^{\lambda_1} a\rho(\lambda_2)da - \left(\frac{1}{\lambda_2} - \frac{1}{\lambda_1}\right)\int_0^{\lambda_2} a\rho(\lambda_2)da \\
&= \frac{\rho(\lambda_2)}{2\lambda_1}\left(\lambda_1^2 - \lambda_2^2\right) - \frac{\lambda_2^2\rho(\lambda_2)}{2}\left(\frac{1}{\lambda_2} - \frac{1}{\lambda_1}\right) \\
&= \frac{\rho(\lambda_2)}{2\lambda_1}\left(\lambda_1^2 - \lambda_1\lambda_2\right) > 0
\end{aligned}$$

根据命题 1，证明了在错分概率随与原始图像的距离单调增加的假设下，一步迭代后的期望噪声幅度也随初始噪声幅度单调增加。基于决策的攻击过程满足无记忆性，即当前噪声仅由最后一步的噪声决定[8]。因此，噪声压缩的单调性满足多步传递性。换言之，当使用相同的基于决策的攻击并对目标模型进行相同次数的查询时，期望的最终噪声幅度与初始噪声幅度正相关。这说明在黑盒场景下，将基于迁移的攻击与基于决策的攻击结合的有效性。边界攻击等基于决策的方法使用随机噪声作为初始噪声，其幅值远大于使用基于迁移的攻击产生的噪声，因此在相同的查询次数下，最终噪声也较大。遵循这个黑盒攻击设置，使用基于迁移的攻击生成的对抗样本作为基于决策的攻击的起点。此外，随着噪声幅度的不断压缩，如果式(7)中球型和源方向的步长 δ 和 ε 保持不变，则新查询的错分概率将逐渐降低。为了弥补查询成功率的下降，引入指数调度来适应性调整两个方向的步长：

$$\delta_s = \delta_0\varphi^s,\quad \varepsilon_s = \varepsilon_0\varphi^s \tag{30}$$

其中，s 表示到目前为止成功的查询数；δ_s 和 ε_s 是 s 个成功查询之后的球型和源方向的步长；δ_0 和 ε_0 是初始步长；$\varphi \in (0,1)$ 是步长调度的衰减因子。由于最近的对抗样本与原始图像之间的距离缩短，新查询的步长也减小，这种指数调度策略平衡了噪声压缩率和查询成功率，实现了对不同图像和同一图像的不同查询阶段的步长的适应性调整。

4.3 CAB 攻击

对于在大输入空间中随机搜索的边界攻击，减少采样空间是提高噪声压缩效率的关键。进化攻击[9]通过双线性插值和将对抗噪声限制在图像中心来减少采样空间。通过相对位置来区分像素对噪声的敏感性对于具有单一结构（例如，人脸识别图像）或小尺寸的图像可能是有效的，对于较大和更复杂的图像则不是有效的。与文献[9]中人工制定的规则相比，当前的噪声 z^* 是像素敏感性的更无偏的表征。因此，只在当前噪声幅度已经

很大的像素上调整噪声：

$$H(z,r)=\underset{\hat{z}\subset z^{*},|\hat{z}|/z^{*}|=r}{\arg\max}\sum_{z\in\hat{z}}|z| \tag{31}$$

$$\text{mask}_i=\begin{cases}1, & z_i^{*}\in H\\ 0, & \text{否则}\end{cases} \tag{32}$$

其中，$\hat{z}$ 是 z^{*} 中具有最大绝对值的像素集合；$r\in(0,1)$ 是 $\hat{z}$ 和 z^{*} 中像素数的比值。具体地说，根据比值 r 选取 z^{*} 中绝对值最大的像素，形成掩模滤除新噪声中敏感度较低的区域。

根据 4.1 节和图 4 的结论，在当前噪声的引导下，CAB 攻击自适应地分配各维正态分布的方差比，并选择对噪声最敏感的区域，进一步缩小采样空间。这两个过程都利用了历史成功的采样。尽管现有的基于决策的攻击直接丢弃失败的采样，但它们实际上包含关于决策边界的信息。修正下一个样本的分布均值，以避免失败的采样：

$$\eta\sim\mathcal{N}\left(-\frac{1}{K}\sum_{j=1}^{K}\tilde{\eta}_j, z^{*2}\right)$$

$$\text{s.t.}F\left(x^{*}+\delta\cdot\frac{\tilde{\eta}}{\|\tilde{\eta}\|_2}+\varepsilon\cdot\frac{x-x^{*}}{\|x-x^{*}\|_2}\right)=F(x)$$

其中，K 是当前对抗样本 x^{*} 上失败采样的总数；$\tilde{\eta}_j$ 是第 j 次失败采样时使用的正态随机向量。维护一个对抗样本的采样记录，并将当前对抗样本 x^{*} 的所有失败采样保存为 $\tilde{x}$。不断更新该记录，直到成功采样，例如，噪声被进一步压缩。由于在基于决策的攻击的下一阶段，采样成功率随着噪声幅度的减小而降低，因此维护该记录可以使新的采样远离历史上的失败采样。算法 3 详细描述了 CAB 攻击。

算法 3 适应性对抗边界攻击

输入：目标 DNN $F(x)$ 以及对抗样本 x^{*}

原始图像 x 及其标签 y

最大查询次数 T，像素保留率 r

球型方向的初始步长 δ_0

源方向的初始步长 ε_0

步长调度的衰减率 φ

1. $W\leftarrow[\,],s\leftarrow 0$
2. for t in 1 to T do

续表

3. if $W \neq \varnothing$ then
4. $z^* \leftarrow x^* - x$
5. // 在自适应调整的正态分布上采样
6. $\eta \sim \mathcal{N}\left(-\frac{1}{|W|}\sum \tilde{\eta}, z^{*2}\right) \text{s.t.} \tilde{\eta} \in W$
7. else
8. $\eta \sim \mathcal{N}\left(0, z^{*2}\right)$
9. end if
10. // 选择 z^* 中具有最大绝对值的像素
11. $H(z,r) = \arg \max\limits_{\hat{z} \subset z^*, |\hat{z}||z^*|=r} \sum\limits_{z \in \hat{z}} |z|$
12. 根据式(32)通过 H 构建 T
13. $x' = x^* + T \cdot \left(\delta_s \cdot \frac{\eta}{\|\eta\|_2} - \varepsilon_s \cdot \frac{z^*}{\|z^*\|_2}\right)$
14. if $y \neq F(x')$ then
15. // 采样成功，噪声已被压缩
16. $x^* \leftarrow x', W \leftarrow [\,]$
17. $s \leftarrow s+1, \delta_s \leftarrow \delta_{s-1}\varphi, \varepsilon_s = \varepsilon_{s-1}\varphi$
18. else
19. // 采样失败，更新失败采样集合
20. $W = W \cup \eta$
21. end if
22. end for
23. return x'

输出：噪声压缩后的对抗样本 x'

5 实验

5.1 实验设置

所有的实验都是在 NIPS 2018 对抗视觉挑战[22]中使用的 Tiny-Imagenet 和 Imagenet[24]上进行的，图像形状分别为 $64 \times 64 \times 3$ 和 $224 \times 224 \times 3$ 。Imagenet 包含 1000 个图像类别。从它的验证集中挑选了 10 000 张图像，这些图像可以被所有目标模型正确分类，每个类

别 10 张图像。对于 200 个图像类别的 Tiny-Imagenet，选择 2000 张图像，每个类别 10 个图像。比较了 8 种不同结构的神经网络模型：resnet-18[39]，resnet-101，inception v3[40]，inception-resnet v2[41]，nasnet[42]，densenet-161[43]，vgg19-bn[44]，senet-154[45]。

在 Foolbox[46]框架上实现了我们的黑盒攻击。为了精确地度量每种方法的攻击效果，在迭代过程之外增加了一个大循环来确定 τ。对于评估标准，选择了 NIPS 2018 对抗性视觉挑战[22]中应用的对抗扰动的中位数和平均值：

$$\text{mid} = \text{median}\left(\left\{\| x' - x \|_2 \mid x \in \boldsymbol{X}\right\}\right) \tag{33}$$

$$\text{avg} = \frac{1}{K}\sum_{i=1}^{K}\left(\left\{\| x' - x \|_2 \mid x \in \boldsymbol{X}\right\}\right) \tag{34}$$

其中，sub 和 F 分别表示替代模型和目标模型；x 是测试集 X 中的原始图像；x' 是发现的最接近 x 的对抗样本。较小的 ℓ_2 距离表示攻击效果更强，生成的对抗样本的迁移性更高。值得注意的是，对于一个更真实的黑盒攻击设置，在将对抗样本输入目标模型之前会进行取整。

5.2 在多个模型上的黑盒攻击

表 1 展示了 Tiny-Imagenet 上对抗干扰的中位数和平均值。在这个 4×4 矩阵中，每个元素表示该行的替代模型相对于该列的目标模型在所有 2000 个图像中的比较结果。对角线上的元素是白盒攻击的结果（用斜体标记）。图 5 展示了当使用 vgg19-bn 作为替代模型时，三个目标模型的扰动的中位数。对于每一对替代模型和目标模型，将我们的方法（Curls&Whey 与仅 Curls）与 FGSM[4]和其他三种基于迁移的攻击——I-FGSM[5]，MI-FGSM[6]和 vr-IGSM[15]进行比较。由于 ℓ_2 范数被用来测量噪声的幅度，不再使用符号函数来更新对抗样本。为了比较的公平性，对于基于迁移的攻击，对目标模型的查询数基本相等。表 2 给出了与查询数相关的参数，包括迭代轮数 T_0、迭代步 T、二分搜索步 bs、Whey 优化中两个挤压步骤的最大尝试数 T_1 和 T_2。我们的方法的总查询数是 $T_0\times(T+\text{bs})\times 2+T_1+T_2$，其他基于迁移的方法的查询数是 $T_0\times T$。初始噪声幅值 τ 和步长 μ 分别为 0.3 和 $T/2$。对于 vr- IGSM 和我们的方法中高斯噪声的方差，设置 $s=1$。Whey 优化中的置零概率 ω 设置为 0.01。

从表 1 可以看出，在黑盒攻击（即非对角元素）中，Curls&Whey 在 ℓ_2 范数下比所有其他方法达到了更小的中位噪声幅度，并且比大多数其他方法达到了更小的平均噪声幅度，如对角线外的元素。随着迭代轨迹的多样化和冗余噪声的挤出，在大多数模型组合中，噪声降低了 20%~30%，有时甚至降低了 40%。仅使用 Curls 迭代的性能也优于现有的绝大多数黑盒攻击方法。由于梯度计算过程中存在高斯噪声，在白盒攻击中，我们

表 1 四个模型两两之间攻击产生的对抗扰动的中值和平均 ℓ_2 距离

	攻击方法	resnet18		inceptionv3		inception resnet v2		nasnet	
		中值	平均值	中值	平均值	中值	平均值	中值	平均值
resnet 18	FGSM	*0.1321*	*0.8893*	4.3085	7.4580	3.6764	5.3257	3.4187	4.5589
	I-FGSM	*0.0800*	***0.0881***	1.9686	2.9287	2.4624	3.3192	2.1865	2.9644
	MI-FGSM	*0.0866*	*0.1029*	2.3220	3.4386	2.9526	3.9267	2.0174	2.9723
	vr-IGSM	*0.0941*	*0.1120*	1.8737	2.8228	2.4803	3.4085	1.7991	2.7645
	Curls	*0.0731*	*0.1182*	1.6443	2.4739	1.8507	2.6290	1.6773	2.4919
	Curls&Whey	***0.0627***	*0.1040*	**1.1942**	**1.7387**	**1.4549**	**1.9450**	**1.3902**	**1.9696**
inception v3	FGSM	0.9944	3.6262	*0.1521*	*1.9010*	2.6171	4.9078	2.8729	4.5217
	I-FGSM	0.6699	1.8883	***0.1132***	***0.1518***	1.3415	1.9095	1.3774	2.1675
	MI-FGSM	0.8124	2.2895	*0.1283*	*0.1989*	1.6248	2.4642	1.6800	2.7336
	vr-IGSM	0.6072	1.7973	*0.1297*	*0.1834*	1.3214	2.0991	1.3569	2.3010
	Curls	0.5760	1.6781	*0.1243*	*0.2194*	1.1163	1.8997	1.2335	2.1067
	Curls&Whey	**0.5140**	**1.4941**	*0.1252*	*0.9200*	**0.9058**	**1.7913**	**0.9398**	**1.9315**
inception resnet v2	FGSM	1.6729	5.0270	4.2482	6.6191	*0.2855*	*4.5974*	4.1107	5.5487
	I-FGSM	0.7019	2.3966	1.3314	2.3834	***0.1293***	*0.3814*	1.3761	2.3732
	MI-FGSM	0.8561	2.8611	1.6342	3.0884	*0.1602*	*0.5419*	1.6594	3.0469
	vr-IGSM	0.6463	2.4453	1.3166	2.6256	*0.1640*	*0.5197*	1.3292	2.6710
	Curls	0.6040	2.0220	1.1325	1.9407	*0.1501*	*0.3450*	1.0978	1.9644
	Curls&Whey	**0.5227**	**1.2404**	**0.8431**	**1.3437**	*0.1485*	***0.3199***	**0.8483**	**1.4403**
nasnet	FGSM	3.7356	6.0550	3.5277	7.2388	3.4829	7.1657	*0.2008*	*6.389*
	I-FGSM	1.5575	4.1401	1.5926	4.3745	1.4180	4.2968	***0.1173***	*1.8225*
	MI-FGSM	0.9518	3.0544	1.8850	3.9685	1.6458	3.7643	*0.1317*	*0.3632*
	vr-IGSM	0.5659	2.4410	1.5006	**3.2440**	1.3066	**3.1112**	*0.1371*	***0.3197***
	Curls	0.5821	2.1520	1.2719	3.9490	1.2048	4.1637	*0.1360*	*2.7497*
	Curls&Whey	**0.5543**	**1.8582**	**1.0003**	3.6760	**0.9599**	3.6069	*0.1354*	*2.5653*

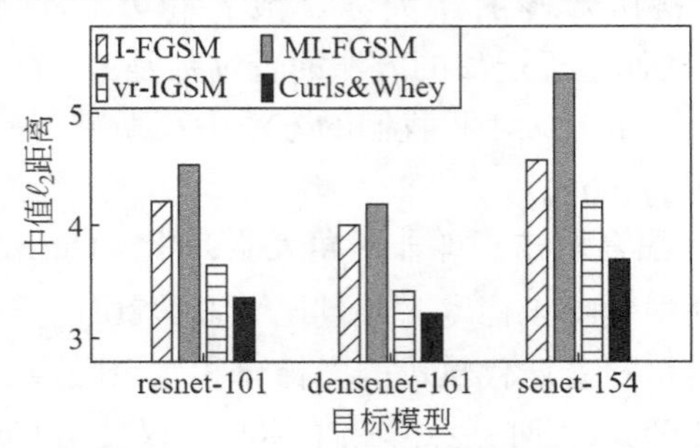

图 5 在 Imagenet 上使用 vgg19-bn 作为替代模型产生的对抗噪声的中值 ℓ_2 距离比较

表 2 两个数据集上的实验的参数集

		T_0	T	bs	T_1	T_2	Total
Tiny-Imagenet	其他方法	20	10	—	—	—	200
	我们的方法	10	4	2	40	40	200
Imagenet	其他方法	24	24	—	—	—	576
	我们的方法	14	7	3	200	100	580

方法的噪声幅度略高于 I-FGSM，此时不考虑迁移性。然而，我们的方法的白盒噪声仍小于 vr-IGSM，这验证了 Whey 优化的有效性。图 6 显示了在两个数据集上构建的对抗样本。Curls&Whey 以几乎不可察觉的噪声实现了针对性与非针对性错分。

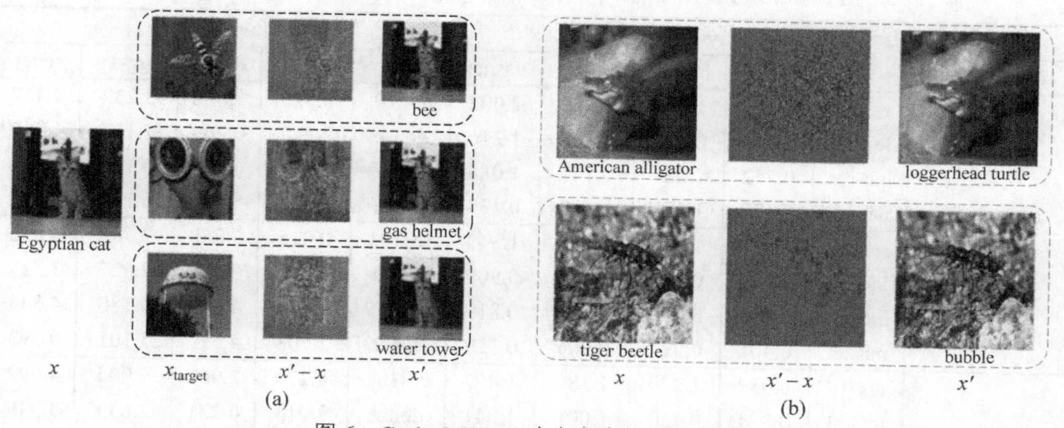

图 6 Curls & Whey 攻击产生的对抗样本

(a) 针对性攻击在 Tiny-Imagenet 上的结果（原始图像 x 、目标类别图像 x_{target} 、噪声 $x'-x$ 及对抗样本 x' 已从左到右列出）；(b) 在 Imagenet 上的非针对性攻击结果（目标模型上的分类结果在底部展示）

5.3 CAB 的噪声压缩性能

表 3 中给出了使用基于决策的不同攻击的中值噪声幅度的比较。第一行展示了初始噪声的 ℓ_2 范数。表中的最后五行表示五种基于决策的不同攻击的中值噪声幅度，两列表示两个数据集，即 Tiny-Imagenet（左）和 Imagenet（右）。针对每个数据集的四个模型，使用 Curls 攻击每个替代-目标模型对，并输入生成的对抗样本作为基于决策的攻击的起点。 6×2 的表中的每个元素都是一个 4×4 的矩阵，其中每行表示 Curls 方法使用的替代模型，每列表示目标模型。因此，对角线上的元素实际上是白盒攻击设置中压缩噪声幅度的结果。每个攻击方法的中值是在相同数量的查询（ $B=300$ ）下计算的。我们的 CAB 攻击的像素保留率 $r=0.2$ 。对于边界、偏置边界、进化和 CAB 攻击，球面方向和源方

向的步长为 $\delta_0 = 0.1$，$\tau_0 = 0.003$。衰减因子 φ 设置为 0.99。对于 BBA[10]，使用在每个步骤中都包含替代模型中的信息的版本。

从表 3 可以看出，对所有相同查询次数的黑盒攻击或 4×4 矩阵中的每个非对角元素，CAB 在不同目标模型上达到了最小的噪声幅度。与边界攻击相比，CAB 对中值噪声幅度有显著的降低，这验证了采用当前噪声和失败采样调整分布的有效性。由于白盒噪声的幅度（每个 4×4 矩阵中的对角元素）往往很小（在 ℓ_2 范数下小于黑盒噪声的 $1/10$），因此边界攻击采用的对每个维度都在标准正态分布采样进行采样的策略更为合适。结果表明，Imagenet 上 CAB 的白盒噪声幅度略大于边界攻击。

表 3　五种基于决策的攻击在两个数据集上对目标模型攻击的中值噪声幅度

		Tiny-Imagenet					Imagenet			
		res-18	inc-v3	inc-res	nasnet		res101	dense	vgg-19	senet
Initial	res-18	0.076	1.617	1.833	2.002	res101	0.325	3.060	3.232	4.367
	inc-v3	0.510	0.124	1.137	1.216	dense	2.777	0.269	3.112	4.241
	inc-res	0.552	1.083	0.147	1.088	vgg-19	6.050	6.034	0.183	5.135
	nasnet	0.577	1.272	1.199	0.134	senet	3.013	5.193	6.141	0.413
Whey	res-18	0.071	1.140	1.250	1.342	res101	0.315	2.868	2.785	3.836
	inc-v3	0.360	0.121	0.865	0.909	dense	2.695	0.262	2.627	3.737
	inc-res	0.369	0.796	0.142	0.819	vgg-19	5.149	4.996	0.180	4.333
	nasnet	0.401	0.927	0.906	0.129	senet	2.708	4.529	5.101	0.393
Boundary	res-18	0.059	1.220	1.386	1.467	res101	**0.279**	2.982	2.904	4.092
	inc-v3	0.384	0.110	1.001	1.041	dense	2.713	**0.231**	2.683	3.949
	inc-res	0.379	0.923	0.134	0.949	vgg-19	5.395	5.161	**0.155**	4.646
	nasnet	0.421	1.024	1.052	0.120	senet	2.738	4.695	5.761	**0.360**
Biaseddary	res-18	0.072	1.129	1.283	1.358	res101	0.318	2.586	2.704	3.745
	inc-v3	0.308	0.122	0.890	0.912	dense	2.441	0.263	2.496	3.542
	inc-res	0.325	0.813	0.144	0.829	vgg-19	4.776	4.402	0.181	4.133
	nasnet	0.332	0.928	0.924	0.132	senet	2.693	4.119	4.953	0.397
Evolutionary	res-18	0.068	0.951	1.112	1.147	res101	0.310	2.518	2.373	3.217
	inc-v3	0.269	0.117	0.881	0.851	dense	2.394	0.256	2.253	3.128
	inc-res	0.292	0.761	0.138	0.797	vgg-19	4.112	4.036	0.176	3.442
	nasnet	0.301	0.849	0.888	0.126	senet	2.569	3.730	4.644	0.386
CAB	res-18	**0.058**	**0.935**	**0.929**	**1.030**	res101	0.290	**2.387**	**1.953**	**3.116**
	inc-v3	**0.263**	**0.109**	**0.692**	**0.733**	dense	**2.294**	0.236	**2.025**	**3.051**
	inc-res	**0.251**	**0.689**	**0.131**	**0.682**	vgg-19	**3.982**	**3.730**	0.157	**3.413**
	nasnet	**0.284**	**0.757**	**0.726**	**0.119**	senet	**2.287**	**3.588**	**4.449**	0.366

5.4 消融分析

本节研究了迭代步 T、二分搜索步长 bs 和高斯噪声方差 var 对黑盒攻击效果的影响。分别使用 inception-resnet v2 和 inception v3 作为替代模型和目标模型。在不同 T、var 和 bs 下，Tiny-Imagenet 数据集上的结果如图 7 所示。如第 3 节所述，尽管 T 与噪声幅度呈负相关，但存在边际效应递减现象。$T=20$ 相对于 $T=16$ 的噪声下降明显不如 $T=8$ 相对于 $T=4$ 的噪声下降多。我们的方法不是简单地增加迭代次数，而是提高迭代轨迹的多样性。因此，Curls&Whey 能够在相同的查询条件下找到具有较小 ℓ_2 范数的对抗样本，并使用部分查询来压缩对抗噪声。

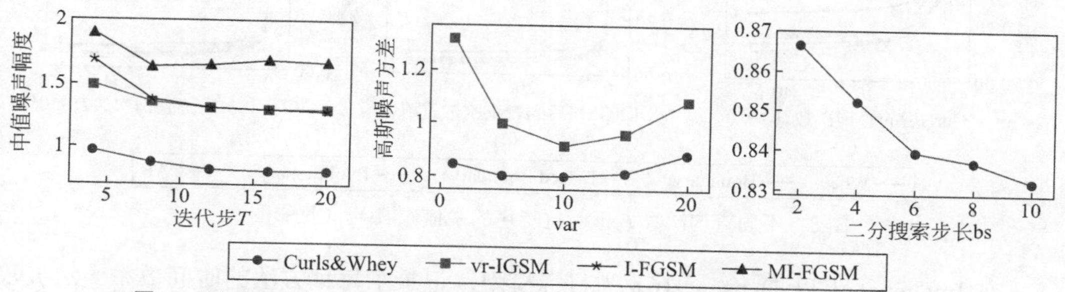

图 7　不同迭代步数下的中值噪声幅度（$T=(4,8,12,16,20)$）、高斯噪声方差（$\text{var}=(1,5,10,15,20)$）及二分搜索步长（$\text{bs}=(2,4,6,8,10)$）

方差 var 与替代模型和目标模型之间的迁移性有关。var 越高，对抗样本从一个模型迁移到另一个非常不同的模型的可能性就越大。然而，随着高斯噪声方差的增大，在梯度计算过程中原始图像所占的比例会逐渐减小，从而导致迁移性下降。因此，在不同 var 的结果中出现局部最小值。如图 7 所示，当使用 inception-resnet v2 攻击 inception v3 时，var 的局部最优值在 10 左右。

对于二分搜索步骤，较大的 bs 意味着在对抗样本和原始图像之间进行更多的二分搜索。作为 Curls 迭代的一个辅助过程，较小的 bs 足以降低噪声。

为了验证我们的攻击方法各部分的有效性，对 Curls&Whey 进行了消融实验。从表 4 可以看出，无论是 Curls 迭代法、二分搜索法（BS），还是 Whey 优化的两步法，每个部分都能有效地降低噪声幅度。

表 4　Curls&Whey 每一部分的增量比较

	Curls	+BS	+Whey(1)	+Whey(2)
中值	1.3138	1.1111	0.9354	**0.8431**
平均值	2.3154	1.9039	1.4723	**1.3437**

对于 CAB，最大查询数 T 是关键参数。使用 inception v3 作为替代模型，resnet-18 作为目标模型，运行三种基于迁移的攻击，包括 Curls，I-FGSM[5]和 VR-IGSM[15]，来生成初始的对抗样本。

图 8 显示了 Tiny-Imagenet 上噪声幅度随 T 增加的曲线。较高的 T 为基于决策的攻击提供了更多的机会来微调对抗噪声。可以看到，在不同的查询次数下，CAB 算法的噪声幅度比其他方法要低。

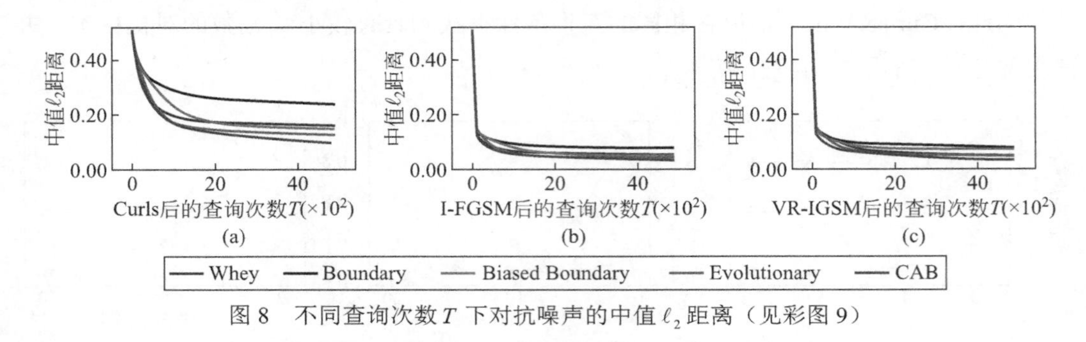

图 8 不同查询次数 T 下对抗噪声的中值 ℓ_2 距离（见彩图 9）

在 Imagenet 上以 Densenet161 为目标模型，记录了每种方法的时间效率。边界攻击和 CAB 攻击查询每幅图像 300 次的平均时间分别为 15.31s 和 13.39s。CAB 利用历史信息避免了使噪声幅度增大的采样空间，从而降低了无效采样的比例，提高了 CAB 的效率。

为了进一步验证 CAB 攻击中每一步的有效性，比较了单独或组合使用几个步骤时的中值和平均对抗噪声[22]。在表 5 中，展示的所有三个步骤，包括方差调整（VC）、步长调度（SS）和利用失败采样（EFS）都有助于提高降噪率。图 9 比较了 Imagenet（前 4 行）和 Tiny Imagenet（后 2 行）上五种不同的基于决策的攻击产生的对抗噪声。每行的第一个图像是 Curls 攻击生成的初始对抗样本 x^*，然后是五种攻击的噪声幅度。从左到右列出了五种方法的压缩噪声（为了更好地展示图像，增强噪声幅度）。由于 CAB 利用当前噪声进行自适应采样，有效地抑制了噪声较大的区域，从而获得了较高的噪声压缩效率。

表 5 CAB 每一步的噪声幅度比较

	x^*	VC	VC+SS	VC+EFS	CAB
中值	0.496	0.314	0.293	0.304	**0.269**
平均值	2.066	1.275	1.264	1.236	**1.202**

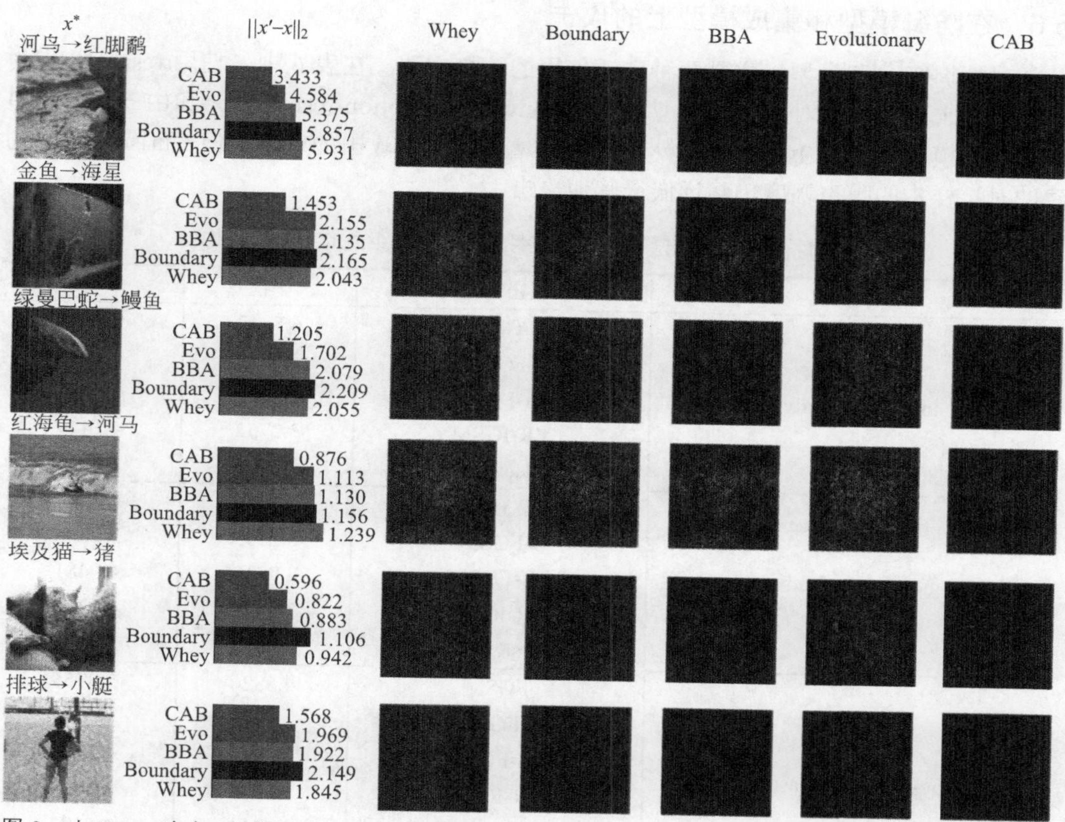

图 9 由 Whey 攻击、边界攻击、有偏边界攻击（BBA）、进化攻击（EVO）和我们的 CAB 攻击产生的对抗噪声的比较（见彩图 10）

标签及错分类别在每一行的 x^* 上方标注为 $y \to F\left(x^*\right)$

5.5 针对性攻击结果

在针对性攻击实验中，为每幅图像分配 5 个不同的目标类别，并计算每个目标类别的原始图像与对抗样本之间的 ℓ_2 距离。如 3.5 节所述，从测试集中选择一幅被分为目标类别的图像进行插值。选择 resnet18 和 inceptionv3 作为替代模型，另外三种模型作为目标模型。在黑盒场景中，三个现有的基于迁移的攻击难以在 ℓ_2 距离较小的条件下实现针对性错分。与边界攻击[8]、逐点攻击和原始插值[46]相比，我们方法的噪声幅度显著降低，这验证了将插值方法集成到 Curls&Whey 攻击中的有效性。

5.6 在防御模型和集成模型上的攻击

对抗训练[25]和模型集成是两种广泛使用的防御方法。在表 6 中，使用 resnet18 作为替代模型来攻击两个对抗训练模型（inceptionv3 和 inception-resnet v2）和由三个模型组成的集成模型。与表 1 相比，虽然防御方法增加了对抗攻击的难度，但 Curls&Whey 构建的对抗样本的噪声幅度仍然远低于其他攻击。

表 6　对抗训练模型与集成模型上对抗扰动的 ℓ_2 距离的中值和平均值

目标模型	攻击方法	中值	平均值
inceptionv3(adv)	FGSM	6.5812	9.1681
	I-FGSM	2.8839	3.76
	MI-FGSM	3.8039	4.6529
	VR-IGSM	3.2752	4.1449
	Curls& Whey	**2.0633**	**2.6349**
inc-resnet v2(adv)	FGSM	4.7029	6.2954
	I-FGSM	3.3195	3.9606
	MI-FGSM	3.9919	4.9481
	VR-IGSM	3.3829	4.2706
	Curls& Whey	**2.2852**	**2.7884**
inceptionv3+ inc-resnet v2+nasnet	FGSM	4.5826	5.9755
	I-FGSM	2.7742	3.595
	MI-FGSM	3.5819	4.5227
	VR-IGSM	3.0785	4.0499
	Curls& Whey	**2.0321**	**2.6187**

6 总结

本文提出了 Curls&Whey 和 CAB 来分别弥补基于迁移和基于决策的黑盒攻击的缺陷。Curls&Whey 使迭代轨迹多样化，并压缩对抗噪声。CAB 使用当前噪声来选择图像的敏感区域和适应性调整采样分布。分析了噪声压缩过程中的单调性，以及方差与当前噪声的关系，从而实现了边界攻击采样过程的自适应。此外，将插值与基于迁移的攻击相结合，显著降低了黑盒场景下针对性攻击的难度。在 Tiny-Imagenet 和 Imagenet 上的实验结果表明，与现有的黑盒攻击相比，Curls&Whey 和 CAB 生成的对抗样本具有较小的 ℓ_2 距离并对各种目标模型具有较强的迁移性。

参考文献

[1] SZEGEDY C, ZAREMBA W, SUTSKEVER I, et al. Intriguing properties of neural networks[C]//International Conference on Learning Representations, 2013.

[2] NGUYEN A, YOSINSKI J, CLUNE J. Deep neural networks are easily fooled: High confidence predictions for unrecognizable images[C]//Proceedings of the IEEE conference on computer vision and pattern recognition. 2015: 427-436.

[3] PAPERNOT N, MCDANIEL P, JHA S, et al. The limitations of deep learning in adversarial settings[C]//IEEE European symposium on security and privacy (EuroS&P).IEEE, 2016:372-387.

[4] GOODFELLOW I J, SHLENS J, SZEGEDY C. Explaining and harnessing adversarial examples[C]//International Conference on Learning Representations, 2015

[5] KURAKIN A, GOODFELLOW I, BENGIO S. Adversarial examples in the physical world[C]//International Conference on Learning Representations Workshop.

[6] DONG Y, LIAO F, PANG T, et al. Boosting adversarial attacks with momentum[C]//The IEEE conference on computer vision and pattern recognition. 2018:9185-9193.

[7] DONG Y, PANG T, SU H, et al. Evading defenses to transferable adversarial examples by translation-invariant attacks[C]//Proceedings of the IEEE/CVF Conference on Computer Vision and Pattern Recognition. 2019: 4312-4321.

[8] BRENDEL W, RAUBER J, BETHGE M. Decision-based adversarial attacks: Reliable attacks against black-box machine learning models[C]//International Conference on Learning Representations, 2017.

[9] DONG Y, SU H, WU B, et al. Efficient decision-based black-box adversarial attacks on face recognition[C]//The IEEE Conference on Computer Vision and Pattern Recognition. 2019: 7714-7722.

[10] BRUNNER T, DIEHL F, LE M T, et al. Guessing smart: Biased sampling for efficient black-box adversarial attacks[C]//The IEEE International Conference on Computer Vision. 2019: 4958-4966.

[11] CHEN P-Y, ZHANG H, SHARMA Y, et al. Zoo: Zeroth order optimization based black-box attacks to deep neural networks without training substitute models[C]//The 10th ACM Workshop on Artificial Intelligence and Security. 2017: 15-26.

[12] LIU L, CHENG M, HSIEH C-J, et al. Stochastic zeroth-order optimization via variance reduction method[Z]. arXiv preprint arXiv:1805.11811.

[13] TU C-C, TING P, CHEN P-Y, et al. Autozoom: Autoencoder-based zeroth order optimization method for attacking black-box neural networks[C]//Proceedings of the AAAI Conference on Artificial Intelligence. 2019: 742-749.

[14] YE H, HUANG Z, FANG C, et al. Hessian-aware zeroth-order optimization for black-box adversarial attack[Z]. arXiv preprint arXiv:1812.11377.

[15] WU L, ZHU Z, TAI C. Understanding and enhancing the transferability of adversarial examples[Z]. arXiv preprint arXiv:1802.09707.

[16] ILYAS A, ENGSTROM L, MADRY A. Prior convictions: Black-box adversarial attacks with bandits and priors[C]//International Conference on Learning Representations, 2019.

[17] CROWD A I. Neurips 2018: Adversarial vision challenge[R/OL]. (2018-11-15)[2021-03-26]. https://www.crowdai.org/organizers/bethgelab/challenges/nips-2018-adversarial-vision-challenge-untargeted-attack-track.

[18] CHENG S, DONG Y, PANG T, et al. Improving black-box adversarial attacks with a transfer-based prior [J]. Advances in Neural Information Processing Systems, 32: 10934-10944.

[19] FAWZI A, FAWZI O, FROSSARD P. Analysis of classifiers' robustness to adversarial perturbations[J]. Machine Learning, 107: 481-508.

[20] LIU Y, CHEN X, LIU C, et al. Delving into transferable adversarial examples and black-box attacks[C]//International Conference on Learning Representations, 2016.

[21] CARLINI N, WAGNER D. Towards evaluating the robustness of neural networks[C]//IEEE symposium on security and privacy (sp).IEEE, 2017:39-57.

[22] BRENDEL W, RAUBER J, KURAKIN A, et al. Adversarial vision challenge[C]//The NeurIPS'18 Competition.Springer. 2020: 129-153.

[23] FINLAY C, POOLADIAN A-A, OBERMAN A. The logbarrier adversarial attack: making effective use of decision boundary information[C]//Proceedings of the IEEE/CVF International Conference on Computer Vision. 2019: 4862-4870.

[24] RUSSAKOVSKY O, DENG J, SU H, et al. Imagenet large scale visual recognition challenge[J]. International journal of computer vision, 115: 211-252.

[25] TRAMèR F, KURAKIN A, PAPERNOT N, et al. Ensemble adversarial training: Attacks and defenses[C]//International Conference on Learning Representations, 2017.

[26] PAPERNOT N, MCDANIEL P, GOODFELLOW I, et al. Practical black-box attacks against machine learning[C]//Proceedings of the 2017 ACM on Asia conference on computer and communications security. 2017: 506-519.

[27] PAPERNOT N, MCDANIEL P, WU X, et al. Distillation as a defense to adversarial perturbations against deep neural networks[C]//2016 IEEE Symposium on Security and Privacy (SP).IEEE, 2016:582-597.

[28] LI X, LI F. Adversarial examples detection in deep networks with convolutional filter statistics[C]// Proceedings of the IEEE International Conference on Computer Vision. 2017: 5764-5772.

[29] MENG D, CHEN H. Magnet: a two-pronged defense against adversarial examples[C]//The ACM SIGSAC Conference on Computer and Communications Security. 2017: 135-147.

[30] CHEN J, JORDAN M I, WAINWRIGHT M J. HopSkipJumpAttack: A query-efficient decision-based attack[C]//2020 IEEE Symposium on Security and Privacy (SP). 2020: 668-685.

[31] CHENG M, LE T, CHEN P-Y, et al. Query-efficient hard-label black-box attack: An optimization-based approach[C]//International Conference on Learning Representation (ICLR), 2019.

[32] WANG B, GAO J, QI Y. A theoretical framework for robustness of (deep) classifiers under adversarial noise[C]//International Conference on Learning Representations Workshop, 2017.

[33] XIE C, ZHANG Z, ZHOU Y, et al. Improving transferability of adversarial examples with input diversity[C]//Proceedings of the IEEE/CVF Conference on Computer Vision and Pattern Recognition. 2019: 2730-2739.

[34] ZHOU W, HOU X, CHEN Y, et al. Transferable adversarial perturbations[C]//Proceedings of the European Conference on Computer Vision (ECCV). 2018: 452-467.

[35] PAPERNOT N, MCDANIEL P, GOODFELLOW I. Transferability in machine learning: from phenomena to black-box attacks using adversarial samples[Z]. arXiv preprint arXiv:1605.07277.

[36] TRAMèR F, PAPERNOT N, GOODFELLOW I, et al. The space of transferable adversarial examples[Z]. arXiv preprint arXiv:1704.03453.

[37] SETHI T S, KANTARDZIC M. Data driven exploratory attacks on black box classifiers in adversarial domains[J]. Neurocomputing, 289: 129-143.

[38] ATHALYE A, ENGSTROM L, ILYAS A, et al. Synthesizing Robust Adversarial Examples[C]//International Conference on Machine Learning. 2018: 284-293.

[39] HE K, ZHANG X, REN S, et al. Deep residual learning for image recognition[C]//Proceedings of the IEEE conference on computer vision and pattern recognition. 2016: 770-778.

[40] SZEGEDY C, VANHOUCKE V, IOFFE S, et al. Rethinking the inception architecture for computer vision[C]//Proceedings of the IEEE conference on computer vision and pattern recognition. 2016: 2818-2826.

[41] SZEGEDY C, IOFFE S, VANHOUCKE V, et al. Inception-v4, inception-resnet and the impact of residual connections on learning[C]//Proceedings of the AAAI Conference on Artificial Intelligence, 2017.

[42] ZOPH B, VASUDEVAN V, SHLENS J, et al. Learning transferable architectures for scalable image recognition[C]//Proceedings of the IEEE conference on computer vision and pattern recognition. 2018:8697-8710.

[43] HUANG G, LIU Z, VAN DER MAATEN L, et al. Densely connected convolutional networks[C]//Proceedings of the IEEE conference on computer vision and pattern recognition. 2017:4700-4708.

[44] SIMONYAN K, ZISSERMAN A. Very Deep Convolutional Networks for Large-Scale Image Recognition[C]//International Conference on Learning Representations, 2015.

[45] HU J, SHEN L, SUN G. Squeeze-and-excitation networks[C]//Proceedings of the IEEE conference on computer vision and pattern recognition. 2018: 7132-7141.

[46] RAUBER J, BRENDEL W, BETHGE M. Foolbox: A python toolbox to benchmark the robustness of machine learning models[Z]. arXiv preprint arXiv:1707.04131.

对抗机器学习：攻击、防御与模型鲁棒性评估

易津锋

（京东科技集团机器学习部，北京 100101）

1 引言

近年来，随着机器学习和人工智能技术在越来越多也越来越复杂的应用场景下落地，产业界对其能力的全面性提出了更高的要求。然而，过去的人工智能往往采用准确率（或与之对应的错误率）作为衡量模型性能好坏的决定性甚至唯一标准。例如，计算机视觉领域的大规模视觉识别挑战赛（ILSVRC）[1]和自然语言处理领域的 SQuAD 机器阅读理解竞赛[2]都分别把衡量准确性的 Top-1/5 分类错误率和 F1-score 作为参赛算法排名的依据。诚然，准确率是衡量模型可用性的一个重要指标，但仅仅在规范的测试集上取得较高的准确率并不一定确保 AI 系统可以胜任真实复杂环境中的任务。近期的多项研究同样表明，模型的准确率与鲁棒性（robustness）之间往往存在一定程度的负相关关系[3-5]。例如，我们[3]通过研究 18 个面向 ImageNet 的经典卷积神经网络分类器，发现准确率越高的模型反而越容易受到对抗样本[6-7]的攻击，如图 1 所示。因此，仅仅追求更高的准确率，反而可能会影响模型的综合性能。

上文提到的对抗样本是 2014 年由 Szegedy 等人[7]首次提出。它是指通过在正常样本中添加（人眼难以察觉的）微小扰动，从而让原本能够正确预测的机器学习模型做出错误判断的样本①。图 2 展示了对抗样本的一个示例。它包含七张看起来几乎一模一样的鸵

① 实际上，Biggio 等人[8]早在 2013 年就提出了针对神经网络分类器的规避攻击（evasion attack），其本质与对抗样本非常相似。甚至早在 2004 年，就已经有工作开展了针对线性分类器的攻击[9-10]。但鉴于对抗样本这个名字已被行业广泛采用，本文仍然使用对抗样本来统称在模型的推理/测试阶段通过对原始样本添加微小扰动而实施的攻击方式。

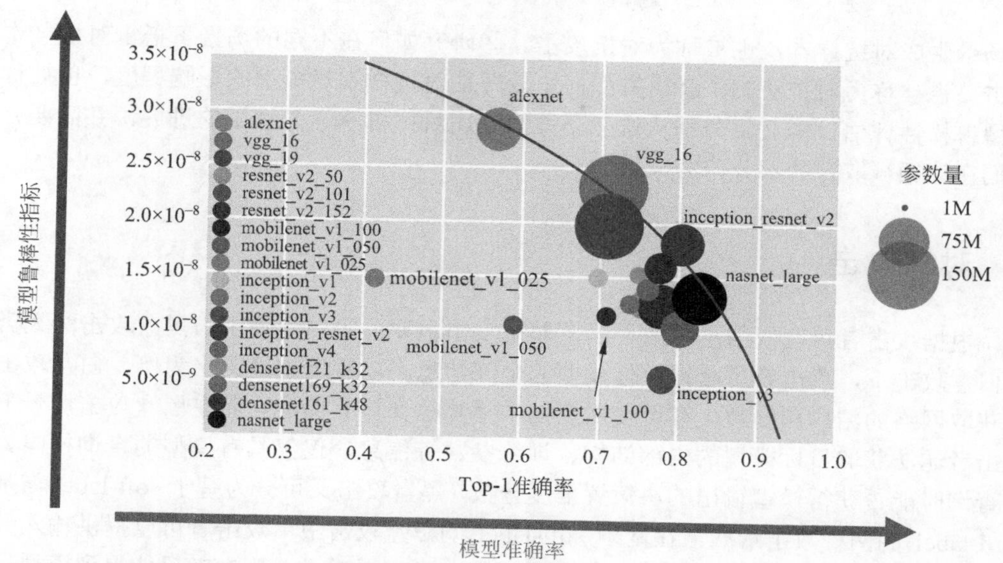

图 1　18 个卷积神经网络的分类准确率和其对抗鲁棒性指标存在的负相关关系（见彩图 11）[3]

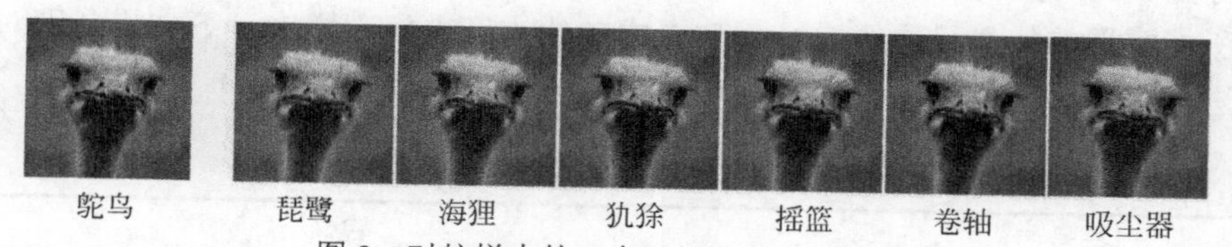

图 2　对抗样本的一个示例（见彩图 12）

在添加不同的对抗扰动后，一个高准确率的 AI 模型可能将鸵鸟图片误识别为多个不同的类别

鸟图片，其中左边第一张是正常的原图，而右边的六张图片都是添加了不同扰动后形成的对抗样本。如果将它们分别输入一个高准确率的图像分类器，除了左边第一张图能被正确识别为鸵鸟之外，右边的六张图都很可能会被错误地判别为海狸、摇篮、吸尘器等完全不相关的类别。由于人工智能已被广泛应用于如自动驾驶、医疗、金融、安防等与安全息息相关的领域，对抗样本的存在很可能会给人们的生命和财产安全带来极大的威胁。例如，类似的图像扰动如果发生在交通标志上，使用 AI 系统的无人驾驶汽车很可能会误判这些标志，从而酿成严重的车祸。同样，对抗样本如果被用于破解人脸识别系统，则可能会造成用户信息与个人财产的损失。

鉴于此，如何保护机器学习模型免受对抗样本威胁成为了一个亟待解决的问题。近年来，相关学者们围绕这一问题开展了大量的研究，形成了对抗机器学习这一领域。一般来说，对抗机器学习的研究内容主要分成三个大的方向：①研究如何针对不同的模型和场景生成对抗样本，也被称为对抗攻击。②研究如何在不同的场景下抵御对抗样本的攻击，也被称为对抗防御。③研究如何定量评估对抗样本对模型的影响程度，也被称为模型鲁棒性评估。本文将分别从这三个方面阐述对抗机器学习领域的进展，并简要介绍我们在该领域取得的部分研究成果。

2 对抗攻击

根据攻击者可获取信息量的不同，对抗攻击主要分为白盒攻击与黑盒攻击两大类。在白盒攻击下，攻击者可获得目标模型的全部信息。以深度神经网络为例，白盒攻击者可获取网络的结构和参数，亦可使用反向传播算法计算其偏导。而在黑盒攻击场景下，攻击者无法获取目标模型的内部结构、训练参数等信息，仅仅具有模型的查询权限。根据查询时能否获得模型输出的决策置信度分数，黑盒攻击又可分为基于 soft-label 与基于 hard-label 的两种攻击形式。在基于 soft-label 的黑盒攻击中，攻击者可以提供输入并得到模型输出的决策置信度分数，而在基于 hard-label 的攻击中，攻击者只能得到模型最后的决策标签。从白盒攻击到基于 soft-label 的黑盒攻击，再到基于 hard-label 的黑盒攻击，攻击者所能掌握的信息量逐级递减，攻击成功的难度也依次增大。下文将分别对各类攻击方式进行介绍。表 1 总结了本文所使用的常用符号及其含义。

表 1 本文使用的常用符号及其含义

符号	含义
x_0	原始正常样本
y	原始样本对应的标签
x'	对抗样本
δ	生成的对抗扰动
t	有目标攻击下的目标类别
ϵ	最大扰动约束
$f(\cdot)$	分类模型
θ	模型参数
L	损失函数

2.1 白盒攻击

白盒攻击是最早被研究的一种对抗攻击方式。针对白盒模型，研究者们提出了多种不同类型的攻击算法，如基于梯度的攻击[6,11-12]、基于决策面的攻击[13]以及基于优化的攻击[14-16]等。早期的白盒攻击主要采用基于梯度的攻击方式，其代表算法是 fast gradient sign method (FGSM)[6]。FGSM 算法的原理十分简单：通过计算模型对输入导数的符号得到具体的梯度方向，再乘以一个步长，就得到了需要添加的对抗扰动，具体形式如下：

$$x' = x_0 + \epsilon \mathrm{sign}\left(\nabla_x L\left(f_\theta\left(x_0\right), y\right)\right) \tag{1}$$

其中，∇_x 表示损失函数 $L\left(f_\theta\left(x_0\right), y\right)$ 在正常样本 x_0 处的梯度。FGSM 只对输入进行一次更新，属于单步攻击算法，而之后提出的 BIM[17]，PGD[11]，MI-FGSM[12]等算法可对扰动进行多次迭代，与单步攻击相比可以产生更强的攻击效果。此外，DeepFool[13]是一种基于决策面的攻击算法，其目标是寻找可使分类器产生误判的最小扰动。JSMA[18]算法则通过计算前向导数构建雅可比显著图（Jacobian saliency map）从而筛选出需要修改的像素点，因此只需改动少量像素即可生成使模型错误分类的对抗样本。

在基于优化的白盒攻击中，C&W 算法[14]是其中的代表性工作。它将对抗样本的生成问题转化为如下的优化问题进行求解：

$$\begin{gathered}\min_x \| x - x_0 \|_2^2 + c \cdot L\left(f_\theta\left(x\right), t\right) \\ \text{s.t. } x \in \left[0,1\right]^p\end{gathered} \tag{2}$$

其中，损失函数 $L\left(f_\theta\left(x\right), t\right)$ 被表示为

$$L\left(f_\theta\left(x\right), t\right) = \max\left\{\max_{i \neq t}\left[Z\left(x\right)\right]_i - \left[Z\left(x\right)\right]_t, -k\right\} \tag{3}$$

其中，$Z\left(x\right)$ 是网络的 logit 层输出；$\left[Z\left(x\right)\right]_i$ 表示 模型将x 判定为类别 i 的 logit；k 则是控制样本迁移性强弱的参数。C&W 使用 $(1+\tanh w)/2$ 代替 x，使其满足式（2）中的约束限制，从而使优化问题变得更加易于求解。相比于上述提到的其他各类攻击方法，C&W 算法通常能够产生更小的对抗扰动，且往往具有更高的攻击成功率。

在此之后，更多基于优化的白盒攻击算法被相继提出。例如，为了进一步增强对抗样本的可迁移性（transferability）[7,19]，即使针对某个模型生成的对抗样本也能以较大的几率成功攻击其他模型，我们提出了一种基于弹性网络正则化的攻击算法 EAD[15]。它采用与 C&W 算法类似的目标函数，并引入了弹性网络正则项 (elastic-net regularization)：

$$\begin{gathered}\min_x c \cdot L\left(f_\theta\left(x\right), t\right) + \beta \cdot \| x - x_0 \|_1 + \| x - x_0 \|_2^2 \\ \text{s.t. } x \in \left[0,1\right]^p\end{gathered} \tag{4}$$

当 $\beta = 0$ 时，C&W 算法即成为 EAD 的一个特例。当 $\beta > 0$ 时，C&W 算法中的 tanh 转换

操作不再适用，因此 EAD 提出使用迭代收缩阈值算法（iterative shrinkage-thresholding algorithm, ISTA）来计算对抗样本 $x^{(k+1)}$：

$$x^{(k+1)} = S_{\beta}\left(x^{(k)} - \alpha_k \nabla g\left(x^{(k)}\right)\right) \tag{5}$$

从实验效果来看，EAD 往往能通过更小的 L_1 扰动生成迁移攻击能力更强的对抗样本。

除此之外，Moosavi-Dezfooli 等人[20]提出通用对抗扰动算法，其生成的单个对抗扰动可以同时影响多个测试样本，并取得较高的攻击成功率。Su 等人[21]提出了基于 L_0 范数约束的单像素攻击方法，通过只修改一个或少量像素实现攻击的目标。为了进一步降低图像的失真程度，Croce 等人[22]提出了 FAB 算法，能够在 L_1，L_2 和 L_∞ 范数约束下提升对抗样本的图像质量并有效防止基于梯度遮蔽（obfuscated gradient）[23]的防御。此外，他们在另一篇工作中[24]提出了基于 PGD 的步长自适应攻击算法 Auto-PGD，可根据攻击情况自动调整攻击步长，从而在减少超参数的情况下进一步提升攻击效果。同时，作者将 Auto-PGD，FAB 和 Square 攻击[25]共同组成一个无参数且相对高效的攻击算法集合 AutoAttack，可用于通过实验方法评估模型的鲁棒性。

2.2 黑盒攻击

与白盒攻击相比，黑盒攻击是一种更为现实的攻击方式。因为在大多数真实场景下，目标模型对攻击者并不透明。由于无法获取目标模型的结构和参数，类似 FGSM 这类直接计算梯度的算法在黑盒攻击中并不适用。在早期，对黑盒模型的攻击主要是利用对抗样本的可迁移性。基于这种特性，攻击者通常会在本地训练一个和黑盒模型近似的代理模型，并利用代理模型生成的对抗样本攻击目标的黑盒模型[26]。虽然这类攻击方法实现起来较为简单，但其攻击成功率往往较低，因此更多的工作主要还是针对黑盒模型本身来进行攻击。它们通常通过查询黑盒模型对给定输入数据的输出来了解模型更多的信息，进而据此生成对抗样本。根据查询时能否获得模型输出的决策置信度分数，这类黑盒攻击又可分为基于 soft-label 与基于 hard-label 的两种攻击形式，下面分别进行介绍。

2.2.1 基于 Soft-label 的攻击

在基于 soft-label 的攻击中，攻击者可以向目标模型输入样本进行查询，而模型返回的结果是对应类别的置信度分数。针对这一问题，我们提出了一种基于零阶优化的黑盒攻击算法 ZOO (zeroth order optimization)[27]。ZOO 的目标函数与 C&W 算法相似。由于在黑盒攻击情况下，攻击者无法获取目标模型对于输入样本的梯度，因此 ZOO 使用对称差商（symmetric difference quotient）[28]来对图像的第 i 个像素点梯度进行估算：

$$\hat{g}_i := \frac{\partial f(x)}{\partial x_i} \approx \frac{f(x + h\boldsymbol{e}_i) - f(x - h\boldsymbol{e}_i)}{2h} \tag{6}$$

其中，$\boldsymbol{e}_i$ 是只有第 i 个分量为 1 的标准基向量。同样，可以得到逐坐标的 Hessian 估计：

$$\hat{h}_i := \frac{\partial^2 f(x)}{\partial x_{ii}^{\ 2}} \approx \frac{f(x + h\boldsymbol{e}_i) - 2f(x) + f(x - h\boldsymbol{e}_i)}{h^2} \tag{7}$$

在得到 $\hat{g}_i$ 和 $\hat{h}_i$ 后，ZOO 可利用随机梯度下降或牛顿法来求解最优扰动。通过实验发现，在 MNIST 和 CIFAR10 数据集上，ZOO 能达到和白盒攻击算法 C&W 相近的 L_2 失真度，并首次在黑盒攻击上实现了接近 100%的攻击成功率。

尽管 ZOO 能实现较高的攻击成功率，但它需要估算图像上几乎每一个像素的梯度，使它的查询复杂度非常高。为了提升查询效率，我们进一步提出了一个通用的优化框架 AutoZoom (autoencoder-based zeroth order optimization method)[29]。与 ZOO 中使用逐坐标估计不同的是，AutoZoom 提出了一种按比例缩放的随机全梯度估计：

$$g = b \cdot \frac{f(x + b\boldsymbol{u}) - f(x)}{\beta} \cdot \boldsymbol{u} \tag{8}$$

其中，$\boldsymbol{u}$ 是从单位欧氏球体空间上随机均匀选取的单位长度矢量；β 是一个平滑参数；b 则用于平衡梯度估计误差中的偏差与方差。AutoZoom 能够在一个更低的维度上求解对抗扰动并降低查询次数。具体来说，它通过一个自编码器 $D: R^d \to R^{d'}$ 在一个 d' 维的低维空间上进行随机梯度估计，并高效地生成对抗样本 $x' = x + D(\delta)$。实验表明，与算法 ZOO 相比，在 MNIST，CIFAR10 及 ImageNet 数据集中，AutoZoom 可以显著减少 93%以上的查询次数，极大地提升了查询效率。

除此之外，众多研究者们不断提出了效率更高且攻击效果更好的黑盒攻击算法。例如，Ilyas 等人[30]给出了在查询量受限、部分信息受限及仅标签信息情况下的三种现实攻击场景，并提出了基于自然演化策略（natural evolutionary strategies）[31]的新型攻击算法，从而可以在这些限制性更强的场景下实现攻击。Bhagoji 等人[32]提出了一种将对抗攻击所需查询次数与输入图像维度解耦的策略，并提出了一种在有目标攻击（targeted attack）和无目标攻击（untargeted attack）情况下都能接近 100%攻击成功率的黑盒攻击算法。Liu 等人[33]设计了一种零阶随机优化算法。该算法与 signSGD 相比，可以利用 $\sqrt{d}$ 倍的迭代次数达到 $O(\sqrt{d} / \sqrt{T})$ 的收敛速率。Ilyas 等人[34]引入梯度先验来优化黑盒攻击，并提出了一种基于 Bandit 优化的方法，可有效地增强黑盒攻击效果。Meunier 等人[35]则指出进化策略（evolution strategy）结合平铺技术可以达到更优的黑盒攻击效果。基于此，他们为黑盒攻击的无梯度优化提出了一个新的目标函数范式，并通过实验验证了该范式下的两个具体实例。Cheng 等人[36]提出了 P-RGF(prior-guided random gradient-free)算法，可利用迁移先验和信息查询来优化黑盒攻击，在一定程度上提高了攻击成功率和查询效率。Abdullah Al-Dujaili 等人[37]提出了一种基于符号的梯度估计算法 SignHunter，可通过对损失函数的有限近似差分，完成对梯度符号信息的高效估计，将查询效率进一步提升。Andriushchenko 等人[25]提出了一种基于随机搜索的攻击方法 Square Attack。在每次迭代中，随机搜索的采样结果都可以最大程度地逼近扰动上限。这种方法一方面可以提高随机搜索的效率，另一方面也可以防止基于梯度遮蔽的防御。除此之外，Shi 等人[38]为了解决基于迭代的黑盒攻击在梯度上升过程中扰动方向单一及无法去除冗余扰动的问题，提出了 Curls & Whey 攻击算法。该算法分为 Curls 和 Whey 两个迭代过程，前者在产生扰动的过程中综合考虑了梯度上升和下降的方向，后者则用于去除冗余扰动。二者结合能让算法生成具有更强迁移性的攻击样本。Du 等人[39]则证明 MetaLearning 也可被用来提升黑盒攻击的查询效率。他们利用 MetaLearning 训练了一个基于 Encoder-Decoder 的生成器。该生成器能通过对目标模型的少量查询来调整自己的参数，从而输出能够成功实现攻击的对抗样本。

2.2.2 基于 Hard-label 的攻击

与基于 soft-label 的攻击不同，在基于 hard-label 的攻击中，攻击者仅能得到模型对输入数据的预测标签。可以想见，由于每次查询能获得的信息更少，基于 hard-label 的攻击比基于 soft-label 的攻击更加困难，也需要更多的查询次数。然而，过多的查询次数容易触发访问量异常，所以现实场景中的机器学习系统可以简单地通过禁止异常的访问量来实现对相关攻击的防御。因此，如何提升查询效率一直是实施 hard-label 攻击时需要解决的关键问题。

Boundary-attack[40]是最早的基于 hard-label 的黑盒攻击算法之一。它首先初始化一个较大的对抗扰动，得到与原始样本差距较大的对抗样本，然后在此基础上通过随机游走逐渐向原始样本靠近，并在移动的同时保持对抗性。这种攻击方式的优点在于几乎不需要设定超参数且不依赖于替代模型，但它需要很高的查询次数且缺乏收敛性的保证，也在一定程度上限制了其适用性。针对该问题，我们提出了一种更为高效的攻击方式 Opt-Attack[41]。它将 hard-label 黑盒攻击转化成一个可以通过零阶优化算法求解的实值优化问题。具体来说，Opt-Attack 引入了一个角度变量 φ，并用 $g(\varphi)$ 表示原始样本 x_0 沿 φ 方向移动生成对抗样本所需的最小距离。通过这种方式，对抗攻击问题被转化为了求解最优角度 φ 的优化问题，从而使沿着 φ 方向生成的对抗样本具有最小的扰动量。我们通过细粒度搜索和二分查找计算不同角度所对应的 $g(\varphi)$ 值，并利用零阶优化算法求解更优的对抗样本。由于 Opt-Attack 不依赖于分类器的梯度，所以也可被用来攻击神经网络之外的其他离散和非连续机器学习模型，如 GBDT[42]等。相比于 Boundary-attack，Opt-Attack 具有更高的查询效率和更低的扰动幅度。

除此之外，Brunner 等人[43]提出了一类基于偏差抽样的算法 Boundary Attacks。该方法通过给采样分布赋予预设偏差，使控制采样的参数不再独立同分布，从而有更大的概率采样到更具攻击性的扰动。Dong 等人[44]则提出了一种进化攻击算法，可对搜索方向的局部进行建模，显著减小了搜索空间的维度，实现了更高的查询效率。Li 等人[45]提出了一种基于边界的黑盒攻击算法 QEBA，从理论上解释了为什么之前在整个梯度空间内进行梯度估计的攻击算法查询效率不高的问题，并为基于降维的梯度估计提供了最优的方案。与之前的算法相比，QEBA 可以在保持 100%攻击成功率的基础上减小扰动幅度并提高查询效率。Chen 等人[46]则提出了 Sign Flip Attack 算法，只需通过随机翻转少数对抗扰动符号即可显著提高黑盒攻击的性能。

2.3 其他对抗攻击

上述攻击算法主要针对图像分类任务。实际上，对抗样本还广泛存在于包括图像描述生成（image captioning）、目标检测、强化学习、机器翻译、语音识别等诸多领域。例如，针对图像描述生成任务，我们提出了两种对抗样本生成方法[47]，可以分别在给定目标标题及给定目标关键词集合时，生成与原图非常相似的对抗样本，从而使目标模型按照给定的标题或关键词集合生成错误描述。此外，还利用梯度投影优化算法为 Seq2Seq 序列模型[48]在离散输入空间上构建出对抗样本[49]，从而可以对机器翻译和文本摘要等任务实施攻击。Gong 等人[50]则通过搜寻词向量的最近邻词语并进行替换，从而在基本不改变文本语义的前提下，诱导神经网络出现误判。针对语义分割任务，Metzen 等人[51]提出了针对该问题的通用性对抗扰动生成方法。该方法可对特定目标类别产生扰动，而不影响其他类别，且该扰动对不同的输入均有效。Xie 等人[52]提出了一种可用于攻击目标检测和语义分割模型的密集对抗生成算法 DAG，并发现生成的对抗样本可以在使用不同训练数据、不同架构，甚至具有不同识别任务的网络间迁移。在语音识别任务中，Carlini 等人[53]通过添加噪声将给定的任意波形转换成人耳不能区分的新波形，成功地攻击了 DeepSpeech 语音识别模型。针对强化学习场景，Huang 等人[54]则发现 A3C[55]，TRPO[56]和 DQN[57]等常用强化学习算法学得的策略均会受到对抗样本的影响从而做出错误的判断。

除了数字域之外，对抗样本同样广泛存在于现实的物理世界。例如，Kurakin 等人[17]通过研究针对物理世界的对抗攻击方法，发现对抗样本被摄像头拍摄后依然具有较强的攻击性。Eykholt 等人[58]提出了一种通用的攻击算法 RP_2 (robust physical perturbation)，当生成的特定图案贴在道路标志上时，可让自动驾驶汽车的标志识别模块产生严重误判。

Brown 等人[59]制作出了通用的“对抗补丁”（adversarial patch），将该补丁打印后贴在原始样本的任意位置皆可对模型实现定向攻击。Athaly 等人[60]则制作出了可在旋转、缩放、光照等变换条件下较为鲁棒的 3D 对抗样本。Wu 等人[61]对目标检测攻击的可转移性进行研究，通过在衣服上打印通用的对抗伪装来绕过 AI 检测模块，并量化了复杂织物变形后的攻击成功率。

3 对抗防御

为了抵御对抗样本的攻击，近年来国内外的研究者们提出了众多的防御方法。根据研究目标的不同，这些方法可分为基于模型的防御方法与基于数据的防御方法两大类。基于模型的防御方法以模型作为出发点，通过训练一个更为鲁棒的模型达到抵御对抗样本的效果。基于数据的防御方法则以输入数据为研究对象，通过对输入数据进行甄别或预处理实现防御的目标。下文将分别对这两类工作进行介绍。

3.1 基于模型的防御

基于模型的防御方法根据对模型的不同处理方式可分为模型蒸馏、鲁棒性结构设计、鲁棒损失函数设计、混淆梯度、对抗训练等多种方式。Papernot 等人[62]在模型蒸馏过程中发现，通过将大网络的知识迁移至小网络，可降低模型对输入扰动的敏感性，从而提升小网络的鲁棒性。Buckman 等人[63]则通过在模型中加入不可微的模块，从而掩盖模型梯度，迫使部分攻击方法失效。此外，通过设计更为鲁棒的网络结构，也可对对抗攻击起到一定的防御效果。Bradshaw 等人[64]注意到现有分类器往往缺乏对自身推理不确定性的估计，因而在遭遇对抗样本或样本分布偏移时可能会表现出不可靠的自信（overconfidence）。为了解决这一问题，研究者们在深度神经网络中引入贝叶斯分类器，可以在对输入进行预测的同时对对抗样本和样本的分布偏移提出警告，从而实现对输出的不确定性估计。此外，一些研究者们还通过在网络结构中加入“非可微层”，从而防止梯度泄漏[65-66]。另一些工作则从损失函数设计入手，构建出“类内紧密，类间分散”的聚类损失以提升模型的鲁棒性[67-68]。虽然研究者们提出了诸多对抗防御方法，但其中的大部分都已被证实在面对某些特定攻击的情况下还不够鲁棒[23,69]。例如，Athalye 等人[23]指出，基于混淆梯度的三类防御方法离散梯度（gradient shattering）、随机梯度（stochastic gradients）和梯度消失&爆炸（vanishing & exploding gradients）都可能被某些攻击方式破解，因此并不可靠。

对抗训练（adversarial training，AT）[11]被普遍认为是现阶段最有效的对抗样本防御方法之一[23,69]。它通过求解如下的鲁棒优化问题，将对抗样本不断加入训练样本集合中以加强模型的鲁棒性：

$$\min_{\theta} \rho(\theta), \quad \rho(\theta) = \mathbb{E}_{(x,y)\sim\mathcal{D}}[\max_{\delta\in\mathcal{S}} L(\theta, x+\delta, y)] \tag{9}$$

具体来说，对抗训练通过内部最大化目标损失生成对抗样本，并在每轮的迭代中注入对抗样本重新训练网络。之后，外部最小化对抗样本带来的损失从而调整网络参数，使模型面对潜在的攻击时能够表现得更为鲁棒。其中，基于投影梯度下降（PGD）的PGD-AT 对抗训练简化流程见算法 1。

算法 1　基于 PGD 的对抗训练（PGD-AT）

输入：批量大小 n，分类网络 f

随机初始化网络 f 的模型参数 θ

重复：

　　从训练集中读取小批量数据 $B = \{x^1, x^2, \cdots, x^n\}$

　　使用投影梯度下降算法（PGD），基于当前模型参数，生成数据 B 对应的对抗样本 B_{adv}

　　使用 B_{adv} 训练网络模型，更新模型参数 θ

直到：训练收敛完成

输出：鲁棒分类网络 f_{rob}

在 MNIST 和 CIFAR-10 数据集上的结果表明，相比基于 FGSM 的对抗训练[6]，PGD-AT 无论针对单步攻击或迭代攻击均可有效提升模型的鲁棒性。在 ImageNet 数据集的实验结果则显示，针对单步攻击，对抗训练可将 Top-1 准确率从 30%显著提升到 70%左右。

虽然对抗训练的防御效果较好，但它仍然面临着一系列的问题，如会导致模型准确率下降[4,70]、计算开销过大等[71-73]。针对上述问题，研究者们提出了一系列的改进方案。例如，针对对抗训练导致模型良性准确率下降的问题，Zhang 等人[4]从理论层面分析了鲁棒性与准确率之间的关系，将鲁棒错误（robust error）拆解为分类错误（natural error）和边界错误（boundary error）两部分。研究者们基于分类校准损失理论（classification-calibrated loss），提供了鲁棒错误与分类错误差异的上界，并提出了一种新的防御方法 TRADES。该方法采用一种带有正则项的替代损失函数，在优化过程中可同时对分类错误和边界错误做出约束，形式如下所示：

$$\arg\min_{\theta} \mathrm{E}\left[L(\theta, x, y) + \max_{\delta\in S} L(\theta, x, x+\delta) / \lambda \right] \tag{10}$$

该目标函数通过最小化第一项的经验风险，最大限度地提高自然精度；同时，使用第二项的正则化促使决策边界远离样本来提高鲁棒性，从而最终达到平衡模型鲁棒性和准确率的目的。

针对对抗训练计算开销大、训练过程慢的问题，Shafahi 等人提出 Free-AT[71]算法提升了对抗训练的速度。具体而言，Free-AT 可以在一次梯度的反向传播中利用计算图同时获得当前损失对于输入图像和模型参数的梯度 $\partial L(x+\delta,\theta)/\partial x$ 及 $\partial L(x+\delta,\theta)/\partial\theta$。在内层循环中，不同于传统的对抗训练，Free-AT 的模型参数和扰动可以同时进行更新，从而既提升了对抗训练的效率，也能获得攻击性更强的对抗样本。在 CIFAR-10 和 CIFAR-100 数据集上的测试结果表明，该方法与标准训练的开销相当，相比其他对抗训练方法可提速 7～30 倍。Zhang 等人提出了 YOPO[72]算法，通过对抗扰动的优化过程将深度神经网络的前向传播和反向传播过程进行解耦，从而达到加速计算的效果。Wong 等人[73]则通过随机化初始扰动点改进了基于 FGSM 的对抗训练，既加速了 PGD-AT，又保证了鲁棒性，取得了较好的防御效果。此外，传统的对抗训练针对的是单一类型的扰动（如 L_∞ 类型），Tramer 等人[74]则进一步证实，即使在对抗训练中简单加入多种不同类型的对抗样本，也无法提升模型面对多种攻击时的鲁棒性。随后，Maini 等人[75]提出了三种多扰动对抗训练的改进策略 MAX，AVG 和 MSD，进一步提升了模型在多种扰动下的对抗鲁棒性。与此同时，Rice 等人[76]发现对抗训练中的过拟合会导致模型鲁棒性的明显下降。他们的研究指出，鲁棒性过拟合作为一种对抗训练中的普遍现象，是对抗训练区别于普通训练的重要标志。实验结果表明，没有任何一种单一的数据增强或正则化方法能够完全消除鲁棒性过拟合。相比于隐式和显式的正则化方法，提早停止训练可以有效提升模型鲁棒性。以基于 PGD 的对抗训练为例，提早停止训练（错误率为 43.2%）可以达到同 TRADES 方法相近的鲁棒性能（错误率为 43.4%）。而后，为了缓解对抗训练中的过拟合现象，一些研究者们也提出了针对性的正则化策略[77-78]。

除了上述工作，我们也对如何进一步提升对抗训练的效率和鲁棒性进行了一些探索。例如，提出了一个名为 FOSC（first-order stationary condition for constrained optimization）[79]的指标来定量评估在求解内部最大化问题过程中生成的对抗样本的收敛质量。该指标定义如下：

$$\text{FOSC}(x') = \max_{\delta\in S}\langle \delta, \nabla_x f(\theta, x')\rangle \tag{11}$$

其中，x' 为内部最大化多轮迭代后生成的对抗样本；$\langle\cdot\rangle$ 为内积操作；$\text{FOSC}(x')\geqslant 0$，且数值越低表明对抗样本 x' 的收敛质量越好。我们发现，为了获得更强的鲁棒性，在对抗训练的早期阶段倾向于使用收敛质量较低的对抗样本，而在后期阶段最好使用收敛质量较高的对抗样本。基于此，我们提出了一种动态训练策略以逐步提高生成对抗样本的收敛质量，从而增强对抗训练模型的鲁棒性。

此外我们还注意到，过去的对抗样本只被定义在最初就被正确分类的样本中，而对于最初被误分类的样本，其对抗样本是未被明确定义的。因此，我们研究了正确分类样本 x^+ 和误分类样本 x^- 在对抗训练下对最终模型鲁棒性的影响。研究发现，误分类样本对模型最终的鲁棒性依然重要，且具有与正确分类样本不同的影响。基于此，我们提出了一种新的防御算法 MART（misclassification aware adversarial training）[80]，对误分类和正确分类样本分别采用不同的优化策略。由于在推导过程中采用的 0-1 损失函数在实际中并不可解，我们构建出了一个更便于优化的替代损失函数，形式如下：

$$L^{\text{MART}}(\theta)=\text{BCE}\left(p\left(x',\theta\right),y\right)+\text{KL}(p\left(x,\theta\right)\parallel p\left(x',\theta\right)\cdot(1-p(x,\theta))) \tag{12}$$

该目标函数使用了增强交叉熵损失(BCE)来替代传统的交叉熵损失，并在最后一项中使用 $1-p(x,\theta)$ 来加强对于误分类样本的学习。

3.2 基于数据的防御

除了直接训练出更为鲁棒的模型，同样可以通过对输入数据进行各种处理来实现防御的目标。根据对输入数据不同的处理方式，这类防御方法又可分为对抗样本检测、数据重建、数据随机化、数据压缩等几个大类。

对抗样本检测旨在对准备输入模型的样本进行甄别，从而找出其中的对抗样本，最终实现防御的目标。Grosse 等人[81]通过设计辅助模型，在原有的 K 个标签下额外增加一个类别，使模型能将对抗样本分配到第 K+1 类。Metzen 等人[82]通过在原始分类器上附加检测子模块，将对抗样本的识别问题转化为一个端到端的二分类问题。该检测子模块将分类器隐藏节点的值作为检测器的输入，判别样本是否来自于真实数据。Lu 等人[83]提出的 SafetyNet 由原始分类器和对抗检测器两部分组成。检测器负责查看原始分类器中后期层的内部状态，从而进行对抗样本的甄别。即使对于训练集中未曾出现过的对抗样本，SafetyNet 仍具有较强的泛化性能。Feinman 等人[84]使用贝叶斯不确定性估计，以及对模型抽取的特征进行密度估计，从而实现模型对对抗样本的置信度检测。Xu 等人[85]则认为原始输入样本和压缩后的样本在模型输出端的差异越大，则输入样本是对抗样本的可能性也越大。基于该思路，他们设计了降低色位深度和空间平滑的两类压缩方法，并对压缩前后的预测差异同选定的阈值进行比较，从而实现对对抗样本的检测。然而，虽然研究者们提出了诸多的检测策略，Carlini 等人[86]在分析多个现有检测方法后发现，通过在攻击中构建新的损失函数，绝大多数现有检测方法仍然可以被攻击算法攻破。

除了直接检测对抗样本，研究者们发现通过对输入数据进行重建，也可能剔除对抗样本中的对抗扰动，从而起到防御的效果。ComDefend[87]作为一种端到端的模型，可以重建出部分对抗样本所对应的原始样本。其整体结构类似于自编码器，由 ComCNN 将输入的对抗样本从通道维度进行压缩并编码成特征向量，然后再使用 RecCNN 将特征向量还原为干净的样本。Mustafa 等人[88]将图像超分技术引入对抗防御，基于图像超分映射函数，将非流形的对抗样本重建至自然图像流形上。实验表明，该方法除了能抵御攻击，还有助于提高图像质量，维持模型对于干净图像的分类性能。Samangouei 等人[89]则提出了一种名为 Defense-GAN 的防御机制。该方法基于生成对抗网络，通过学习生成与已有样本分布近似的替代样本，作为网络的真实输入代替原有的输入进行预测。当遭遇对抗样本时，生成对抗网络会模拟未受干扰的图像分布，从而在生成替代样本时将扰动去除。然而，该方法虽然具有一定的防御能力，但已被证实可被基于流形投影的梯度下降法攻破[23]。

除了以上介绍的方法外，输入数据随机化和数据压缩等方式都有助于降低对抗样本的攻击能力。例如，Xie 等人[52]发现通过随机调整对抗样本的大小或在对抗性样本中加入一些随机纹理都可以降低对抗样本的攻击效果。此外，当模型的训练或防御方法被暴露给攻击者时，模型会有更大的可能性被攻破。为了解决这一问题，Wang 等人[90]通过使用与网络模型分离的基于 DNN 的数据转换模块，在数据被输入神经网络前交由该模块对数据进行处理。该转换模块既可以增加模型的复杂度，也能够有效阻断梯度的反向传播，降低攻击成功率。Dziugaite 等人[91]还发现，通过 FGSM 产生的对抗样本可以通过 JPEG 图像压缩显著降低其攻击有效性。此外，Das 等人[92]使用相似的 JPEG 压缩方法和显示压缩技术研究了针对 FGSM 和 DeepFool 攻击的防御方法。然而该方法仍存在一定的局限性：少量的压缩往往不足以消除对抗扰动的影响，而大量的压缩会显著降低原始图像的分类准确率。

4 模型鲁棒性评估

除了对抗样本的生成与防御，对抗机器学习关注的另一个重要问题是如何定量评估机器学习模型对对抗样本防御能力的强弱，也即本文要讨论的模型鲁棒性评估问题。给定分类器 $f(\cdot)$ 和测试样本 x，鲁棒性评估问题的核心是计算出能够使分类器预测结果异于测试样本真实标记的最小对抗扰动（minimal adversarial perturbation），也即下述优化问题的最优解：

$$\begin{gathered}\min \|\delta\| \\ \text{s.t. } f\left(x_0+\delta\right) \neq y\end{gathered} \tag{13}$$

显然地，在给定的测试数据分布上，模型的最小对抗扰动越大，模型的对抗鲁棒性就越强。然而，对于许多机器学习模型，精确求解其最小对抗扰动是一个 NP 难问题[93]，因此，许多研究者转而使用其上界或下界来评估模型的对抗鲁棒性。上文提到生成对抗样本所需的对抗噪音就是最小对抗扰动的一个上界，因为对抗样本已经成功地改变了模型的预测。基于此，对抗攻击常常被用来评估模型的对抗鲁棒性，而通过这种方式评估出的鲁棒性被称为经验鲁棒性（empirical robustness）。经验鲁棒性通常借助对抗攻击的成功率或所需施加的扰动大小来评估模型鲁棒性的强弱。然而，这种方法存在先天的局限：一个模型即使在实验中能够成功地抵御某一个攻击算法，也并不意味着它一定可以抵御其他的攻击。因此，在经验鲁棒性下表现优秀的模型并不能说明它是足够鲁棒和安全的。

基于上述原因，一些研究者转而开始研究验证鲁棒性（certified robustness）问题。验证鲁棒性计算的是最小对抗扰动的下界。任何小于该下界的扰动，都不能使分类器做出错误的标记。因此，这一下界仍旧可以用于衡量模型的对抗鲁棒性：最小对抗扰动的下界越大，则代表模型的对抗鲁棒性越强。图 3 给出了最小对抗扰动与经验鲁棒性和验证鲁棒性的关系。

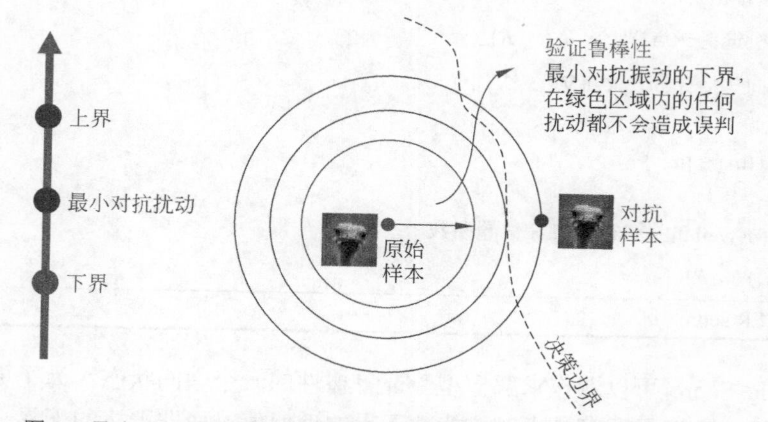

图 3　最小对抗扰动与经验鲁棒性和验证鲁棒性的关系（见彩图 13）

值得注意的是，虽然对抗鲁棒性因为深度神经网络而受到广泛的关注，但是对抗鲁棒性缺失的问题并不是深度神经网络模型所独有，传统机器学习模型同样存在着对抗鲁棒性问题。下面我们将从深度学习与其他机器学习模型两个方面分别介绍相关的模型鲁棒性评估工作。

4.1　面向深度神经网络的鲁棒性评估

早在 2014 年，Szegedy 等人[7]就已经利用全局 Lipschitz 常数对某些神经网络模型的鲁棒性进行理论分析。Hein 等人[94]则基于交叉 Lipschitz 定理计算出可让分类器改变输出所需的最小扰动范数，从而为单隐藏层神经网络的鲁棒性提供形式化的保证。为了能够评估具有更多隐层的大型神经网络，我们在 2018 年提出了 CLEVER（cross lipschitz extreme value for network robustness）Score[95]，可用于定量评估任意神经网络分类器的对抗鲁棒性强弱。此外，由于该指标与特定的攻击方式无关，因此更能充分反映模型本身的鲁棒性。它将鲁棒性评估问题转化为一个估算模型局部 Lipschitz 常数的问题，同时使用极值理论极大地减少了评估过程的计算量，从而可用于评估面向 ImageNet 等大型数据集的复杂分类网络。面向定向攻击的 CLEVER Score 计算流程见算法 2。

算法 2 面向定向攻击的 CLEVER Score 计算方法

输入：分类器 F，输入 x 及其类别 c，目标类别 t，批大小 N_b，批的样本个数 N_s，扰动范式 p，最大扰动 R

$S \leftarrow \{\varnothing\}, g(x) \leftarrow f_c(x) - f_t(x), q \leftarrow p / p-1$

For $i \leftarrow 1$ to N_b：

  For $k \leftarrow 1$ to N_s：

    随机选择一点 $x^{(i,k)} \in B_p(x_0, R)$

    反向传播计算 $b_{ik} \leftarrow \| \nabla g(x^{(i,k)}) \|_q$

  End

  $S \leftarrow S \cup \{\max_k \{b_{ik}\}\}$

End

$\hat{a}_w \leftarrow S$ 上逆向 Weibull 分布的 MLE 位置参数

$\mu \leftarrow \min(g(x_0) / \hat{a}, R)$

输出：CLEVER score μ

CLEVER Score 可以看作求解模型验证鲁棒性的一个中间状态：为了更加高效地评估模型鲁棒性，它选择计算最小对抗扰动下界的近似值，而非严格的下界。在此之后，多种针对神经网络的鲁棒性验证方法被相继提出。针对验证鲁棒性下界较低或计算开销较高的问题，Weng 等人[96]针对基于 ReLU 激活函数的神经网络，提出了两个高效的验证方法 Fast-Lin 和 Fast-Lip。它们分别通过计算线性逼近约束和局部 Lipschitz 常数的约束获取更优的最小扰动下界。为了使鲁棒性验证的框架更为通用，Zhang 等人提出了 CROWN 算法[97]，使用线性和二次函数对给定的激活函数进行约束，因此放宽了对激活函数的限制。除了常见的 ReLU 激活函数外，它还可用于评估基于 Tanh，ArcTan，Sigmoid 等激活函数的网络。此外，Boopathy 等人[98]提出一个通用且高效的框架 CNN-Cert，能够处理包括卷积层、最大池化层、批量归一化层、残差块及通用激活函数等多种架构。

通过利用卷积层的特殊结构，显著提升了计算效率，与 Fast-Lin 和 CROWN 等方法相比，分别实现了高达 17 倍和 11 倍的提速效果。

为了对前馈、卷积等不同类型的架构进行精准的评估，DeepPoly 系统[99]在解决鲁棒性验证的可扩展性和精度方面做了进一步探索。该系统不像传统的验证方法那样遍历可能的扰动图片来进行验证，而是通过构建扰动空间并利用专门设计的 Transformer 对神经网络特有的组成成分进行分析，从而完成对整个扰动空间的鲁棒性分析与验证。一系列的实验结果表明，当面对大型网络时，DeepPoly 相比过去的评估方法更为准确。

4.2 面向其他机器学习模型的鲁棒性评估

除了深度神经网络，很多其他机器学习模型同样存在着对抗鲁棒性问题。针对这些模型，研究者们进行了诸多的探索，已对包括贝叶斯模型、决策树、最近邻分类器、度量学习等多种模型开展了验证鲁棒性研究。

Dalvi 等人[9]将朴素贝叶斯模型的最小对抗扰动问题转化为一个整数规划问题，并利用动态规划算法在接近线性的时间复杂度内进行求解。Lowd 等人[10]进一步限制了攻击者能掌握的信息并将最小对抗性扰动的求解推广到一般的线性模型。对于决策树模型，现有的验证方法通常利用混合整数线性编程（MILP）来求解最小对抗扰动，但该方法需要指数级时间，并不适用于大型树模型[100]。基于此，Chen 等人[101]将验证鲁棒性问题转化为在 K-边图上计算最大团（max-cliques）的问题，并提出了一种高效的多层次验证算法，可以针对大规模树群给出鲁棒性的严格下界。针对 10 个数据集训练的 RF/GBDT 模型结果表明，该方法相比之前需要求解 MILP 的方法快了数百倍，且能够在大型 GBDT 模型上给出严格的验证鲁棒性保证。Ranzato 等人[102]利用决策数的鲁棒性设计了一种验证算法 Meta-Silvae。该算法对决策树分类器执行基于抽象理解的静态分析，能够精确计算决策树中可被对抗扰动攻击到的叶子集合。此外，对于增强型决策树，Andriushchenko 等人[103]提出了一种验证鲁棒性的高效方法，可在 $O(T\log T)$ 时间内计算出确切的鲁棒损失和测试误差。

针对最近邻分类器（1-NN）这一离散阶梯函数，Dubey 等人[104]通过攻击其可微的近似模型间接攻击最近邻分类器，从而评估出模型的鲁棒性。Sitawarin 等人[105]则尝试使用简单的启发式策略，在测试样本和异类中心的连线上搜索对抗样本。然而，这类方法无法实现最优的对抗攻击，也无法给出具有理论保证的验证鲁棒性结果。为了解决包含最近邻分类器在内的 K-近邻分类器的验证鲁棒性问题，我们将该类问题形式化为一组凸二次规划问题[106-107]，通过求解该凸二次规划问题间接实现对模型最优鲁棒性的评估。具体而言，针对最近邻问题，我们在原始对偶框架下给出一个高效求解器，使其能在与进行经验鲁棒性验证（对抗攻击）相近的时间内完成验证鲁棒性的评估工作。针对 K 近邻问题，为了解决凸二次规划问题复杂度随 K 的增大呈指数级增长的问题，我们对原始对偶框架进行了约束放松，从而可以在多项式时间内求解最小对抗扰动的下界。然而，该下界会随着 K 值的增大而变得不严格。因此，我们引入随机平滑法对 K 近邻分类器进行鲁棒性验证：先将基分类器转化为平滑分类器，再验证平滑分类器的对抗鲁棒性。这两种方法互为补充，使得当 K 值较大和较小时都能取得较为理想的验证鲁棒性结果。

在研究 K -近邻分类器验证鲁棒性问题的基础上，我们进一步对基于度量学习的马氏 K -近邻分类器进行了研究[108]。马氏距离下的 K 近邻分类器的最小对抗扰动可以形式化为如下的凸二次规划问题：

$$\begin{gathered}\min \|\delta\| \\ \text{s.t.} d_{\mathrm{M}}\left(x+\delta, x^{-}\right) \leqslant d_{\mathrm{M}}\left(x+\delta, x^{+}\right)\end{gathered} \tag{14}$$

该问题的闭式解为 $\dfrac{[d_{\mathrm{M}}(x,x^-)-d_{\mathrm{M}}(x,x^+)]_+}{2\sqrt{(x^+-x^-)^{\mathrm{T}}\boldsymbol{M}^{\mathrm{T}}\boldsymbol{M}(x^+-x^-)}}$，即马氏 K -近邻分类器的验证鲁棒性。进一步地，通过最大化每一个训练样本点的最小对抗扰动，可以学习出一个由半正定矩阵 $\boldsymbol{M}$ 描述的马氏距离，从而显著提升 K -近邻分类器的鲁棒性。此外，不同于传统的对抗训练方法，我们的方法可以保证在提升鲁棒性的同时不会影响模型本身的准确性。

5 总结与展望

本文从对抗攻击、对抗防御及模型鲁棒性评估三个方面对对抗机器学习领域的部分工作进行了梳理和介绍。值得注意的是，对抗机器学习是一个非常活跃的研究领域，近年来相关论文的总量更是以接近指数的速度高速增长①。因为篇幅的限制，许多有价值的研究工作并未纳入本文的讨论范畴，但它们仍然对该领域的发展起到了重要的推动作用。可以预见，随着对抗防御与鲁棒性评估体系的不断完善，未来的人工智能系统将会变得越来越安全和可靠，也会在更多与安全相关的应用场景下发挥出巨大的实用价值。

从产业界实际应用需求的角度出发，我们对对抗机器学习的研究提出一些展望：

（1）即插即用 vs. 端到端模型。目前大多数的对抗防御研究都集中在训练更加鲁棒的端到端模型上。然而在工业界中，很多已经部署的系统中存在大量相互依存的模块，替换其中的 AI 模型会带来巨大的成本。因此，相比于直接替换现有系统中的模型，开发出可即插即用的鲁棒性模块（如前置的对抗样本检测模块）往往会更加符合大规模系统

① https://nicholas.carlini.com/writing/2019/all-adversarial-example-papers.html。

部署的需求。

（2）模型鲁棒性的科学评估。到目前为止，模型的对抗鲁棒性仍然缺乏科学统一的评估标准。虽然最小对抗扰动是目前主流的鲁棒性评估指标，但这一指标仍与样本点的具体位置相关，并没有充分体现出模型自身的鲁棒性，此外，最小对抗扰动在很多情况下难以求解，因此只能转而计算其上界（对抗攻击）或下界（验证鲁棒性）。对前者而言，一个模型能够成功抵御某些特定类型的对抗攻击并不意味着真正鲁棒，因为它仍然可能被其他攻击算法攻破。而对后者而言，验证鲁棒性方法推导得出的边界往往不够精准，与最优的理论边界存在不小的差距，用来评估模型的鲁棒性仍然可能会有失偏颇。因此，建立科学、普适的模型鲁棒性评估体系仍然是现有对抗机器学习领域亟待解决的核心问题之一。

（3）鲁棒模型的“鲁棒性”问题。这里的两个鲁棒代表不同的含义。前者是指具有对抗防御能力的 AI 模型，后者则是指这些模型在不同环境下的适应能力。目前很多防御算法都基于一些特定的假设，如数据分布、攻击者的攻击方式等。然而，现实世界中的攻防是一个动态博弈的过程，真实的攻击很可能会与事先的预想不符。甚至现实世界中如摄像头的分辨率、拍摄角度、光线强度等因素都可能会对算法的效果产生重大的影响。因此，不断提升鲁棒模型对不同环境的鲁棒性，使它们在各种环境和条件下都能正常工作，也是防御算法能在实际应用中落地的关键。

（4）以更全面的视角综合评估 AI 模型。如本文开篇所述，仅仅通过准确率这个单一指标来衡量 AI 模型的优劣很可能会有失偏颇。然而，虽然本文着重探讨的是模型鲁棒性问题，但只关注鲁棒性这个指标亦同样不可取。一个真正实用的 AI 模型必定是全面的，需要同时满足准确、鲁棒、高效等一系列特性。甚至在某些应用中，还要求模型具有公平、可解释、可扩展等优点。因此，这样的要求也需要我们以更加全面的视角去综合看待不同模型的长处与不足，从而设计和开发出更加全面且实用的 AI 模型。

参考文献

[1] RUSSAKOVSKY O, DENG J, SU H, et al. Imagenet large scale visual recognition challenge[J]. International Journal of Computer Vision, 2015, 115(3): 211-252.

[2] RAJPURKAR P, ZHANG J, LOPYREV K, et al. SQuAD: 100 000+ Questions for machine comprehension of text[C]//Conference on Empirical Methods in Natural Language Processing (EMNLP). 2016: 2383-2392.

[3] SU D, ZHANG H, CHEN H. Is robustness the cost of accuracy? A comprehensive study on the robustness of 18 deep image classification models[C]//Proceedings of the European Conference on Computer Vision (ECCV). 2018: 631-648.

[4] ZHANG H, YU Y, JIAO J, et al. Theoretically principled trade-off between robustness and accuracy[C]// International Conference on Machine Learning (ICML). 2019: 7472-7482.

[5] TSIPRAS D, SANTURKAR S, ENGSTROM L, et al. Robustness may be at odds with accuracy[C]// International Conference on Learning Representations (ICLR), 2019.

[6] GOODFELLOW I J, SHLENS J, SZEGEDY C. Explaining and harnessing adversarial examples[C]// International Conference on Learning Representations (ICLR), 2015.

[7] SZEGEDY C, ZAREMBA W, SUTSKEVER I, et al. Intriguing properties of neural networks[C]// International Conference on Learning Representations (ICLR), 2014.

[8] BIGGIO B, CORONA I, MAIORCA D, et al. Evasion attacks against machine learning at test time[C]// Joint European Conference on Machine Learning and Knowledge Discovery in Databases. 2013: 387- 402.

[9] DALVI N, DOMINGOS P, SANGHAI S, et al. Adversarial classification[C]//InSIGKDD International Conference On Knowledge Discovery and Data Mining (KDD). 2004: 99-108.

[10] LOWD D, MEEK C. Adversarial learning[C]//SIGKDD International Conference On Knowledge Discovery in Data Mining (KDD). 2005: 641-647.

[11] MADRY A, MAKELOV A, SCHMIDT L, et al. Towards deep learning models resistant to adversarial attacks[C]//International Conference on Learning Representations (ICLR), 2017.

[12] DONG Y, LIAO F, PANG T, et al. Boosting adversarial attacks with momentum[C]// Proceedings of the IEEE Conference on Computer Vision and Pattern Recognition. 2018: 9185-9193.

[13] MOOSAVI-DEZFOOLI S M, FAWZI A, FROSSARD P. DeepFool: A simple and accurate method to fool deep neural networks[C]//Proceedings of the IEEE Conference on Computer Vision and Pattern Recognition. 2016: 2574-2582.

[14] CARLINI N, WAGNER D. Towards evaluating the robustness of neural networks[C]//2017 IEEE Symposium on Security and Privacy. 2017: 39-57.

[15] CHEN P Y, SHARMA Y, ZHANG H, et al. EAD: Elastic-net attacks to deep neural networks via adversarial examples[C]//Proceedings of the AAAI Conference on Artificial Intelligence (AAAI). 2018, 32(1).

[16] CHEN J, ZHOU D, YI J, et al. A Frank-Wolfe framework for efficient and effective adversarial attacks[C]//Proceedings of the AAAI Conference on Artificial Intelligence. 2020, 34(04): 3486-3494.

[17] KURAKIN A, GOODFELLOW I, BENGIO S. Adversarial examples in the physical world[C]//ICLR Workshop, 2017.

[18] PAPERNOT N, MCDANIEL P, JHA S, et al. The limitations of deep learning in adversarial settings[C]// 2016 IEEE European Symposium on Security and Privacy (EuroS&P). 2016: 372-387.

[19] XIE C, ZHANG Z, ZHOU Y, et al. Improving transferability of adversarial examples with input diversity[C]// Proceedings of the IEEE/CVF Conference on Computer Vision and Pattern Recognition. 2019: 2730-2739.

[20] MOOSAVI-DEZFOOLI S M, FAWZI A, FAWZI O, et al. Universal adversarial perturbations[C]//

Proceedings of the IEEE Conference on Computer Vision and Pattern Recognition. 2017: 1765-1773.

[21] SU J, VARGAS D V, SAKURAI K. One pixel attack for fooling deep neural networks[J]. IEEE Transactions on Evolutionary Computation, 2019, 23(5): 828-841.

[22] CROCE F, HEIN M. Minimally distorted adversarial examples with a fast adaptive boundary attack[C]// International Conference on Machine Learning (ICML). 2020: 2196-2205.

[23] ATHALYE A, CARLINI N, WAGNER D. Obfuscated gradients give a false sense of security: Circumventing defenses to adversarial examples[C]//International Conference on Machine Learning (ICML). 2018: 274-283.

[24] CROCE F, HEIN M. Reliable evaluation of adversarial robustness with an ensemble of diverse parameter-free Attacks[C]//International Conference on Machine Learning (ICML). 2020: 2206-2216.

[25] ANDRIUSHCHENKO M, CROCE F, Flammarion N, et al. Square attack: A query-efficient black-box adversarial attack via random search[C]//European Conference on Computer Vision (ECCV). 2020: 484-501.

[26] PAPERNOT N, MCDANIEL P, GOODFELLOW I, et al. Practical black-box attacks against machine learning[C]//Proceedings of the Asia Conference on Computer and Communications Security. 2017: 506-519.

[27] CHEN P Y, ZHANG H, SHARMA Y, et al. ZOO: Zeroth order optimization based black-box attacks to deep neural networks without training substitute models[C]//Proceedings of the 10th ACM Workshop on Artificial Intelligence and Security. 2017: 15-26.

[28] LAX P D, TERRELL M S. Calculus with applications[M]. New York, NY: Springer, 2014.

[29] TU C C, TING P, CHEN P Y, et al. Autozoom: Autoencoder-based zeroth order optimization method for attacking black-box neural networks[C]//Proceedings of the AAAI Conference on Artificial Intelligence(AAAI). 2019, 33(01): 742-749.

[30] ILYAS A, ENGSTROM L, ATHALYE A, et al. Black-box adversarial attacks with limited queries and information[C]//International Conference on Machine Learning (ICML). 2018: 2137-2146.

[31] WIERSTRA D, SCHAUL T, GLASMACHERS T, et al. Natural evolution strategies[J]. The Journal of Machine Learning Research, 2014, 15(1):949-980.

[32] BHAGOJI A N, HE W, LI B, et al. Practical black-box attacks on deep neural networks using efficient query mechanisms[C]//Proceedings of the European Conference on Computer Vision (ECCV). 2018: 154-169.

[33] LIU S, CHEN P Y, CHEN X, et al. SignSGD via zeroth-order oracle[C]//International Conference on Learning Representations, 2018.

[34] ILYAS A, ENGSTROM L, MADRY A. Prior convictions: black-box adversarial attacks with bandits and priors[C]//International Conference on Learning Representations, 2018.

[35] MEUNIER L, ATIF J, TEYTAUD O. Yet another but more efficient black-box adversarial attack: Tiling and evolution strategies[Z]. arXiv preprint arXiv:1910.02244, 2019.

[36] CHENG S, DONG Y, PANG T, et al. Improving black-box adversarial attacks with a transfer-based prior[C]//Advances in Neural Information Processing Systems (NeurIPS), 2019.

[37] AL-DUJAILI A, O'REILLY U M. There are no bit parts for sign bits in black-box attacks[Z]. arXiv preprint arXiv:1902.06894, 2019.

[38] SHI Y, WANG S, HAN Y. Curls & whey: Boosting black-box adversarial attacks[C]//Proceedings of the IEEE/CVF Conference on Computer Vision and Pattern Recognition (CVPR). 2019: 6519-6527.

[39] DU J, ZHANG H, ZHOU T, et al. Query-efficient meta attack to deep neural networks [C]// International Conference on Learning Representations (ICLR), 2020.

[40] BRENDEL W, RAUBER J, BETHGE M. Decision-based adversarial attacks: Reliable attacks against black-box machine learning models[C]//International Conference on Learning Representations (ICLR), 2018.

[41] CHENG M, LE T, CHEN P Y, et al. Query-efficient hard-label black-box attack: An optimization-based approach[C]//International Conference on Learning Representations (ICLR), 2019.

[42] FRIEDMAN J. H. Greedy function approximation: A gradient boosting machine[J]. Annals of Statistics, 2001, 1189-1232.

[43] BRUNNER T, DIEHL F, LE M, et al. Guessing smart: Biased sampling for efficient black-box adversarial attacks[C]//Proceedings of the IEEE/CVF International Conference on Computer Vision. 2019: 4958-4966.

[44] DONG Y, SU H, WU B, et al. Efficient decision-based black-box adversarial attacks on face recognition[C]//Proceedings of the IEEE/CVF Conference on Computer Vision and Pattern Recognition. 2019: 7714-7722.

[45] LI H, XU X, ZHANG X, et al. QEBA: Query-efficient boundary-based blackbox attack[C]// Proceedings of the IEEE/CVF Conference on Computer Vision and Pattern Recognition. 2020: 1221-1230.

[46] CHEN W, ZHANG Z, HU X, et al. Boosting decision-based black-box adversarial attacks with random sign flip[C]//European Conference on Computer Vision (ECCV). 2020: 276-293.

[47] CHEN H, ZHANG H, CHEN P Y, et al. Attacking visual language grounding with adversarial examples: A case study on neural image captioning[C]//Conference of the Association for Computational Linguistics (ACL). 2018: 2587-2597.

[48] SUTSKEVER I, VINYALS O, LE Q V. Sequence to sequence learning with neural networks [C]// Advances in Neural Information Processing Systems (NIPS), 2014.

[49] CHENG M, YI J, CHEN P Y, et al. Seq2sick: Evaluating the robustness of sequence-to-sequence models with adversarial examples[C]//Proceedings of the AAAI Conference on Artificial Intelligence (AAAI). 2020, 34(04): 3601-3608.

[50] GONG Z, WANG W, LI B, et al. Adversarial texts with gradient methods[Z]. arXiv preprint arXiv:1801.07175, 2018.

[51] METZEN J, KUMAR M, BROX T, et al. Universal adversarial perturbations against semantic image segmentation[C]//Proceedings of the IEEE International Conference on Computer Vision (ICCV). 2017: 2755-2764.

[52] XIE C, WANG J, ZHANG Z, et al. Adversarial examples for semantic segmentation and object detection[C]//Proceedings of the IEEE International Conference on Computer Vision (ICCV). 2017: 1369-1378.

[53] CARLINI N, WAGNER D. Audio adversarial examples: Targeted attacks on speech-to-text[C]//IEEE Security and Privacy Workshops. 2018: 1-7.

[54] HUANG S H, PAPERNOT N, GOODFELLOW I J, et al. Adversarial attacks on neural network policies[C]// International Conference on Learning Representations (ICLR), 2017.

[55] MNIH V, PUIGDOMENECH A, MIRZA M, et al. Asynchronous methods for deep reinforcement learning[C]//International Conference on Machine Learning(ICML), 2016

[56] SCHULMAN J, LEVINE S, MORITZ P, etc. Trust region policy optimization[C]//International Conference on Machine Learning(ICML), 2015.

[57] MNIH V, KAVUKCUOGLU K, SILVER D, etc. Playing atari with deep reinforcement learning[C]// Advances in Neural Information Processing Systems (NIPS), 2013.

[58] EYKHOLT K, EVTIMOV I, FERNANDES E, et al. Robust physical-world attacks on deep learning visual classification[C]//Proceedings of the IEEE Conference on Computer Vision and Pattern Recognition. 2018: 1625-1634.

[59] BROWN T B, MANÉ D, AURKO R, et al. Adversarial patch[Z]. arXiv preprint arXiv:1712.09665.

[60] ATHALYE A, ENGSTROM L, ILYAS A, et al. Synthesizing robust adversarial examples[C]// International Conference on Machine Learning (ICML). 2018: 284-293.

[61] WU Z, LIM S N, DAVIS L S, et al. Making an invisibility cloak: Real world adversarial attacks on object detectors[C]//European Conference on Computer Vision (ECCV). 2020: 1-17.

[62] PAPERNOT N, MCDANIEL P, WU X, et al. Distillation as a defense to adversarial perturbations against deep neural networks[C]//IEEE Symposium on Security and Privacy (SP). 2016: 582-597.

[63] BUCKMAN J, ROY A, RAFFEL C, et al. Thermometer encoding: One hot way to resist adversarial examples[C]//International Conference on Learning Representations (ICLR), 2018.

[64] BRADSHAW J, MATTHEWS A G G, GHAHRAMANI Z. Adversarial examples, uncertainty, and transfer testing robustness in gaussian process hybrid deep networks[Z]. arXiv preprint arXiv:1707.02476, 2017.

[65] BUCKMAN J, ROY A, RAFFEL C, et al. Thermometer encoding: One hot way to resist adversarial examples[C]//International Conference on Learning Representations (ICLR), 2018.

[66] GUO C, RANA M, CISSE M, et al. Countering adversarial images using input transformations[C]// International Conference on Learning Representations (ICLR), 2017.

[67] PANG T, XU K, DONG Y, et al. Rethinking softmax cross-entropy loss for adversarial robustness[C]// International Conference on Learning Representations (ICLR). 2020.

[68] MUSTAFA A, KHAN S, HAYAT M, et al. Adversarial defense by restricting the hidden space of deep neural networks[C]//Proceedings of the IEEE/CVF International Conference on Computer Vision. 2019: 3385-3394.

[69] TRAMER F, CARLINI N, BRENDEL W, et al. On adaptive attacks to adversarial example defenses[C]//Advances in Neural Information Processing Systems (NeurIPS), 2020.

[70] CARMON Y, RAGHUNATHAN A, SCHMIDT L, et al. Unlabeled data improves adversarial robustness[C]//Advances in Neural Information Processing Systems (NeurIPS), 2019.

[71] SHAFAHI A, NAJIBI M, GHIASI A, et al. Adversarial training for free[C]//Advances in Neural Information Processing Systems (NeurIPS), 2019.

[72] ZHANG D, ZHANG T, LU Y, et al. You only propagate once: Accelerating adversarial training via maximal principle[C]//Advances in Neural Information Processing Systems (NeurIPS), 2019.

[73] WONG E, RICE L, KOLTER J Z. Fast is better than free: Revisiting adversarial training[C]//International Conference on Learning Representations (ICLR), 2020.

[74] TRAMER F, BONEH D. Adversarial training and robustness for multiple perturbations[C]//Advances in Neural Information Processing Systems (NeurIPS), 2019.

[75] MAINI P, WONG E, KOLTER Z. Adversarial robustness against the union of multiple Perturbation Models[C]//International Conference on Machine Learning (ICML). 2020: 6640-6650.

[76] RICE L, WONG E, KOLTER Z. Overfitting in adversarially robust deep learning[C]//International Conference on Machine Learning (ICML). 2020: 8093-8104.

[77] CHEN T, ZHANG Z, LIU S, et al. Robust overfitting may be mitigated by properly learned Smoothening[C]//International Conference on Learning Representations (ICLR), 2021.

[78] WU D, XIA S, WANG Y. Adversarial weight perturbation helps robust generalization[C]// Advances in Neural Information Processing Systems (NeurIPS), 2020.

[79] WANG Y, MA X, BAILEY J, et al. On the convergence and robustness of adversarial training[C]//International Conference on Machine Learning (ICML), 2019.

[80] WANG Y, ZOU D, YI J, et al. Improving adversarial robustness requires revisiting misclassified examples[C]//International Conference on Learning Representations (ICLR), 2020.

[81] GROSSE K, MANOHARAN P, PAPERNOT N, et al. On the (statistical) detection of adversarial examples[Z]. arXiv preprint arXiv:1702.06280, 2017.

[82] METZEN J H, GENEWEIN T, FISCHER V, et al. On detecting adversarial perturbations[C]//International Conference on Learning Representations(ICLR), 2017.

[83] LU J, ISSARANON T, FORSYTH D. Safetynet: Detecting and rejecting adversarial examples robustly[C]//Proceedings of the IEEE International Conference on Computer Vision. 2017: 446-454.

[84] FEINMAN R, CURTIN R R, SHINTRE S, et al. Detecting adversarial samples from artifacts[Z]. arXiv preprint arXiv:1703.00410, 2017.

[85] XU W, EVANS D, QI Y. Feature squeezing: Detecting adversarial examples in deep neural networks[C]//Network and Distributed System Security Symposium (NDSS), 2018.

[86] CARLINI N, WAGNER D. Adversarial examples are not easily detected: Bypassing ten detection methods[C]//Proceedings of the ACM Workshop on Artificial Intelligence and Security. 2017: 3-14.

[87] JIA X, WEI X, CAO X, et al. Comdefend: An efficient image compression model to defend adversarial examples[C]//Proceedings of the IEEE/CVF Conference on Computer Vision and Pattern Recognition. 2019: 6084-6092.

[88] MUSTAFA A, KHAN S H, HAYAT M, et al. Image super-resolution as a defense against adversarial attacks[J]. IEEE Transactions on Image Processing, 2019, 29: 1711-1724.

[89] SAMANGOUEI P, KABKAB M, CHELLAPPA R. Defense-GAN: Protecting classifiers against adversarial attacks using generative models[C]//International Conference on Learning Representations(ICLR), 2018.

[90] WANG Q, GUO W, ZHANG K, et al. Learning adversary-resistant deep neural networks[Z]. arXiv preprint arXiv:1612.01401, 2016.

[91] DZIUGAITE G K, GHAHRAMANI Z, ROY D M. A study of the effect of JPG compression on adversarial images[Z]. arXiv preprint arXiv:1608.00853, 2016.

[92] DAS N, SHANBHOGUE M, CHEN S T, et al. Keeping the bad guys out: Protecting and vaccinating deep learning with JPEG compression[Z]. arXiv preprint arXiv:1705.02900, 2017.

[93] KATZ G, BARRETT C, DILL D L, et al. Reluplex: An efficient SMT solver for verifying deep neural networks[C]//International Conference on Computer Aided Verification. 2017: 97-117.

[94] HEIN M, ANDRIUSHCHENKO M. Formal guarantees on the robustness of a classifier against adversarial manipulation[C]//Advances in Neural Information Processing Systems (NIPS), 2017.

[95] WENG T W, ZHANG H, CHEN P Y, et al. Evaluating the robustness of neural networks: An extreme value theory approach[C]//International Conference on Learning Representations (ICLR), 2018.

[96] WENG L, ZHANG H, CHEN H, et al. Towards fast computation of certified robustness for ReLU networks[C]//International Conference on Machine Learning (ICML). 2018: 5276-5285.

[97] ZHANG H, WENG T W, CHEN P Y, et al. Efficient neural network robustness certification with general activation functions[C]//Advances in Neural Information Processing Systems (NeurIPS), 2018.

[98] BOOPATHY A, WENG T W, CHEN P Y, et al. CNN-Cert: An efficient framework for certifying robustness of convolutional neural networks[C]//Proceedings of the AAAI Conference on Artificial Intelligence (AAAI). 2019, 33(1): 3240-3247.

[99] SINGH G, GEHR T, PÜSCHEL M, et al. An abstract domain for certifying neural networks[J]. Proceedings of the ACM on Programming Languages. 2019, 3(POPL): 1-30.

[100] CHEN H, ZHANG H, BONING D, et al. Robust decision trees against adversarial examples[C]//International Conference on Machine Learning (ICML). 2019: 1122-1131.

[101] CHEN H, ZHANG H, SI S, et al. Robustness verification of tree-based models[C]//Advances in Neural Information Processing Systems (NeurIPS). 2019: 12317-12328.

[102] RANZATO F, ZANELLA M. Genetic adversarial training of decision trees[Z]. arXiv preprint arXiv: 2012.11352.

[103] ANDRIUSHCHENKO M, HEIN M. Provably robust boosted decision stumps and trees against adversarial attacks[C]//Advances in Neural Information Processing Systems (NeurIPS), 2019.

[104] DUBEY A, MAATEN L, YALNIZ Z, et al. Defense against adversarial images using web-scale nearest-neighbor search[C]//Proceedings of the IEEE/CVF Conference on Computer Vision and Pattern Recognition. 2019: 8767-8776.

[105] SITAWARIN C, WAGNER D. On the robustness of deep K-nearest neighbors[C]//IEEE Security and Privacy Workshops. 2019: 1-7.

[106] WANG L, LIU X, YI J, et al. Evaluating the robustness of nearest neighbor classifiers: A primal-dual perspective[Z]. arXiv preprint arXiv:1906.03972, 2019.

[107] WANG L, LIU X, YI J, et al. Robustness verification of nearest neighbor classifiers[C]//Workshop on Formal Methods for ML-Enabled Autonomous Systems, 2020.

[108] WANG L, LIU X, JIANG Y, et.al. Provably robust metric learning[C]//Advances in Neural Information Processing Systems (NeurIPS), 2020.

基于城市视觉大数据的交通预测与调度

余正旭　魏　龙　金仲明　黄建强　华先胜

（阿里巴巴达摩院城市大脑实验室，杭州）

1　引言

城市交通治理作为城市治理的重要一环，其中的重点业务问题一直受到学界和业界的广泛关注。《城市公共交通“十三五”发展纲要》提到，从2014年起，我国的机动车保有量增长率保持在5%以上，2020年机动车保有总量已达到了3.6亿辆，机动车保有量超过100万的城市有69个，超过300万的城市有12个。另一方面，交通部《“十三五”现代综合交通运输体系发展规划》提到，到2020年末，我国公路总里程达到500万公里。该目标对应的新增里程为公路总里程新增42.27万公里，年均增长8.45万公里，年均增长率小于2%。从这两组数据可以看出，随着经济的快速发展，城市车辆保有率增速与道路扩建率之间的供需不平衡，这也一方面导致了重点交通问题如交通拥堵、事件事故频发等问题的突出显现。因此，研究并提出这些重点交通问题的解决方案具有重要的经济和社会价值。

近年来，以深度学习为主的人工智能算法在城市场景中的应用广度和深度不断发展，逐渐发展出了一系列系统化、产业化的解决方案，形成了从数据获取、感知、推理预测到研判决策的完整业务链路。其中一个典型的、系统化的解决方案就是2016年阿里巴巴提出的“城市大脑”，致力于打造城市级的人工智能系统。从非常宏观的角度来看，“城市大脑”基于海量的城市多模态数据，结合云计算提供的强大算力和先进的人工智能算法来服务城市，帮助城市管理者解决“城市病”，包括保障公众安全、提升公众出行效率、促进经济繁荣，从而提升人们的幸福感等。以“城市大脑”在城市交通领域下的应用场景为例，“城市大脑”将人工智能算法应用于城市交通场景下的一系列应用场景。图1

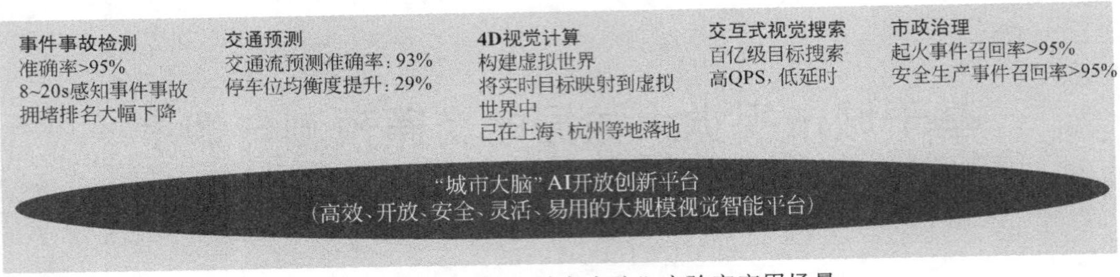

图 1　阿里云达摩院“城市大脑”实验室应用场景

所示是“城市大脑”在城市交通领域的主要方向，既包括城市级人工智能平台计算底座的构建，也包括在事故事件检测、交通预测、四维可视化、交互式视觉搜索、市政管理等重点问题上的探索。在各个方向上，“城市大脑”通过以深度学习为主的人工智能算法解决了传统方法在复杂多变的交通场景中低时效性、低精度的问题。

深度学习作为一种数据驱动的方法，对训练数据的数量、质量有很强的依赖性。其中最为重要的数据就是视频数据，此外也包括 GPS 等其他数据。视频数据的特点之一就是，在用于城市场景下的交通推理、预测、干预等环节前，需要先进行分割、物体识别、异常事件检测等感知、理解层面上的分析。另外，为了能够快速检索出目标视觉对象，感知、识别出的视觉对象还需要建立索引，并通过挖掘视觉对象间的关系和模式建立如阿里巴巴“拍立淘”式的快速检索功能。通过快速检索功能，就能知道目标对象的历史和实时状态信息，如位置、行为等。然后，可以根据所有过去的数据和当前的情况进行预测，例如，十字路口的交通量、购物中心某个区域的人数、根据城市的天气和事件估计出的交通事故数量等。有了这种预测，就能更好地完成干预决策层面的功能，例如，可以通过提前做出反应动作来阻止城市中一些坏事的发生，如控制交通流量以避免天气不好时发生事故等。近年来，视频设备、计算资源等硬件成本的逐渐降低有效地扩充了城市场景下视觉数据源的数据丰富度和覆盖率，减轻了缺乏训练数据的问题，并使基于城市视觉大数据的大规模深度学习应用成为可能。并且，较大规模的交通场景数据集的出现使学界和业界在交通预测、控制问题上取得了较为长足的进展[1-3]。

另外，在数据质量方面，可以形象地说，数据驱动的深度学习方法就像是“盲人摸象”。要解决的问题是如何让“盲人”（网络模型）通过接触“象”的局部（训练数据）而学习“象”的本质规律（真实数据分布）。为了让“盲人”了解“象”的更多特征，需要更多“有用”的数据。因此，在深度学习模型被应用到现实生活中的同时，研究学者们很快意识到单纯增加数据量面临的数据“以偏概全”的瓶颈问题。这不仅体现在数据源种类常常不是完备的，更体现在数据获取场景的地域、时间局限性导致的数据生成（获取）偏差（data generation bias）上。例如，在城市交通场景中，能够获得一定区域、一段时间内的车流数据。但由于信号灯控制策略改变、事故事件发生等原因，在一定时空范围内获取的数据常常具有与其他时空范围不同的数据分布，即数据是非独立同分布的。而目前大部分深度学习方法和“盲人摸象”学习法所遵循的一个基本假设就是独立同分布（independent and identically distributed, I.I.D）假设[4]，即获得更多的数据就能了解“象”的全貌。在符合该基本假设的理想条件下，模型可以通过增加训练数据量，并以在训练数据集上求得最低损失的方式获得具有良好泛化性能的模型。然而，在真实场景下，由于数据生成过程中的特征、对象等多层次的耦合关系（如上述车辆与时间、空间、事件等因素间的耦合关系），数据往往是非独立同分布（non-I.I.D）的。由于异常情况发生规律刻画难度大、可用样本少、难以区分等特点，在与正常数据样本混杂后常呈现出数据分布长尾、高噪声等性质，使城市交通场景下的预测与控制任务具有很大的挑战性。这使得即使坐拥海量数据，模型在测试场景下时常会取得与训练数据集上相差甚远的效果。针对该问题，学界与业界基于视觉大数据对异常情况的异常检测方法进行了一系列的研究，如 Zhao 等人[5]提出的时空异常检测算法。然而，在许多场景中，异常事件的影响往往是区域性的，与正常规律混杂的。因此，研究混杂着异常事件的场景中如何进行稳定的预测是目前亟待解决的重要问题。

针对上述问题，如图 2 所示，阿里巴巴“城市大脑”提出了整体的交通优化框架，包括多模态数据的数据源层、感知推理层、挖掘城市交通运行规律的预测层及减轻车和路之间时空不平衡性的干预层。接下来，本文分别从交通预测和控制中的具体任务出发，

干预层	接驳车调度	出租车调度	交通信号控制	……
预测层	交通流量预测	交通速度预测	交通需求预测	……
感知推理层	目标识别	事件检测	轨迹还原	……
数据源层	GPS	环境	轨迹	……

图 2　阿里巴巴“城市大脑”交通优化框架

介绍异常事件、事故场景下感知推理层、预测层和干预层三个不同层面上的算法进展和成果：

（1）在感知推理层面上，介绍了一种时空异常检测算法（STAE），该算法基于视觉数据对数据中的异常事件进行检测。由于异常类型的样本较少，在数据分布中处于长尾分布的尾部。因此，还介绍了一种基于因果推理理论的长尾分类方法来有效地减轻长尾分布对分类的影响。该方法针对长尾分布下 SGD-momentum 优化器中动量（momentum）引入数据分布导致模型偏向于样本量多的头部类的问题，提出了一种因果干预方法，通过去除动量对模型训练的影响进行稳定的分类。

（2）在预测层面上，针对批采样（mini-batch）训练策略在非独立同分布场景下存在的批数据间分布偏移的问题，介绍了一种基于卷积长短时记忆网络的渐进学习方法(PTL)，通过渐近地将各批数据的分布信息收集到记忆单元中，并利用记忆单元中汇聚的分布信息对每个批数据的特征嵌入进行校正，以提高模型的稳定性。

（3）在干预层面上，针对城市交通场景中的重要的非 I.I.D 场景——交通信号灯控制场景，介绍了一种基于强化学习的预测与干预融合的区域交通信号灯稳定控制方法（MaCAR）。MaCAR 通过预测智能体动作下的未来交通态势并建立智能体间状态信息、决策信息的通信机制，提高模型在多智能体协同干预情况下的稳定性。

2 视频异常检测及长尾分类方法

2.1 背景介绍

视频异常在真实应用中十分复杂。以交通事故监控检测为例，正常的视频片段和异常的视频片段不能通过一帧来区分。但是在给定多个连续帧的情况下，可以从不规则的车流中检测出异常，这些异常往往表现为车辆在十字路口的不同运动状态。如图 3 所示是真实场景下的视频异常案例。在这个例子中，两辆车相撞并停留在十字路口，而另一辆车从它们旁边经过。这种全局性的时空异常增加了这项任务的难度。

针对视频异常任务，本节介绍一种基于时空自编码器的视频异常检测方法（STAE）。该工作 *Spatio-Temporal AutoEncoder for Video Anomaly Detection*[5] 发表在 ACM Multimedia 2017 上。典型的自编码器模型只有一个解码器分支来重构输入数据，而本节介绍的方法设计了一个预测分支来预测未来帧。因为之前的一些研究已经证明，预测网络有利于学习视频表示。

通过异常检测方法，能够检测出视频中能够直接观测到的异常事件。然而在真实场景中，常常会有大量出现频次少的正常事件被当作异常事件检测出，因此无法很好地将

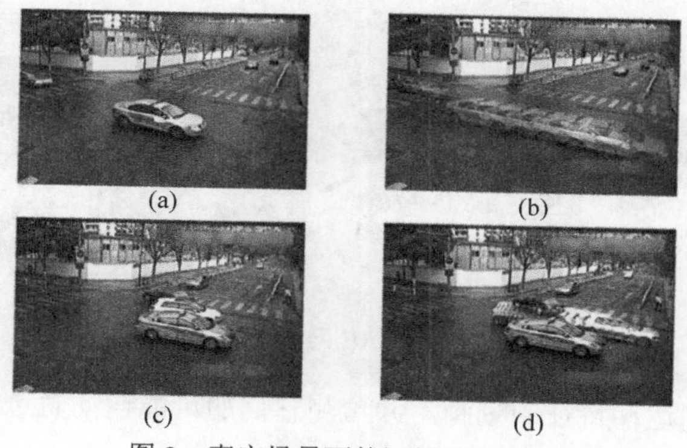

图 3　真实场景下的视频异常案例

（a）和（b）正常情况；（c）和（d）车辆撞击

正常事件和异常事件分开。通常情况下，检测出的事件分布是一个长尾分布，正常事件的类别属于长尾分布的头部类别，样本频次高。而异常事件类别属于长尾分布的尾部类，样本数较少。在这种长尾分布数据集上，常用的分类模型倾向于将分类结果靠近头部类，导致尾部类的分类效果较差。这种情况在异常检测任务中是不可容忍的，因此，针对长尾分布问题，本节还介绍了一个基于因果推理的稳定学习框架（TDE）。首先，对先前工作在长尾分布上分类效果变差的原因进行了分析。理论分析表明 SGD 动量（momentum）本质上是长尾分类中的因果关系混杂（confounder）。一方面，它有一个有害的因果效应，误导尾部类别的样本预测结果偏向头部类别。另一方面，它的诱导方式也有利于提高表征学习和头部类别预测准确率。针对上述问题，TDE 通过探索输入样本造成的直接因果效应，解释了 SGD 中动量的矛盾效应。特别是在训练中该框架使用因果干预，在推理中使用反事实推理，以去除“坏的”而保留“好的”。在三个长尾视觉识别基准上的实验结果表明，该框架能够达到最前沿的性能。本节中介绍的方法 *Long-tailed classification by keeping the good and removing the bad momentum causal effect*[6]已发表在 NeurIPS’2020 上。

2.2 STAE 网络

STAE 网络的框架如图 4 所示，在编码器阶段使用三维卷积，在解码器阶段使用三维反卷积。并且对每个卷积层进行批处理归一化，加速了训练阶段的收敛。此外，通过利用 Leaky ReLU，解决了 ReLU 导致的特征图高稀疏性的问题。

如图 4 所示，解码器阶段的两个分支有不同的损失函数。重构分支的损失函数是输入和输出间的欧几里得损失，其中预测视频片段的每一帧对总损失具有相同的权重。然而，恒权损失不适合用于异常检测任务，因为在大多数视频异常检测场景中，视点是固定的，新事物的出现是很难预测的。因此，在预测分支，STAE 应用预测损失函数来加

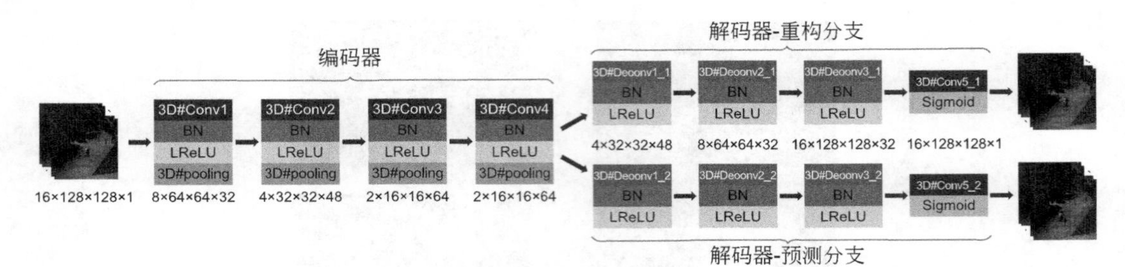

图 4　STAE 网络的框架

强模型对现有物体的运动特征的刻画，并预测它们的可能运动轨迹。

2.3　STAE 网络实验验证

STAE 在多个异常检测数据集上进行了大量实验，实验效果表明该方法能够有效地提高异常检测准确率。在 UCSD 和 CUHKAvenue 数据集上的实验结果见表 1。

表 1　UCSD 和 CUHKAvenue 数据集上的实验结果

Algorithm	Ped1		Ped2		Avenue	
	AUC	EER	AUC	EER	AUC	EER
MPPCA[7]	59.0	40.0	69.3	30.0	-	-
SF[8]	67.5	31.0	55.6	42.0	-	-
SF+MPPCA[9]	66.8	32.0	61.3	36.0	-	-
MDT[9]	81.8	25.0	82.9	25.0	-	-
SCL[10]	91.8	15.0	-	-	-	-
AMDN[11]	92.1	16.0	90.8	17.0	-	-
ConvAE[12]	81.0	27.9	90.0	21.7	70.2	25.1
STAE-grayscale	92.3	15.3	91.2	16.7	77.1	33.8
STAE-optflow	87.1	18.3	88.6	20.9	80.9	24.4

2.4　TDE 框架

图 5 是为了系统地研究长尾分布及 SGD 中动量如何影响预测效果提出的因果图，其中包含四个变量：动量(M)、物体特征(X)、头部方向的投影(D)和模型预测输出(Y)。因果图是一种有向无环图，用来表示感兴趣的变量$\{M, X, D, Y\}$如何通过因果联系相互作用。节点 M 和 D 分别是混杂因子和中介因子。混杂因素是影响相关变量和自变量的变量，造成虚假的统计相关性。以一个锻炼、年纪、是否患癌症的因果图为例：$\{$锻炼 $\leftarrow$ 年纪 $\rightarrow$ 是否患癌症$\}$。

老人退休后，虽然他们花更多的时间锻炼，但是由于年纪大，他们同样容易患上癌症。在该例中，年纪因素导致了锻炼导致患癌风险增加的伪相关性。此外，图 5 中中介因子的例子是$\{$药物$\rightarrow$安慰剂$\rightarrow$治愈$\}$，安慰剂（中介因子）是指服用药物的副作用，阻止了对药物和治愈率间的直接因果关系的分析。

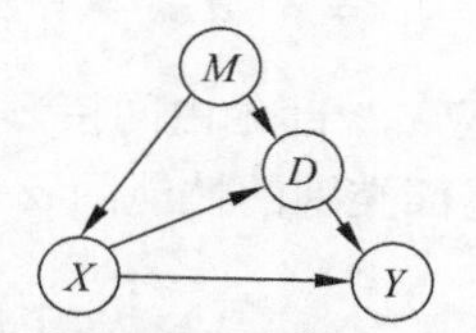

图 5　解释动量因果效应的因果图

以 Pytorch 中实现的版本为例来简单回顾一下 SGD-momentum：

$$v_t = \mu \cdot v_{t-1} + g_t,\quad \theta_t = \theta_{t-1} - lr \cdot v_t \tag{1}$$

其中，$\mu \cdot v_{t-1}$ 是动量；θ_t 是模型参数；g_t 是梯度；v_t 是速度；μ 是动量的衰减率；lr 是训练率参数。动量的使用在很大程度上抑制了由单个样本引起的振荡。在图 5 所示的因果图中，动量 M 是 $\mu \cdot v_{T-1}$ 在收敛时（$t = T$）的总效应，是衰减速率为 μ 的所有历史样本的梯度的指数滑动平均。如式 1 所示，给定超参 μ 和 lr，每个样本的动量 $M = m$ 是模型初始化和批数据训练策略的函数，即 M 有无限个可能的采样值。在一个平衡的数据集中，动量是由每个类共同贡献的。然而，当数据集是长尾时，它将被头部样本所主导，出现以下三种因果联系：①$M \rightarrow X$，这个关系表示用于生成特征向量 X 的主要参数在 M 的作用下进行了训练。这在式（1）中很明显，如图 6 所示，可以看到 X 的大小从头到

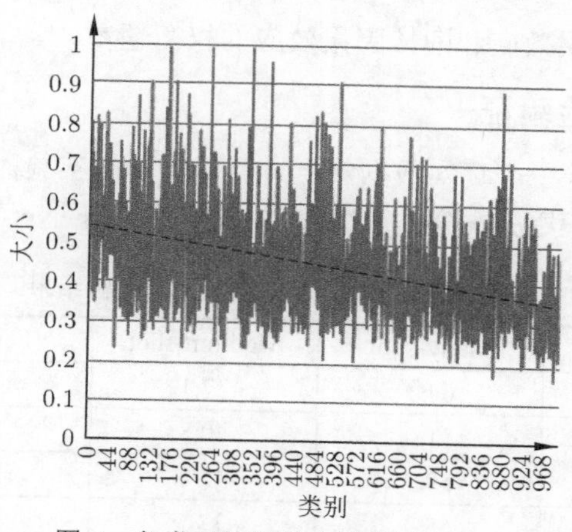

图 6　各类别的特征向量的平均大小

尾是如何变化的。②$(M,X)\to D$，该关系表示动量也会导致特征向量 X 向头部方向 D 偏移，这也是由 M 决定的。在长尾数据集中，少数头部类别占据了大部分的训练样本，它们的方差小于数据少但类丰富的尾部，因此滑动平均动量将指向一个稳定的头部方向。③$X\to D\to Y$ 和 $X\to Y$，这些因果关系表明，X 的影响可以分为间接(中介)效应和直接效应。对任意特征向量 x，可以将 x 分解为 $x=\ddot{x}+d$，其中 $D=d=\hat{d}\cos\left(x,\hat{d}\right)\|x\|$。根据这样的正交分解，$X$ 和 Y 间的影响效应 x 一部分由间接影响 d 决定，一部分由直接影响 $\ddot{x}$ 决定。当改变 d 的尺度参数 α 时，尾部类别的性能随 α 单调增加，这启发了下面介绍的去除 D 中介效应的算法框架。

基于图 5 中提出的因果图，可以对长尾分类任务目标做如下描述，即求解沿着 $X\to Y$ 关系的直接因果效应。根据因果推理理论，可以用总直接影响（TDE）表示：

$$\arg\max_{i\in C}\mathrm{TDE}\left(Y_i\right)=\left[Y_d=i|\mathrm{do}\left(X=x\right)\right]-[Y_d=i\mid\mathrm{do}(X=x_0)] \tag{2}$$

其中，x_0 表示空输入。将因果关系定义为对第 i 类的逻辑预测值 Y_i。下标 d 表示在因果图模型图 2.3 中，而 do 算子表示去除 M 和 X 间的因果关系的因果干预操作。在因果干预操作 $\mathrm{do}(X=x)$ 下，对所提出的因果图模型进行优化，目的是保持从动量中学习“好的”特征，并切断其“坏的”混淆效果。结合后门调整方法（backdoor adjustment）和能量模型，本节提出去混杂模型：

$$\mathrm{TDE}\left(Y_i\right)=\frac{\tau}{K}\sum_{k=1}^{K}\left(\frac{\left(w_i^k\right)^{\mathrm{T}}x^k}{\left(\|w_i^k\|+\gamma\right)\|x^k\|}-\alpha\cdot\frac{\cos\left(x^k,\hat{d}^k\right)\cdot\left(w_i^k\right)^{\mathrm{T}}\hat{d}^k}{\|w_i^k\|+\gamma}\right) \tag{3}$$

其中，α 是调解直接因果效应和间接因果效应的权衡超参。

2.5 TDE 框架的实验验证

在 ImageNet-LT[13]等三个真实数据集上进行了大量的实验。实验证明，TDE 方法的效果超过了之前最优的长尾分布分类算法。表 2 介绍了 ImageNet-LT 数据集上的实验结果。

表 2 ImageNet-LT 数据集上的实验结果

方法	Many-shot	Medium-shot	Few-shot	Overall
Focal Loss[14]	64.3	37.1	8.2	43.7
OLTR[13]	51.0	40.8	20.8	41.9
Decouple-OLTR[13-14]	59.9	45.8	27.6	48.7
Decouple-Joint[13]	65.9	37.5	7.7	44.4

续表

方法	Many-shot	Medium-shot	Few-shot	Overall
Decouple-NCM[13]	56.6	45.3	28.1	47.3
Decouple-cRT[13]	61.8	46.2	27.4	49.6
Decouple- τ -norm[13]	59.1	46.9	30.7	49.4
Decouple-LWS[13]	60.2	47.2	30.3	49.9
Baseline	66.1	38.4	8.9	45.0
Cosine[15-16]	67.3	41.3	14.0	47.6
Capsule[13,17]	67.1	40.0	11.2	46.5
(Ours)De-confounder	67.9	42.7	14.7	48.6
(Ours)Cosine-TDE	61.8	47.1	30.4	50.5
(Ours)Capsule-TDE	62.3	46.9	30.6	50.6
(Ours)De-confound-TDE	62.7	48.8	31.6	51.8

另外，值得注意的，本节工作中使用 Grad-CAM 可视化 TDE 方法训练得到特征图后发现，De-confound-TDE 事实上使特征图更紧致，即更关注少数区分度高的区域，而非整体结构。比如图 7 中“长牙野猪”的例子，传统的算法关注整个身体，而这部分其实和“猪”这个大类没什么区别，唯一的区别在于“长牙”，TDE 方法则明显关注到了这些区分度高的紧致区域。

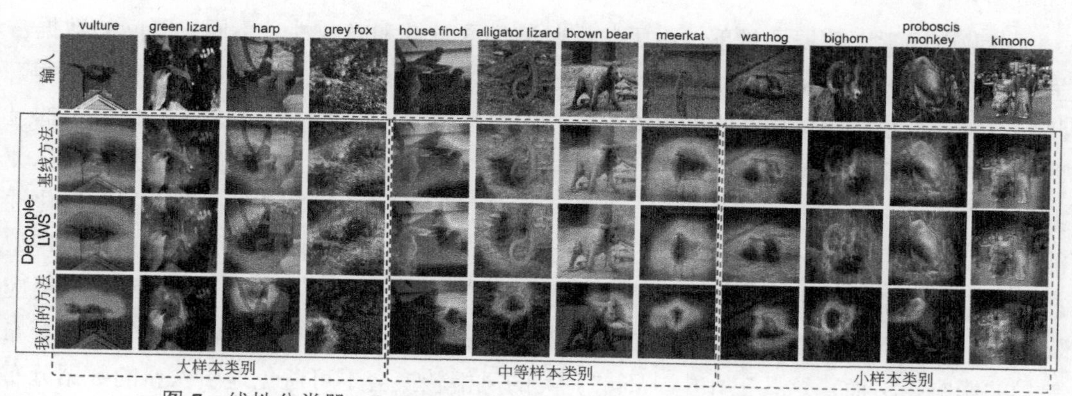

图 7 线性分类器、Decouple-LWS 和 De-confound-TDE 的特征图可视化

2.6 实际应用示例

如图 8 所示，在实际落地过程中，阿里巴巴“城市大脑”通过将基于视频数据的异常检测方法和长尾分类方法相结合，形成了事故事件检测分类的业务闭环。

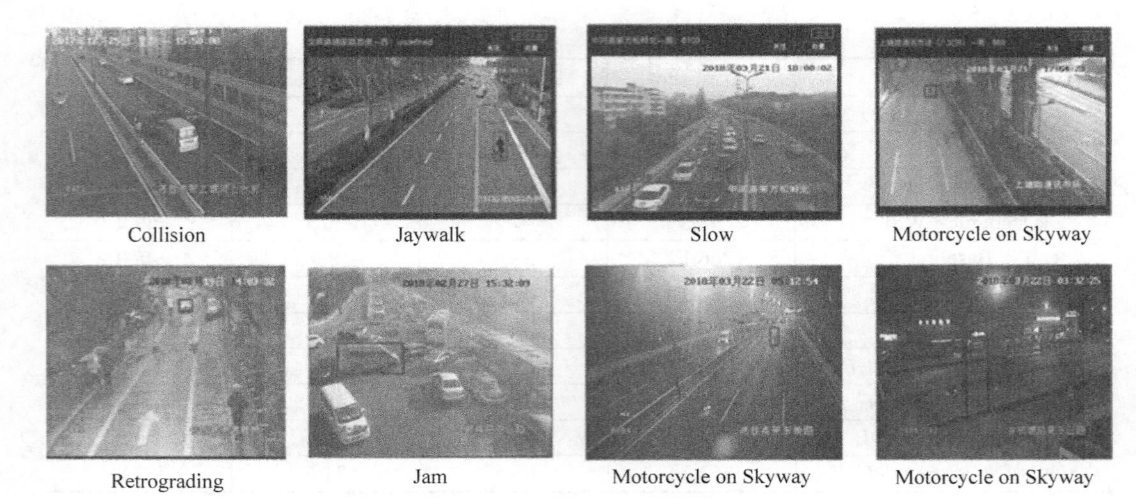

图 8 视频异常事件检测及分类实际应用示例

3 基于卷积长短时记忆网络的渐进学习方法

3.1 背景介绍

在城市视觉大数据场景下，由于数据收集的区域性和时效性，不同场景间的数据常呈现非独立同分布（non-I.I.D），即不同场景下采样得到的数据集间存在数据生成偏差。此外，由环境变化、异常事件事故等引起的不同场景间（空间）、不同时刻（时间）的显著数据分布变化同样使数据呈现时空上的非独立同分布性质，因此，如果直接在目标数据集上应用一个预先训练好的模型，常常会出现预测效果的明显下降，即模型的泛化性能较差。目前，大部分相关工作[18-19]主要关注如何利用目标数据集的先验知识减轻模型在数据集间迁移时数据分布差异的影响。其中大部分工作是基于生成对抗网络的域适应（domain adaptation）方法，通过将目标域和源域的风格转换为相近风格来减轻数据分布差异的影响。然而，这些工作中，不够完美的风格转换会不可避免地引入新的数据分布偏移，这同样会导致模型的泛化性能差的问题。而上面提到的同一场景下的环境变化导致的数据分布偏移同样对模型的泛化性能具有重要影响，但在最近提出的工作中很少提及，这正是本节内容所关注的。这些 non-I.I.D 数据间的显著差异会导致每个批数据（mini-batch）与整个数据集之间的分布偏移，并且很大程度上会影响模型的泛化性能。

因为这种差异会导致模型训练过程中梯度估计的偏差，从而影响模型微调的效果。缓解该问题最直接的方法是增加批数据的数据量（batch size）。然而，Keskar 等人的研究成果[20]显示，使用大的批数据量会导致模型收敛到更为狭小的局部最优解，并导致更差的泛化性能。

针对上述问题，本节研究了如何通过收集数据全局信息的方式，利用该信息减轻批数据间数据分布偏移的影响。首先，提出了一种新的 CNN 模块，称为批相关卷积单元（BConv-Cell）。BConv-Cell 逐批、渐进地收集数据集的分布信息，并保存在单元状态中。然后将 BConv-Cell 中收集的数据集信息用于修正下一个批的数据分布。在 BConv-Cell 的基础上，本节进一步提出了渐进学习方法（PTL），通过将待训练的骨干网络与 BConv-Cell 结合，BConv-Cell 组成旁枝网络，对预训练的模型进行微调。在 MSMT17[21]，Market-1501[22]，CUHK03[23]和 DukeMTMC-reID[24]数据集上进行了大量实验，表明本节提出的方法能够有效提升模型泛化性能。最后，进一步将提出的方法拓展到图片分类基准碑数据集 CIFAR-10、CIFAR-100[25]上。实验结果表明，本节提出的方法训练的模型在所有实验中都优于基线模型，这表明本节提出的方法对提高模型泛化性能的有效性。本节中介绍的方法已经以论文 *Progressive Transfer Learning for Person Re-identification*[26]发表在 IJCAI'2019 上。

3.2 批相关卷积单元

批相关卷积单元的出发点是基于一个简单的想法，即利用数据集的全局信息来减轻批数据间分布差异对模型训练效果的不利影响。BConv-Cell 的结构是受到卷积长短时记忆网络（Conv-LSTM）启发，但 BConv-Cell 与 Conv-LSTM 存在本质上的区别。首先，BConv-Cells 输入数据间的时序关系与 Conv-LSTM 不同。其次，BConv-Cells 与 Conv-LSTM 的网络结构明显不同，BConv-Cell 只收集并聚合数据的隐层信息，并且该信息不会回传到同一批数据的特征提取过程中，而是在下一批数据输入时用于修正下一批数据的输出。而 Conv-LSTMs 同时保留了隐层信息和数据输入的输出信息，并且这些信息会在同一批数据的前向传播过程中被回传到网络中。更为重要的是，BConv-Cell 的作用并不是根据输入做出预测，而是收集和汇聚数据分布信息。利用 LSTM 的存储机制，BConv-Cells 可以逐步收集全局信息，并在模型训练过程中使用它来修正模型参数的优化过程。与其他基于 LSTM 的元学习方法不同，BConv-Cells 的输出是直接作用于提出特征的修正。同时，由于 BConv-Cells 的本质是卷积神经网络层的堆叠，因此，BConv-Cells 可以十分方便、简单地与大多先有的深度特征提取网络模型相结合。BConv-Cells 的关键公式如下：

$$\begin{cases} i_b = \sigma\left(W_{xi} * x_b + b_i\right) \\ f_b = \sigma\left(W_{xf} * x_b + b_f\right) \\ o_b = \sigma\left(W_{xo} * x_b + b_o\right) \\ C_b = f_b \circ C_{b-1} + i_b \circ \tanh(W_{xc} * x_b + b_c) \\ y_b = o_b \circ \tanh(C_b) \end{cases} \tag{4}$$

其中，$*$表示卷积算子；$\circ$表示哈德曼积（Hadamard product）；σ表示 sigmoid 函数；x_b表示 BConv-Cell 在第b个批数据的输入；i_b，f_b，o_b分别是输入门i、遗忘门f、输出门o的输出；C_b表示第b个批输入后 BConv-Cell 收集、汇集得到的隐层状态；W是该 BConv-Cell 中的卷积神经网络的参数；y_b是 BConv-Cell 在该批次的最终输出。x_b，C_b，i_b，f_b，o_b均是三维数组对象。如式（4）所示，输出y_b是由隐层状态C_b和输入x_b决定的。而C_b是由x_b和上一个批数据输入时更新的C_{b-1}决定的，可以知道C_b保存了过去所有历史输入的部分数据分布信息。

3.3 渐进学习方法

给定一个基于深度神经网络的骨干网络如 DenseNet，渐进学习方法中将 BConv-Cells 与骨干网络中的卷积块（Conv-block）进行配对，多个 BConv-Cell 模块形成了一个旁枝网络，命名为渐进学习(PTL)网络。图 9 给出了 PTL 网络的结构示例，其中黑色虚线框表示卷积神经网络骨干网络。x_b^i表示骨干网络中第i个卷积块的输出，y_b^i表示对应的 BConv-Cell 的输出，x_b^0表示第b批输入图片，C_{b-1}^i表示上一批训练之后 BConv-Cells 中存储的隐层状态。红色虚线框表示一个 PTL 网络的基本构成模块，包括 Conv-block 和 BConv-Cell 构成的模块对。在每一个对中，首先将x_b^i和y_b^{i-1}串联后输入到一个核大小（kernel size）为1×1的卷积层中，然后再输入到对应的 BConv-Cell 中。特别地，最后一个 BConv-Cell 的输出会和骨干网络的输出串联之后输入到一个核大小（kernel size）为1×1的卷积层中完成特征图的融合。对所有的 BConv-Cell，都保存下其中收集、更新后

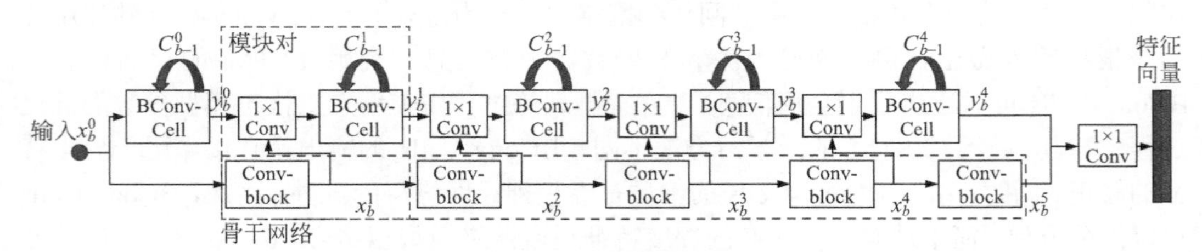

图 9 PTL 网络的结构（见彩图 14）

的隐层状态直到下一个批输入。用以下公式（5）定义该模块对：

$$\begin{cases} x_b^i = F_{\text{conv}}\left(x_b^{i-1}\right) \\ y_b^i = F_{\text{bconv}}\left(F_{1\times1}\left(x_b^i, y_b^{i-1}\right), C_{b-1}^i\right) \\ C_b^i = g(x_1^i, x_2^i, \cdots, x_b^i) \end{cases} \tag{5}$$

其中，x_b^i 表示第 i 个 Conv-block（$i \geqslant 1$）在第 b 个批时的输出；y_b^i 表示第 i 个 BConv-Cell 在第 b 个批时的输出。$i=0$ 时，x_b^0 表示输入图片，因此 C_b^0 没有意义，函数 g 表示式（4）中的计算过程。符号 F_{conv} 和 F_{bconv} 分别表示 Conv-block 和 BConv-Cell 学习的特征映射，$F_{1\times1}$ 表示图 9 中核大小为 1×1 的卷积层。本方法中适用的骨干网络种类很多，具有比较强的灵活性，大部分现有的卷积神经网络和图神经网络都可以结合。如式（5）所示，BConv-Cell 在收集全局信息和更新潜在状态的同时，学习从输入到特征空间的映射函数。值得注意的是，在式（5）中函数 g 将历史输入数据的辨别性信息渐进地收集、更新到 BConv-Cell 的潜层状态中。

3.4 实验验证

本节在四个有说服力的 personReID 数据集上进行了验证，包括 Market-1501, DukeMTMC-reID, MSMT17 和 CUHK03。在 Market-1501，DukeMTMC-reID 和 CUHK03 上的所有实验中使用 Open-ReID①提供的评价代码。这里仅介绍部分核心实验结果。在 Market-1501 数据集上的实验结果见表 3。从该表中可以看出，使用 PTL 方法可以有效提升骨干网络模型的效果。表 3 中 DenseNet-161*表示使用批数据量参数为 90 训练的模型，而 DenseNet-161 模型使用批数据量参数 32。可以注意到，使用更大的批数据量并不能有效提高模型的泛化性能，这与前面提到的 Keskar 等人的研究成果[20]相映证。从表 3 可以看出，本节提出的 PTL 方法可以有效提高模型的泛化性能。此外，STD 方法是本节中使用的模型蒸馏方法，其作用是利用 PTL 方法训练出大模型来帮助训练一个与骨干网络完全相同的小模型，以去除参数量对实验比较公平性的影响。

表 3　MSMT17 数据集上的实验结果

方法	mAP	CMC-1
DML[27]	70.51	89.34
HA-CNN[28]	75.70	91.20

① https://github.com/cysu/open reid。

续表

方法	mAP	CMC-1
PCB+RPP[29]	81.60	93.80
MGN[30]	86.90	95.70
DenseNet-161*	69.90	88.30
DenseNet-161	76.40	91.70
DenseNet-161+PTL	77.50	92.50
DenseNet-161+PTL+STD	77.50	92.50
MGN(reproduced)	85.80	94.60
MGN+PTL	87.34	94.83

为了验证渐进学习方法对提高模型泛化性能的有效性，本节进一步使用 Li 等人[31]提出的可视化工具，在 CIFAR-100 数据集上可视化骨干网络 DenseNet-100 使用和不使用我们提出的 PTL 方法的泛化能力并进行比较。如图 10 所示，DenseNet-100+PTL-v2 的损失面比 DenseNet-100 宽，根据 Li 等人[31]的研究成果，损失面越大，模型的泛化能力越好。因此，可以得出一个实验性的结论，即 PTL 方法有助于提高模型的泛化性能。

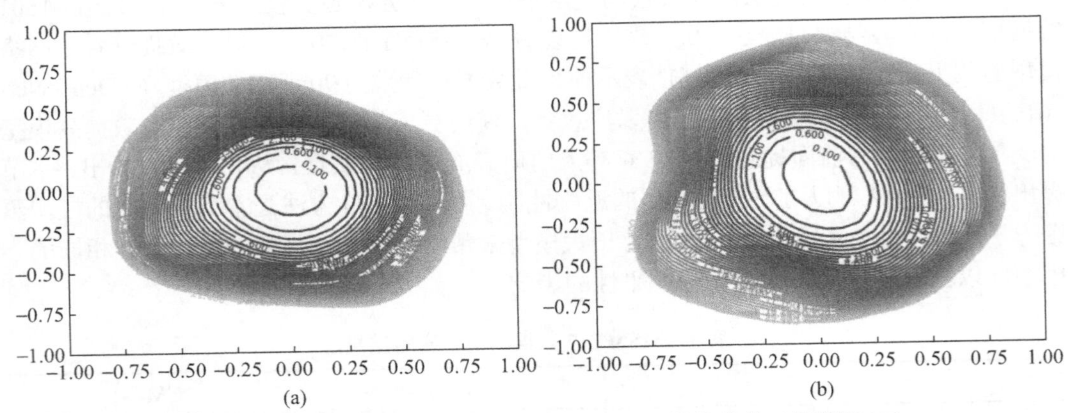

图 10　骨干网络使用和不使用 PTL 方法的泛化性能比较（见彩图 15）

（a）DenseNet-100；（b）DenseNet-100+PTL

其中，横、纵坐标均表示梯度值

4 预测与干预融合的区域交通信号灯稳定控制方法

4.1 背景介绍

本节介绍针对交通领域中十分重要的交通信号灯控制任务提出的稳定干预方法。城市交通信号灯控制和优化（TLC）是一个十分重要并且具有挑战性的真实业务问题。基于强化学习（reinforcement learning）的干预方法多采用在线训练策略，其基本流程如图11所示。首先在城市信号灯系统中部署独立或协作式的强化学习智能体（agent），用于控制单个或多个交叉路口。然后，利用历史的环境观测作为输入做出干预动作。最后，利用环境对智能体干预动作的反馈值设计动作评价指标，并对智能体进行训练。而常见的环境观测量包括路口监控摄像头等视觉数据来源，以及感应线圈等脉冲信号数据来源。通过物体识别等技术，可以将视觉数据转化为结构化数据，结合路网拓扑结构等信息整合成城市路网交通态势图数据，提高数据的聚合度。近年来，随着摄像头、存储设备等硬件成本的降低，基于城市视觉大数据的方法取得了一定的进展。

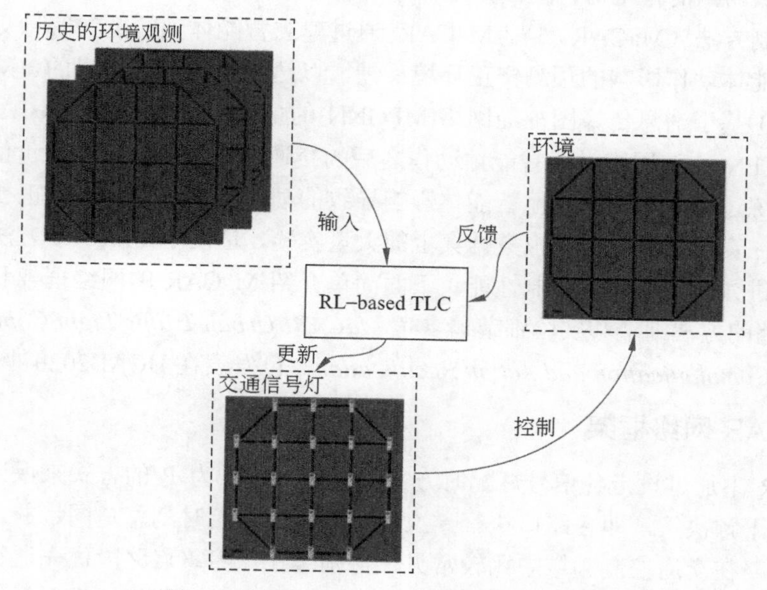

图11 基于强化学习的控制方法基本流程

目前的主流方法以独立智能体控制单个交叉路口[32-34]为主。这些方法中，智能信号灯系统能够根据实时路况做出决策动作，并实时动态地调节信号灯控制方案。然而，在城市信号灯控制任务中实际落地应用时，可以观察到独立的智能体间的动作常常会存在互相冲突的情况。例如，上游路口的信号灯为了保持所控制路口的通畅，会倾向于将车流尽量排向下游路口，并导致下游交叉口的溢流或道路资源浪费。在这种情况下，交叉路口（智能体）的状态数据受到其他智能体动作的影响，这是一个典型的 non-I.I.D 场景。

针对独立智能体优化时智能体间可能的动作干扰问题，近年出现了一些拓展智能体的可观测环境范围的方法[35-37]，一定程度上缓解了上述智能体动作互相干扰的问题。这一类方法所采用的智能体间信息交互机制可以看作一种被动的智能体间信息交互方式。这里的被动是指智能体间并不直接交互动作决策信息、历史状态信息、未来动作信息等直接影响未来交通态势的信息，而是让智能体观测更大范围的具有很大噪声和动作时延的路网环境。由于路网上复杂的交通态势和智能体动作对环境影响的时延效应，在这些工作中智能体同时动作时，环境反馈的动作价值依然是有偏的。并且由于智能体间交互的信息量不足，智能体间并不能真正地达到协同控制的效果。

针对区域路网的稳定协同控制问题，本节将介绍一种预测与干预融合的区域交通信号灯稳定控制方法（MaCAR）[3]。MaCAR 通过建立智能体间的主动信息交互机制并基于对其他智能体动作影响的预测修正环境反馈，以达到稳定学习的目的。MaCAR 由两部分组成：①基于消息传播图神经网络(MPGNN)的主动通信决策网络（CAN）；②交通预测网络（TFN），该网络学习智能体动作之后的交通态势和对应的动作价值，并利用预测信息在训练过程中减轻受其他智能体影响导致的动作价值偏差，以修正智能体的未来动作。通过在真实数据集和模拟数据集上的大量实验，证明了 MaCAR 方法可以获得超过目前最先进方法的信号灯控制性能。下面简单介绍 MaCAR 的网络框架和部分实验验证结果，完整的方法细节和实验细节请参考 *MaCAR: Urban Traffic Light Control via Active Multi-agent Communication and Action Rectification*，已发表在 IJCAI’2020 上。

4.2 MaCAR 网络框架

MaCAR 中通过设定各信号灯配时方案的周期总时长为 P 的方式来保证各个信号灯会同时生成干预决策，即各路口执行一次配时方案的起点时刻总是同步的。这种使用相同周期的策略在交通控制领域中比较常见，特别是在保障重点路段优先通行时，常在固定周期长度相同的同时设置控制方案，以获得重点路段上的“绿波带”效应（即车辆可以一路绿灯地通过整个重点路段）。而信号灯优化控制的目标就是最小化整个路网上的车辆拥堵指标，如路口车辆排队长度等。为了适应真实应用中各路口对交通控制策略的定制化需求，同时避免造成不良的驾驶体验和交通事故，MaCAR 通过修改各信号灯配时方案中各相位（一段时间内允许通行的车道列表组成的控制状态）占周期总长的比例来调解路口的交通态势。

MaCAR 的网络框架如图 12 所示，交通预测网络（TFN）用于预测智能体动作后可能的交通态势和对应的动作价值。而基于消息传播图神经网络(MPGNN)的主动通信决策网络（CAN）利用 TFN 提供的交通态势预测值，生成各信号灯下一周期使用的相位比例。

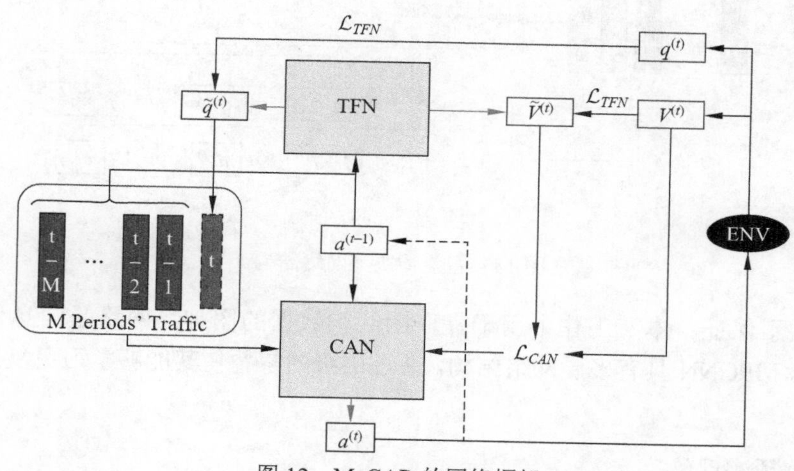

图 12　MaCAR 的网络框架

4.2.1　智能体决策网络

MaCAR 的第一部分是带主动通信机制的智能体决策网络（CAN），其输出是路网上各智能体在下一周期将要执行的信号灯配时方案。CAN 由消息传播图网络（MPGNN）组成的中心通信网络和多个可自定义的智能体子网络组成，如图 13 所示是 CAN 的网络框架图。在 CAN 中，将过去 M 个周期的路网上各路口车辆排队长度和路网拓扑结构图组成的图结构数据和对应时刻的历史动作信息分别提取特征嵌入后，输入中心通信网络完成信息在智能体间的传播、汇集和更新，然后将更新后的信息提供给智能体子网络。值得注意的是，决策信息的引入促使模型在学习路网车流传播规律的同时加入对其他智能体动作的考量。各智能体子网络分别控制路网上的一个信号灯。智能体子网络生成相位比例后，再根据信号灯的配置信息等先验信息结合信号灯安全运行规则将相位比例翻译成可执行的配时方案，以完成对信号灯的调节。这种规则约束的加入为路口交通安

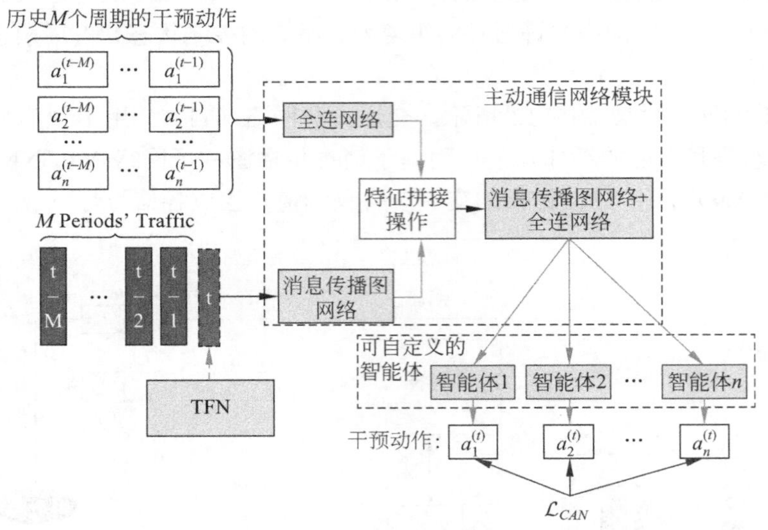

图 13 智能体决策网络结构

全运行提供了保证。本节工作中用于提取图结构数据的图神经网络是消息传播图网络（MPGNN）。MPGNN 具有多层网络结构，通过层级的累叠使其能够学习更复杂的信息传播模式。

4.2.2 交通预测网络

MaCAR 的第二部分是交通预测网络(traffic forecasting network, TFN)，其目的是在给定的动作下预测未来可能的交通态势及给定动作的环境反馈的动作价值。通过将这些交通态势预测信息输入 CAN 网络中，补充模型对不改变动作情况下的未来交通状态的感知。另一方面，利用预测的动作价值修正实际从环境中获得的动作价值，从而修正智能体受到其他智能体动作的影响。图 14 是 TFN 的网络框架。TFN 首先使用一个 MPGNN（k=2）从历史 M 个周期的交通态势数据中提取特征嵌入，然后将其与各周期对应动作经过神经网络提取得到的特征嵌入串联起来。然后将串联之后的特征嵌入件输入到两个分支（预测器和评价器）中。其中预测器分支用于预测整个路网的未来交通态势 $\tilde{q}^{(t+1)}$，评价器用于预测当前动作 $a^{(t)}$ 下的动作价值 $\tilde{v}^{(t)}$。如图 14 所示，预测器分支到评价器分支额外建立一个短接，通过这种方式为评价器引入对未来交通态势的预测信息，以减轻其他智能体对 $\tilde{v}^{(t)}$ 的影响，提高稳定性。

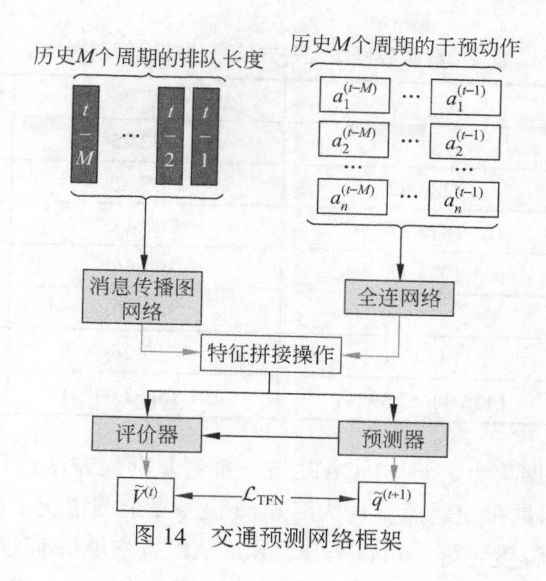

图 14 交通预测网络框架

4.3 实验验证

本节在两个最常用的开源交通信号灯控制模拟器上进行了实验：CityFlow① 和 SUMO②。此外，在三个从真实世界中采集到的真实数据集上进行了一系列实验，包括 D_{Hangzhou}，D_{Jinan} 和 D_{NewYork} 数据集。这三个数据集都来自最前沿方法 CoLight[35]。NewYork, Hangzhou 和 Jinan 数据集分别有 196 个、16 个和 12 个路口。其中，在三个真实数据集上的实验结果见表 4，表中数值表示车辆到达目的地所花费的平均旅行时长。可以看出，MaCAR 在所有三个真实数据集上取得了低于所有其他方法的平均旅行耗时。与 CoLight 相比，MaCAR 取得了平均 2.78%的效果提升。另外，可以看到 MaCAR 的效果明显优于基于多智能体协同动作的 CGRL 方法。在 CGRL 中，CGRL 使用一个统一的模型同时控制相邻的两个交叉口的动作。在多路口情况下，CGRL 需要搜索很大的动作空间，并且可能会面临拓展性的问题。与 CGRL 相比，由于仍然使用独立的代理，因此我们的方法具有较小的搜索空间。例如，在 New York 数据集上的实验中，MaCAR 可以在 196 个路口的路网上取得超过目前最好方法的效果，这说明 MaCAR 具有良好的大规模应用能力。

① https://cityflow-project.github/。
② http://sumo.dlr.de/index.html。

表 4 真实数据集上各方法平均旅行耗时对比

方法	$D_{Hangzhou}$	D_{Jinan}	$D_{NewYork}$
CGRL	2187.12	1582.25	1210.70
NeighborRL	2280.92	1053.45	1168.32
GCN	1876.37	768.43	625.66
OneModel	1973.11	394.56	728.63
Individual RL	—	345.00	325.56
CoLight	1459.28	297.26	291.14
MaCAR	**1425.00(+2.30%)**	**291.18(+2.04%)**	**279.49(+4.00%)**

此外，在模拟数据集上，将 MaCAR 与一系列基准线方法进行了多种交通路况下的性能比较，图 15 所示是在交通流量较大且车流趋势复杂的情况下的性能可视化图，排队长度越小，路网通畅程度越好。可以看出，MaCAR 方法能够有效地减少路网排队长度。同时，整体上 MaCAR 的性能优于去掉 TFN 网络的 MaCAR-noTFN 模型。

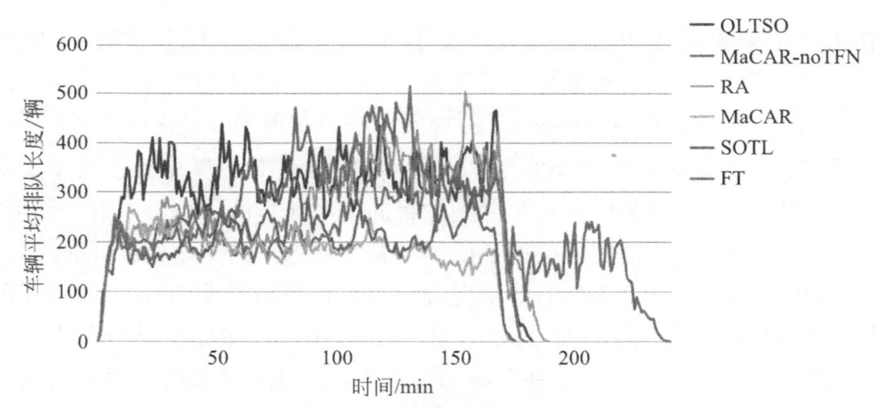

图 15 复杂交通流下的路网排队长度值随时间变化图（见彩图 16）

4.4 实际应用示例

本节提出的交通控制模型也可被应用于出租车调度系统。图 16 所示案例是阿里巴巴“城市大脑”在成都市交通运行监测调度中心（TOCC）的一个实践。基于成都核心区域约 300km^2 的车辆、枢纽（包括成都东站火车站、成都双流机场）客流数据，系统实时预测枢纽未来客流、出租车等候时长、出租车空车量、出租车 OD，准确率都稳定在 90%

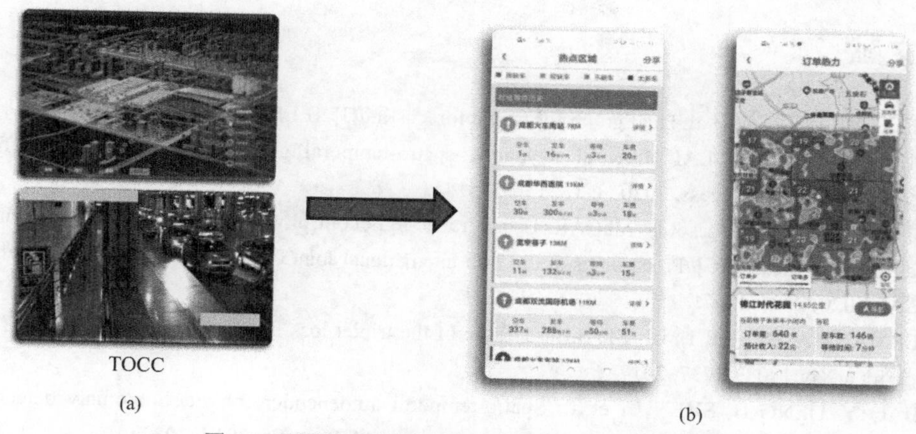

图 16 预测与干预融合的车辆调度系统应用示例

（a）预测干预融合的出租车调度系统；（b）生态合作伙伴派尔的信息发布平台

以上。基于预测，交通控制模型实时生成调度信息传递给生态合作伙伴派尔（成都本地一家出租车 APP 公司，主要产品为司机小秘书软件）。派尔 APP 将相关信息发布给出租车司机，出租车司机会根据 APP 上的信息决定自己的下一步接驳动作，这样就形成了整个系统的闭环反馈。

5 总结

近年来，以深度学习为代表的人工智能技术在城市大数据场景下的应用引起了学术界和工业界的广泛关注并快速发展。城市交通场景中异常事件事故的发生对预测模型的稳定性提出了挑战。目前大部分机器学习方法遵循的一个基本假设是独立同分布（independent and identically distributed, IID）假设。在符合基本假设的理想条件下，模型可以通过在训练数据上求得最低损失的最优模型来获得在训练数据集上具有良好泛化性能的模型。然而，在城市大数据场景下，由于数据生成过程中的某些因果关系，数据往往是非独立同分布（non-IID）的，这使得即使坐拥海量数据，模型在测试场景下时常会取得与训练数据集上相差甚远的效果。因此，本文就城市交通场景如何学习具有良好泛化性能的稳定模型的一系列问题展开了研究。特别地，着重对交通场景下基于视觉大数据的一系列基础问题进行了研究，包括异常检测任务、长尾图片分类任务、信号灯控制任务等。

参考文献

[1] WEI L, YU Z, JIN Z, et al. Dual graph for traffic forecasting[J]. IEEE Access, 2019,99: 1-1

[2] CHEN K, CHEN F, LAI B, et al. Dynamic spatio-temporal graph-based CNNs for traffic flow prediction[J]. IEEE Access, 2020, 8: 185136-185145.

[3] YU Z, LIANG S, WEI L, et al. MaCAR: Urban traffic light control via active multi-agent communication and action rectification[C]//Proceedings of the 20th International Joint Conference on Artificial Intelligence. IJCAI, 2020: 2491-2097.

[4] HERMANS A, BEYER L, LEIBE B. In defense of the triplet loss for person re-identification[Z]. arXiv preprint arXiv:1703.07737, 2017.

[5] ZHAO Y, DENG B, SHEN C, et al. Spatio-temporal autoencoder for video anomaly detection[C]// Proceedings of the 25th ACM international conference on Multimedia. 2017: 1933-1941.

[6] TANG K, HUANG J, ZHANG H. Long-tailed classification by keeping the good and removing the bad momentum causal effect[J]. Advances in Neural Information Processing Systems, 2020, 33.

[7] KIM J, GRAUMAN K. Observe locally, infer globally: a space-time MRF for detecting abnormal activities with incremental updates[C]//Proceedings of the IEEE Conference on Computer Vision and Pattern Recognition. 2009: 2921-2928.

[8] MEHRAN R, OYAMA A, SHAH M. Abnormal crowd behavior detection using social force model[C// Proceedings of the IEEE Conference on Computer Vision and Pattern Recognition,2009: 935-942.

[9] MAHADEVAN V, LI W, BHALODIA V, et al. Anomaly detection in crowded scenes[C]//Proceedings of the IEEE Conference onComputer Vision and Pattern Recognition. 2010: 1975-1981.

[10] LU C, SHI J, JIA J. Abnormal event detection at 150 fps in matlab[C]//Proceedings of the IEEE International Conference on Computer Vision. 2013: 2720-2727.

[11] XU D, RICCI E, YAN Y, et al. Learning deep representations of appearance and motion for anomalous event detection[Z]. arXiv preprint arXiv:1510.01553, 2015.

[12] HASAN M, CHOI J, NEUMANN J, et al. Learning temporal regularity in video sequences[C]// Proceedings of the IEEE Conference on Computer Vision and Pattern Recognition. 2016: 733-742.

[13] LIU Z, MIAO Z, ZHAN X, et al. Large-scale long-tailed recognition in an open world[C]// Proceedings of the IEEE Conference on Computer Vision and Pattern Recognition. 2019: 2537-2546.

[14] LIN T Y, GOYAL P, GIRSHICK R, et al. Focal loss for dense object detection[C]//Proceedings of the IEEE International Conference on Computer Vision. 2017: 2980-2988.

[15] GIDARIS S, KOMODAKIS N. Dynamic few-shot visual learning without forgetting[C]// Proceedings of the IEEE Conference on Computer Vision and Pattern Recognition. 2018: 4367-4375.

[16] QI H, BROWN M, LOWE D G. Low-shot learning with imprinted weights[C]// Proceedings of the IEEE Conference on Computer Vision and Pattern Recognition. 2018: 5822-5830.

[17] SABOUR S, FROSST N, HINTON G E. Dynamic routing between capsules[C]//Proceedings of the 31st

International Conference on Neural Information Processing Systems. Red Hook, NY, USA: Curran Associates Inc., 2017: 3859-3869.

[18] DENG W, ZHENG L, YE Q, et al. Image-image domain adaptation with preserved self-similarity and domain-dissimilarity for person re-identification[C]//Proceedings of the IEEE Conference on Computer Vision and Pattern Recognition. 2018: 994-1003.

[19] MA L, SUN Q, GEORGOULIS S, et al. Disentangled person image generation[C]//Proceedings of the IEEE Conference on Computer Vision and Pattern Recognition. 2018: 99-108.

[20] KESKAR N S, MUDIGERE D, NOCEDAL J, et al. On large-batch training for deep learning: Generalization gap and sharp minima[C]//Proceedings of the 5th International Conference on Learning Representations, 2017.

[21] WEI L, ZHANG S, GAO W, et al. Person transfer gan to bridge domain gap for person re-identification[C]// Proceedings of the IEEE Conference on Computer Vision and Pattern Recognition. 2018: 79-88.

[22] ZHENG L, SHEN L, TIAN L, et al. Scalable person re-identification: A benchmark[C]// Proceedings of the IEEE International Conference on Computer Vision. 2015: 1116-1124.

[23] LI W, ZHAO R, XIAO T, et al. Deepreid: Deep filter pairing neural network for person re-identification[C]// Proceedings of the IEEE Conference on Computer Vision and Pattern Recognition. 2014: 152-159.

[24] ZHENG Z, ZHENG L, YANG Y. Unlabeled samples generated by gan improve the person re-identification baseline in vitro[C]//Proceedings of the IEEE International Conference on Computer Vision. 2017: 3754-3762.

[25] KRIZHEVSKYA , HINTON G. Learning multiple layers of features from tiny images[M]. Handbook of Systemic Autoimmune Diseases, 2009.

[26] YU Z, JIN Z, WEI L, et al. Progressive transfer learning for person re-identification[C]// Proceedings of the 28th International Joint Conference on Artificial Intelligence. 2019: 4220-26.

[27] ZHANG Y, XIANG T, Hospedales T M, et al. Deep mutual learning[C]//Proceedings of the IEEE Conference on Computer Vision and Pattern Recognition. 2018: 4320-4328.

[28] LI W, ZHU X, GONG S. Harmonious attention network for person re-identification[C]// Proceedings of the IEEE Conference on Computer Vision and Pattern Recognition. 2018: 2285-2294.

[29] SUN Y, ZHENG L, YANG Y, et al. Beyond part models: Person retrieval with refined part pooling (and a strong convolutional baseline)[C]//Proceedings of the European Conference on Computer Vision (ECCV). 2018: 480-496.

[30] WANG G, YUAN Y, CHEN X, et al. Learning discriminative features with multiple granularities for person re-identification[C]//Proceedings of the 26th ACM international conference on Multimedia. 2018: 274-282.

[31] LI H, XU Z, TAYLOR G, et al. Visualizing the loss landscape of neural nets[C]//Proceedings of the 32nd International Conference on Neural Information Processing Systems. 2018: 6391-6401.

[32] WEI H, ZHENG G, YAO H, et al. Intellilight: A reinforcement learning approach for intelligent traffic

light control[C]//Proceedings of the 24th ACM SIGKDD International Conference on Knowledge Discovery& Data Mining. 2018: 2496-2505.

[33] NISHI T, OTAKI K, HAYAKAWA K, et al. Traffic signal control based on reinforcement learning with graph convolutional neural nets[C]// Proceedings of the 21st International Conference on Intelligent Transportation Systems (ITSC). IEEE, 2018: 877-883.

[34] CASAS, N. Deep deterministic policy gradient for urban traffic light control[Z]. arXiv Preprint arXiv:1703.09035, 2017.

[35] WEI H, XU N, ZHANG H, et al. Colight: Learning network-level cooperation for traffic signal control[C]//Proceedings of the 28th ACM International Conference on Information and Knowledge Management. 2019: 1913-1922.

[36] VAN DER POL E, OLIEHOEK F A. Coordinated deep reinforcement learners for traffic light control[C]// Proceedings of Learning, Inference and Control of Multi-Agent Systems, 2016.

[37] CHU T, WANG J, CODECÀ L, et al. Multi-agent deep reinforcement learning for large-scale traffic signal control[J]. IEEE Transactions on Intelligent Transportation Systems, 2019, 21(3): 1086-1095.

基于深度学习的命名实体识别

黄萱菁　桂　韬　李孝男　马若恬

（复旦大学计算机科学技术学院，上海 200433）

1　引言

命名实体识别（named entity recognition，NER）是信息抽取（information extraction，IE）的重要子任务，其目的是识别文本中具有特定意义的实体，主要包括人名、地名、机构名、专有名词等。例如，在“Democrat Biden replaces Trump as President of the United States”这句话中，可以识别出四个命名实体，其中“Democrat”是组织名，“Biden”和“Trump”都是人名，“United States”是地名。识别的过程包括两个步骤：实体边界的判断和实体类别的判断。其中实体边界的判断要确定实体在文本中的开始和结束位置，实体类别的判断则要完成实体的分类。

命名实体识别的应用十分广泛，在自然语言处理任务中起到承上启下的作用。一些底层任务如中文分词、词性标注、句法分析等对命名实体识别有帮助作用，同时命名实体识别也对下游任务起到关键作用。比如，在机器翻译任务中，待翻译文本中包含机构名时，翻译结果往往较差，当加入命名实体标注后，就能够有效地提升机器翻译系统的性能[1]。更重要的是，命名实体是信息抽取、信息检索、构建知识图谱和问答系统的基本单元。命名实体作为信息的主要载体，是实现信息抽取的第一步，也是信息抽取中最有实用价值的一项关键技术。命名实体的识别可以改善系统检索文档的相关度，并提高检索系统的召回率和准确率；将命名实体识别的技术应用在问答系统上，可以对文本中的关键信息做出更准确的分析，使问答系统给出更精确、更简洁的短语级答案。

近几年来，基于神经网络的深度学习方法在命名实体识别的研究中取得了巨大的成功。与传统机器学习方法相比，其高度非线性表示能力及端到端的训练方式有利于学习到更复杂的特征表示。对输入表示而言，以词向量表示词语的方法作为一种代表性的深度学习技术，一方面解决了高维向量空间带来的数据稀疏问题，另一方面词向量本身也比人工选择的特征包含更多的语义信息，而且该方法可以从非结构文本中获取统一向量空间下的特征表示，对于 NER 这种典型的序列化标注问题，俨然能够带来强大的发展动力。对模型结构而言，经典深度学习方法使用 LSTM，CNN 等非线性模型结构，在命名实体识别任务上已经能取得较好的效果[2-4]。

然而，在深度学习时代，命名实体识别仍然涉及一些需要解决的问题。首先，命名实体识别需要对大量未登录词进行处理，大多数命名实体本身属于未登录词，且命名实体识别是未登录词中数量最多、识别难度最大、对分词效果影响最大的问题。其次，命名实体识别存在语义依赖问题。在不同领域、场景下，命名实体的外延有差异，存在分类模糊的问题；不同命名实体之间存在大量的交叉和互相包含现象，且部分命名实体常常容易与普通词混淆。这些实体的识别需要更好地对上下文依赖进行建模。此外，中文命名实体识别存在边界歧义问题。相对英文命名实体而言，中文中未分词的句子导致实体边界的识别更加困难。词级别的中文命名实体识别会导致汉语分词和命名实体识别互相影响，而字级别的命名实体识别缺少词的信息，导致性能下降。最后，命名实体识别需要更好的解码方式。命名实体识别通常采用序列标注的解决方案，序列标签之间具有非常强的相互依赖关系，而现有的 CRF 解码方式无法对长距离依赖进行建模，且维特比算法的解码效率较低。

围绕上述挑战性问题，我们在深度命名实体识别的通用框架下开展了一系列研究，并取得一定进展。该框架可以被拆解为三个部分：输入表示层、上下文编码器和标签解码器。在输入表示层，针对未登录词问题，提出了一个 Teacher-Student 框架来学习未登录词的输入表示；在上下文编码器层，针对语义依赖问题，将 Transformer 模型应用于命名实体识别。进一步地，针对中文命名实体识别的边界歧义问题，提出了一种基于 Transformer 的中文命名实体识别模型来全方位优化编码模型；在标签解码器层，提出了一种两阶段的解码方式，既能实现并行预测，也能对长距离标签依赖进行建模。最后，提出了一种深度融合实体知识与预训练语言模型的预训练方式。

本文其余部分内容如下：第 2 节介绍基于深度学习的命名实体识别的相关工作，第 3 节介绍基于深度学习的 NER 通用框架下的基础工作及针对上述四个问题的改进，第 4 节进行总结。

2 相关工作

基于深度学习的命名实体识别已经涌现了许多成熟的工作。下面将分别从文本的编码方式、解码框架、预训练与 NER 三个方面介绍基于深度学习的 NER 工作。

2.1 文本的编码方式

词语的分布式表示（distributed representation）被广泛应用在自然语言处理的深度学习模型中。相较于独热编码（one-hot），分布式表示虽然能够对不同词语之间的语义关系进行建模[5-7]，但无法对未登录词进行有效的表示。在英语 NER 中，由于词语的形态信息（如大小写、词根、词缀等）有助于判断一个词语是否是实体，Santos[8]和 Ling 等人[9]提出基于神经网络模型的字符嵌入（character embedding）模型，同时验证了字符嵌入可以增强未登录词（out-of-vocabulary word, OOV）的表示[10-12]。但字符嵌入方式仍然无法解决未登录词的上下文语义。因此，提出了基于 Student-Teacher 的学习范式来训练未登录词表示预测网络，通过这种方式有效提高了未登录词的实体识别率。

获得词语的词嵌入与字符嵌入之后，需要充分融合词语及上下文的信息。Chiu 等人[11-13]尝试了基于循环神经网络（recurrent neural network, RNN）的模型，但 RNN 模型无法充分并行，因此 Strubell 等人[14-15]使用了空洞卷积神经网络（convolution neural network，CNN）的结构。但是 RNN 和 CNN 都无法有效对长距离依赖进行建模，因此具备长距离依赖建模能力的 Transformer 结构[16]在提出之后就被广泛应用在各类 NLP 任务之中，但 Transformer 结构在 NER 任务中的性能并不理想[17]。这是由于原始的 Transformer 结构无法像 RNN，CNN 网络一样较好地学习局部上下文信息，因此提出了基于相对距离的注意力计算机制并获得了更优秀的性能。

由于汉字之间没有明显的边界信息，并且汉字也不存在形态变化，英文中的字符嵌入方式并不能直接应用到中文 NER 中。同时，由于中文以词语为基本语素，在模型之中忽略词语信息可能带来性能损失。因此在中文 NER 中，一般会使用汉字嵌入和中文词典。由于词典中的词语一般包含多个汉字，并且一个汉字可能被多个词语包含，因此需要特殊的手段来编码这两类信息。针对中文 NER 的编码器主要有词典增强嵌入和改良模型架构两种范式。Liu 等人[18]和 Peng 等人[19]直接将词语的信息编码到字的表示中，这种将词语信息添加到词语输入层的方式与架构无关，具备较强的迁移性，但这类词典增强嵌入的方法需要额外的网络来编码词典信息，并且字与词之间的交互不够充分。Zhang 等人[20]通过 Lattice-LSTM 的结构来融合字与包含该字的词语，Gui 等人[21]通过使用 CNN 缓解 Lattice-LSTM 无法充分并行及无法处理词汇信息冲突的问题。但以上方法存在交互不够充分、解码效率不高和无法对长距离依赖进行建模的问题，通过修改 Transformer 注意力的计算方式将栅格结构展平，使不同文本之间的栅格结构可以并行处理，不但有效提高了推理效率，还取得了较好的性能。

2.2 解码框架

由于命名实体识别任务不仅包含了对实体边界的识别，还包含了对实体类型的分类。为了达成这个目的，通常将 NER 视为一个序列标注问题，通过将 BIO 或者 BILOU 体系（scheme）与实体类型做笛卡尔积来为每个词语分配一个标签[22]。过去的大部分工作都是使用条件随机场（conditional random field，CRF）作为解码器，并且在推断的时候利用维特比算法得到最终的标签序列[4,12,15,17]。由于一阶 CRF 只能对相邻词语之间的标签依赖进行建模，所以 Cui 等人[23-24]尝试通过引入标签嵌入的方式对更长范围的标签依赖进行建模。除此之外，Wang 等人[25-26]还尝试了通过序列到序列（sequence-to-sequence）的框架来进行解码，但需要使用自回归的方式生成，在长句场景下效率较低。为了解决过去解码方法中解码效率不高、标签依赖较短及标签错误传播等问题，提出基于不确定度量的两阶段解码来缓解这些现象。这种两阶段解码方式不但可以并行解码和对长距离标签依赖进行建模，同时也可以在文档级别 NER 任务上保持相同实体之间类别的一致性[27]。

除了将 NER 视为一个序列标注问题，Li 等人[28-30]尝试直接枚举句中所有的片段，然后对片段进行实体分类。通过这种方式，可以使用相同的模型架构同时解决非嵌套 NER 和嵌套 NER 任务。但这种方式需要枚举的片段数量将是 $O(n^2)$，一般在使用这种方式建模 NER 时，通常会限定最大实体长度[31-32]。

2.3 预训练与 NER

静态的预训练词向量无法处理一词多义的问题，但是当它通过神经网络与上下文交互之后便可以在不同语境下得到不同的表示。因此，Peters[33]等人尝试直接对带有 RNN 结构的模型进行预训练，在下游任务直接使用预训练模型对文本进行编码，通过这种预训练模型可以大幅提高 NER 模型的性能。为了获取更好的表示，FlairNLP 团队尝试了不同粒度的预训练[34-35]并在序列标注任务上取得了显著提升。由于 ELMo 等[33-35]预训练模型都是基于 RNN 网络进行的，一方面无法充分并行计算，另一方面也无法在较大规模的数据上有效训练。具备高并行能力的 Transformer 结构通过在大量语料上进行训练，如 BERT, RoBERTa 等[36-38]，在 NER 任务上取得了较好的性能和效率。这种通用领域的预训练模型学会的主要是词语依赖等浅层语义，研究发现 BERT 等预训练模型中缺乏实体知识[39]。因此，ERNIE[39]和 KnowBERT[40]等模型尝试为实体生成静态词向量表示并将其融入到预训练模型中。但这种方式并没有充分考虑实体的上下文信息，也无法对实体的表示与语言模型联合训练，因此提出了深度融合实体知识和预训练语言模型的 CoLAKE 模型，以进一步提升 NER 性能。

3 基于深度学习的命名实体识别

如第 1 节中所讨论，基于深度学习的命名实体识别存在的主要问题包括未登录词问题、语义依赖问题、中文命名实体识别的边界歧义问题、标签解码问题。下面将介绍在目前深度学习命名实体识别的通用框架（由输入表示层、上下文编码器、标签解码器构成）下针对这四个问题开展的研究工作及成果，包括在输入表示层增强未登录词表示、在上下文编码器层全方位优化编码模型、在标签解码层高效率解码标签序列及融合实体知识的预训练。

3.1 未登录词表示增强

在命名实体识别的深度学习方法中，对于未登录词，最常见的表示方法是用一个统一的向量进行表示。这一做法显然是不合理的，因为不同的未登录词的含义是不同的，一个统一的向量无法对它们之间的区别进行建模。很多工作表明，在遇到未登录词时，深度命名实体识别模型的效果会显著下降[41-42]。

为此，提出通过对未登录词的词形特征及上下文信息进行建模来学习未登录词的表示。为了训练这样一个未登录词表示模型，使用了 Teacher-Student 学习框架。这一框架的核心思想是基于训练数据集构建一个未登录词的表示预测任务。其中，Teacher 模型基于训练集和任务标注进行训练，得到一个包含词表示的命名实体识别模型；Student 模型随机挑选训练集中的一些词作为未登录词，然后基于这些词的词形和上下文信息训练得到它们在 Teacher 模型中的表示。

具体地，首先挑选合适的深度学习框架（不需要特殊处理未登录词）训练一个命名实体识别模型，称为 Teacher 模型。采用当前常用的深度命名实体识别框架 CNN+BiLSTM+CRF 来实现 Teacher 模型。CNN+BiLSTM+CRF 使用一个稠密向量 $\boldsymbol{e}_w(w)$ 和一个基于 CNN 网络和词的字符序列得到的向量 $\boldsymbol{e}_c(w)$ 来表示一个词：

$$\boldsymbol{e}(w) = [\boldsymbol{e}_w(w) \oplus \boldsymbol{e}_c(w)]$$

其中，$\oplus$ 表示向量的拼接操作。句子 s 对应的词序列的向量表示被送入 BiLSTM 中，得到每个词对应的隐层向量表示 $\boldsymbol{h}(w|s)$。在此基础之上，使用序列化的条件随机场（CRF）[43] 对序列标签进行预测：

$$p(\boldsymbol{Y}|\boldsymbol{X};\boldsymbol{\theta}_T)=\frac{\prod_{t=1}^{m}\phi_t(y_{t-1},y_t|s)}{\sum_{Y'\in\mathcal{Y}}\prod_{t=1}^{m}\phi_t(y'_{t-1},y'_t|s)}$$

其中，$\phi_t(y',y|\boldsymbol{X})=\exp(\boldsymbol{w}_{y',y}^{\mathrm{T}}\boldsymbol{h}(w|s)+\boldsymbol{b}_{y',y})$。模型的训练损失定义如下：

$$\mathcal{L}=\sum_{i=1}^{N}\ln p(\boldsymbol{Y}_i|\boldsymbol{X}_i;\boldsymbol{\theta}_T)$$

其中，N 为训练样本（句子）数目。Teacher 模型使用所有的训练数据进行训练，对于未登录词，使用统一的随机向量进行初始化表示，并在模型训练过程中进行微调。之后，基于以上获得的 Teacher 模型训练 Student 模型。Student 模型基于词的词形特征和上下文信息来学习词的表示。

词的词形特征来源于词对应的字符序列 $w=[c_1,c_2,\cdots,c_n]$。具体地，参考文献[44]，首先对词的字符序列进行填充，进而得到该序列的包含至多 k 个字符的 ngram 集合：

$$G(w)=\bigcup_{m=1}^{k}\bigcup_{i=0}^{n+2-m}\{c_i,\cdots,c_{i+m-1}\}$$

基于此，使用平均池化的方法得到 $G(w)$ 的向量表示：

$$\boldsymbol{v}_{\mathrm{form}}(w)=\frac{1}{G(w)}\sum_{g\in G(w)}\boldsymbol{W}_{n\mathrm{gram}}(g)$$

其中，$\boldsymbol{W}_{n\mathrm{gram}}(g)$ 返回的是 g 的可训练稠密向量。

为了获得词的上下文信息，用 BiLSTM 对词 w_i 的上下文 $s_{\backslash i}$ 进行建模。考虑两种典型的 Student 模型应用场景：① 整个输入序列只有一个词为未登录词；② 整个输入序列有多个词为未登录词。在第一种场景下，其他词在 Teacher 模型中的表示已知，所以可以直接用 BiLSTM 对 $s_{\backslash i}$ 进行建模。具体地，使用前向 LSTM 建模 $s_{<i}$，得到词的前向上下文表示 $\boldsymbol{h}(w_i|s)$；使用后向 LSTM 建模 $s_{>i}$，得到词的后向上下文表示 $\boldsymbol{h}(w_i|s)$。这两个表示拼接起来得到词的上下文表示：

$$v_{\mathrm{context}}(w)=[\boldsymbol{h}(w_i|s)\oplus\boldsymbol{h}(w_i|s)]$$

在第二种场景下，对于未登录词 w_i，序列中的一些其他词也是未登录词，所以无法直接用 BiLSTM 对 w_i 的上下文进行建模。因此，使用迭代方法得到 w_i 的表示。令

$\boldsymbol{v}_{\text{context}}^{t}(w\_i|s)$ 表示 w_i 在第 t 次迭代时的表示，且 $\boldsymbol{v}_{\text{context}}^{0}(w\_i|s)$ 为零向量。在第 $t+1$ 次迭代时，基于每个词在第 t 时刻的表示，使用 BiLSTM 对词的上下文进行建模，得到 $\boldsymbol{v}_{\text{context}}^{t+1}(w\_i|s)$。将该更新过程应用于序列中的每个词，得到每个词在 $t+1$ 时刻的表示。在 K 次迭代后，得到词的最终表示。在本项工作中，$K=2$。

将词的词形和上下文表示拼接起来得到词基于 Student 模型的表示：

$$\boldsymbol{v}_{w}(w)=\alpha\boldsymbol{v}_{\text{form}}(w)+(1-\alpha)\boldsymbol{v}_{\text{context}}(w)$$

其中，

$$\alpha=\sigma\left(\boldsymbol{w}^{\mathbf{T}}\left[\boldsymbol{v}_{\text{form}}(w)\oplus\boldsymbol{v}_{\text{context}}(w)\right]+\boldsymbol{b}\right)$$

为了训练 Student 模型，随机采样训练数据中的部分词作为未登录词。之后，使用 Student 模型预测这些词的表示，并用这些表示替换其在 Teacher 模型中的表示。最后，将替换后的表示输入到 Teacher 模型的 BiLSTM 和 CRF 模块中，得到 Teacher 模型的损失，并基于这一损失训练 Student 模型。

在表 1 所示的 4 个数据集上进行测试，表 2 给出了模型在未登录实体上的表现。其中，RandomUNK 表示用统一的随机初始化向量对未登录词进行表示；SingleUNK 表示用统一的微调向量对未登录词进行表示；Proposed 表示我们提出的方法；Lazaridou 等人[45]提出的方法在测试阶段使用平均池化的方法由词的上下文得到词的向量表示；Khodak 等人[44]提出的方法在文献[44]的基础上用一个线性变换对池化的向量进行线性变换；Schick 等人[46]提出的方法和我们所提方法类似，但是其 Student 模型的训练目标是重构出筛选词在 Teacher 模型中的表示而不是直接使 Teacher 模型表现更好。

表 1　测试数据集及其对应的未登录实体数目

数据集	Dev		Test	
	#OOV	OOVRate	#OOV	OOVRate
CoNLL02-Spanish	2,216	50.91%	1,544	43.38%
CoNLL02-Dutch	1,819	69.53%	2,564	65.05%
Twitter-English	1,266	79.15%	4,131	79.13%
CoNLL03-German	3,928	81.27%	2,685	73.10%

从表 2 可以看出：①在绝大多数任务上，RandomUNK 的表现最差。②使用词形和上下文信息的 Schick 等人提出的方法表现得比只使用上下文信息的 Khodak 等人提出的方法更好，表明词形信息和上下文信息的互补性。③在绝大多数任务上，Schick 等人提出的方法表现得比 SingleUNK 更好，表明了采样思路的正确性。④我们提出的方法表现

得比 Schick 等人提出的方法更好，一方面验证了 Teacher 模型和 Student 模型之间存在的误差传递问题，另一方面验证了该方法在解决未登录词方面的有效性。

表 2 模型在未登录实体上的表现

Arch	模型	CoNLL02-Spanish		CoNLL02-Dutch		Twitter-English		CoNLL03-German	
		Dev	Test	Dev	Test	Dev	Test	Dev	Test
LSTM	RandomUNK	69.36	72.06	64.23	64.08	56.88	56.38	55.92	56.89
	SingleUNK	68.79	71.59	67.83	66.39	56.82	56.39	59.69	60.16
	Lazaridou 等人(2017 年)	68.61	69.08	65.99	65.43	47.72	47.20	47.87	49.17
	Khodak 等人(2018 年)	68.74	69.53	66.34	65.70	48.22	47.28	47.97	49.33
	Schick 等人(2018 年)	70.84	72.88	68.88	67.51	59.18	57.21	55.83	58.42
	Akbik 等人(2018 年)	61.78	64.06	60.49	62.09	49.68	50.22	55.06	53.01
	我们提出的方法	**73.91**	**74.63**	**70.33**	**70.12**	**60.14**	**58.32**	**60.55**	**61.79**
CNN	RandomUNK	61.07	61.61	53.48	57.31	44.82	43.72	56.23	56.94
	SingleUNK	56.87	58.30	**60.68**	**60.46**	**57.13**	**57.34**	62.33	62.07
	Lazaridou 等人（2017 年）	54.97	60.39	53.73	56.99	42.91	46.99	43.38	43.54
	Khodak 等人(2018 年)	55.12	60.41	54.20	57.00	48.21	47.78	53.19	53.50
	Schick 等人(2018 年)	61.23	61.54	53.60	57.48	46.79	46.59	56.16	57.44
	我们提出的方法	**63.38**	**63.02**	59.24	60.33	57.06	57.32	**62.42**	**63.01**

3.2 文本编码模型优化

如第 2 节所述，基于循环神经网络的模型运行速度有限，且基于 CNN 和 LSTM 的模型都难以对长距离上下文信息进行建模。因此，提出了 TENER，将 Transformer 应用于 NER[4]，以解决以上两个问题。

Transformer 的计算没有循环结构，因此运算效率较高。Transformer 主要由 Self-attention 和 Feed-forward(FFN)这两个模块构成，得益于前者，它能够直接对任意两个符号之间的依赖进行建模。Self-attention 的计算如下：

$$\mathrm{Att}(\boldsymbol{A},\boldsymbol{V})=\mathrm{softmax}(\boldsymbol{A})\boldsymbol{V}$$

$$\boldsymbol{A}_{ij}=\left(\frac{\boldsymbol{Q}_i\boldsymbol{K}_j^{\mathrm{T}}}{\sqrt{d_{\mathrm{head}}}}\right)$$

其中，$\boldsymbol{A}$ 代表注意力分数矩阵；$\boldsymbol{K},\boldsymbol{Q},\boldsymbol{V}$ 代表自注意力计算的三个部分——键、查询、值，均来自于上一层隐层表示的线性变换。我们发现原 Transformer 的绝对位置编码难以对符号的相对位置和方向进行建模，而这对 NER 任务来说是很关键的。我们的 TENER 模型将原始 Transformer 针对 NER 任务进行适配，以达到更好的性能和运行效率。具体地，在 TENER 中，对 Transformer 在以下两个方面进行了改进：

（1）将绝对位置编码方式替换为更适合 NER 任务的相对位置编码方式，具体地，Self-attention 的计算修改如下：

$$\boldsymbol{Q},\boldsymbol{K},\boldsymbol{V}=\boldsymbol{H}\boldsymbol{W}_q,\boldsymbol{H},\boldsymbol{H}\boldsymbol{W}_v$$

$$\boldsymbol{R}_{t-j}=\left[\cdots\sin\left(\frac{t-j}{10000^{2i/d_k}}\right)\cos\left(\frac{t-j}{10000^{2i/d_k}}\right)\cdots\right]^{\mathrm{T}}$$

$$\boldsymbol{A}_{t,j}^{\mathrm{rel}}=\boldsymbol{Q}_t^{\mathrm{T}}\boldsymbol{K}_j+\boldsymbol{Q}_t^{\mathrm{T}}\boldsymbol{R}_{t-j}+\boldsymbol{u}^{\mathrm{T}}\boldsymbol{K}_j+\boldsymbol{v}^{\mathrm{T}}\boldsymbol{R}_{t-j}$$

$$\mathrm{Attn}(\boldsymbol{Q},\boldsymbol{K},\boldsymbol{V})=\mathrm{softmax}(\boldsymbol{A}^{\mathrm{rel}})\boldsymbol{V}$$

其中，t 是目标符号的位置；j 是上下文符号的位置；$\boldsymbol{Q}_t,\boldsymbol{K}_j$ 是第 t 个符号的查询向量和第 j 个符号的键向量；$W_q,W_v\in\boldsymbol{R}^{d\times d_k},u\in\boldsymbol{R}^{d_k},v\in\boldsymbol{R}^{d_k}$ 是可学习的参数；R_{t-j} 是相对距离编码。这里 Self-attention 的计算借鉴 Transformer-XL[4]，但由于 NER 的训练数据一般是小规模的，避免了两个可学习的矩阵直接相乘的情况，因为它们的结果可以被一个矩阵代替。

（2）在 Transformer 的原始文献中，对 Self-attention 中的注意力分数做了除以模型维度的缩放，然而通过实验发现，在 NER 任务中，去掉缩放之后模型性能会更好，这可能是因为这样做使注意力分数分布更加尖锐，更符合 NER 任务中实体的稀疏特性。

表 3 给出了 TENER 在两个常用英文 NER 数据集上的性能，实验结果显示使用更强编码器的 TENER 模型的性能要明显胜过 LSTM。

表 3　TENER 在常见英文数据集上的性能

模型	CoNLL2003	Ontonotes5.0
BiLSTM-CRF[2]	88.83	—
CNN-BiLSTM-CRF[11]	90.91	86.12
BiLSTM-BiLSTM-CRF[12]	90.94	—

续表

模型	CoNLL2003	Ontonotes5.0
ID-CNN[14]	90.54	86.84
TENER	91.43	88.43

进一步地，针对中文命名实体识别的边界歧义问题，提出了一个基于 Transformer 的中文 NER 模型 FLAT[4]，它能实现并行地处理汉字与词典信息。具体来说，给每个字一个位置标签，然后给每个匹配的词对应两个位置标签，分别为头和尾。比如在句子“重庆人和药店”中，词语“人和药店”的头和尾是“人”和“店”的位置标签，也就是 3 和 6。这样就能将字-词联合结构转化为一个三元组的集合，每个三元组包括字或者词及其头、尾位置，比如(“人”，3，3)、(“人和药店”，3，6)，称这个集合为 Flat-Lattice。

FLAT 模型也使用了类似 TENER 的相对位置编码方式，但是它的位置编码是针对栅格结构（lattice）特殊设计的，具体计算如下：

$$R_{ij} = \mathrm{ReLU}\left[W_r\left(p_{d_{ij}^{(hh)}} \oplus p_{d_{ij}^{(th)}} \oplus p_{d_{ij}^{(ht)}} \oplus p_{d_{ij}^{(tt)}}\right)\right]$$

其中，d_{ij}^{hh}，d_{ij}^{ht}，d_{ij}^{th}，d_{ij}^{tt} 是第 i 个符号和第 j 个符号间的头和头、头和尾、尾和头、尾和尾的相对距离。用上面的式子替代 TENER 中 R_{ij} 的计算，就得到了 FLAT 中注意力分数的计算。

图 1 给出了模型 FLAT 整体的结构图，具体来说，同时将栅格中的字和词传入 Transformer 模型，然后设计一种针对栅格结构的位置编码以更好地编码栅格结构信息，这样 FLAT 就不需要像之前的模型一样动态修改结构来表征输入的结构了，同时，FLAT 也采用了和 TENER 类似的相对位置编码方法来加入栅格结构信息。由于 Transformer 各符号之间互相都能够注意到，输入中的字能够直接接收输入中的词信息，所以在运行得更快的同时，信息传递效率也非常高。表 4 给出了 FLAT 模型在四个常见数据集上的性能，它不仅超过了基线模型和其他结合词典的模型，且在大数据集上的性能提升尤为明显。图 2 给出了 FLAT 和其他基于词汇的中文 NER 模型的速度对比，可以看出 FLAT 在速度及 GPU 利用率上较其他模型有较大提升。

3.3 高效率标签序列解码

标签解码器的目的是为每一个字符预测一个对应的标签，常见的解码器包括 Softmax 和条件随机场（CRF）。Softmax 可以并行解码，但是无法对标签间的依赖关系进行建模。CRF 可以捕获相邻标签的依赖性，但是依赖于效率低下的维特比（Viterbi）解码。许多最近提出的方法试图引入标签嵌入对更长范围的依赖性进行建模，例如，两阶段标

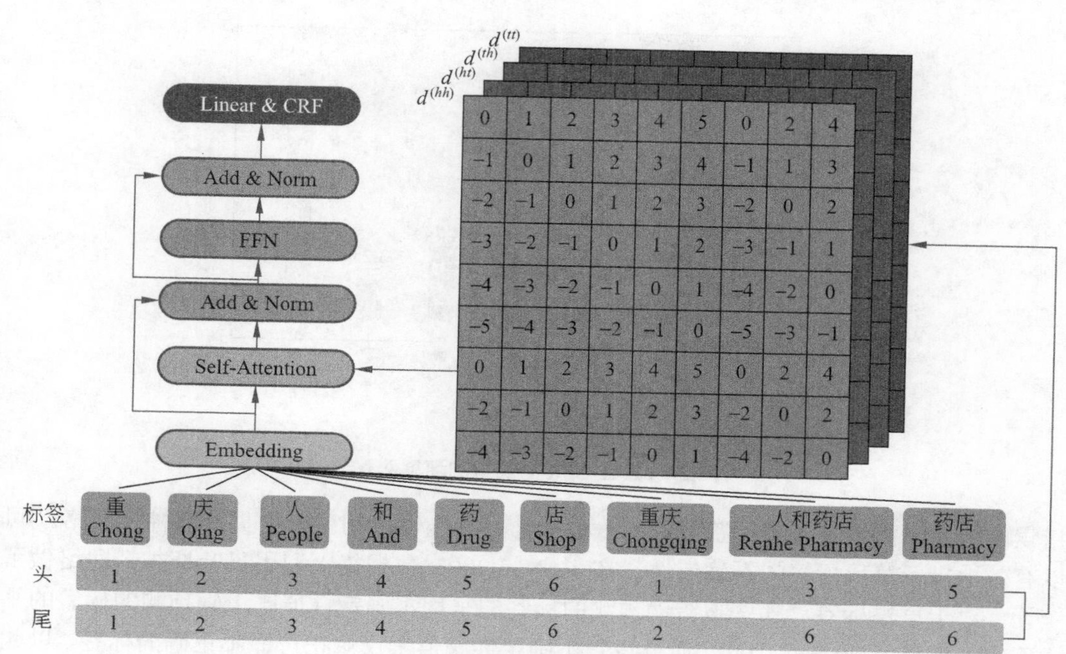

图 1　FLAT 模型详细结构

表 4　FLAT 在四个常见数据集的性能

	Lexicon	Ontonotes	MSRA	Resume	Weibo
BiLSTM[2]	—	71.81	91.87	94.41	56.75
TENER[12]	—	72.82	93.01	95.25	58.39
Lattice LSTM[45]	YJ	73.88	93.18	94.46	58.79
CNNR[21]	YJ	74.45	93.71	95.11	59.92
LGN[47]	YJ	74.85	93.63	95.41	60.15
FLAT	YJ	76.45	94.12	95.45	60.32
FLAT_msm	YJ	73.39	93.11	95.03	57.98
FLAT_mld	YJ	75.35	93.83	95.28	59.63
CGN[48]	LS	74.79	93.47	94.12	63.09
FLAT	LS	75.70	94.35	94.93	63.42

签修正框架[23-24]和 Seq2seq 框架[48-49]。尽管这些方法可以对更长的标签依赖关系进行建模，但它们容易受到错误传播的影响，即如果在推理过程中错误地预测了标签，则错误将被传播，且以此为条件的其他标签将受到影响[50]。因此，模型应选择性地更正更可能

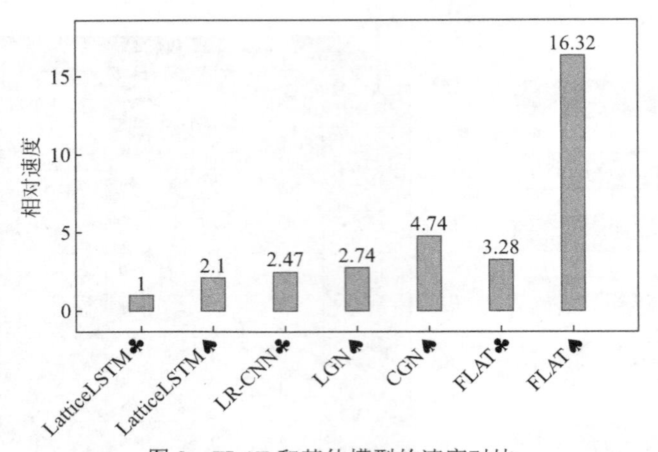

图 2 FLAT 和其他模型的速度对比

错误的标签，而不是全部标签。作为一种估计不确定性值的工具，贝叶斯神经网络[51]可以有效地指示草稿标签的不确定性。在 CoNLL2003 数据集上利用贝叶斯神经网络做了统计，对于草稿标签，那些被错误预测的标签的平均不确定性值比正确预测的标签的平均不确定性值大 29 倍。因此，通过设置不确定性阈值来仅修正可能不正确的标签，以此防止对正确标签产生负作用。

具体地，提出一种基于不确定性的两阶段标签修正网络 UANet（图 3）。在第一阶段，贝叶斯神经网络将句子作为输入，并产生所有的草稿标签及对应的不确定性数值。在第二阶段，为了充分利用已经预测的草稿标签，提出一个双流自注意力模型对单词到单词

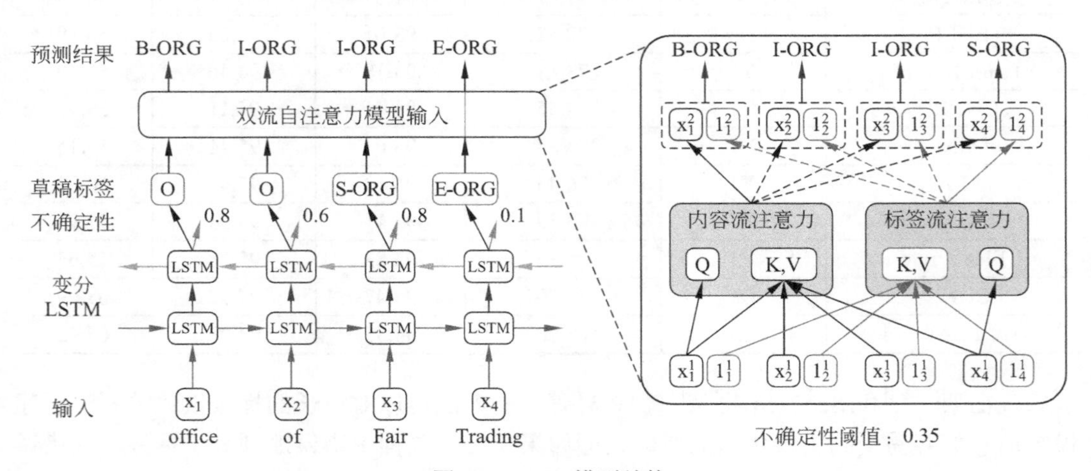

图 3 UANet 模型结构

的自注意力交互和单词到标签的自注意力交互进行建模，然后将两种形式的表示融合起来进行第二阶段的标签预测，以此对可能错误的草稿标签进行修正。以上的所有步骤都是可以并行处理的，相比维特比解码可以实现更快的预测。在三个序列标注的数据集上的实验结果显示，我们提出的模型不仅在预测准确性上超过了基于 CRF 的模型，而且可以极大地加快模型推断的速度。

在具体模型结构上，UANet 在第一阶段使用变分 LSTM 对输入序列进行预测，得到草稿标签及对应的不确定性。在不确定性的计算上，采用 Monte Carlo Dropout[51]先对贝叶斯网络进行采样：

$$\boldsymbol{p}_i(y=c|S,\mathcal{D})\approx\frac{1}{M}\sum_{j=1}^{M}\text{softmax}\left(\boldsymbol{h}_i\,|\,\boldsymbol{W}_j\right)$$

其中，M 表示采样次数；$\boldsymbol{W}_j\sim q_\theta^*(\boldsymbol{W})$；$q_\theta^*(\boldsymbol{W})$是用 Dropout 实现的提议分布。为了使模型多次采样操作后的时间与标准 LSTM 的时间接近，复制了同样的输入 M 并组成一个批次的输入数据，这样使多次采样的数据可以直接在 GPU 上并行。因此，M 次采样可以在前向计算时并发进行，最终导致与标准 Dropout 相似的运行时间。类似于经典的序列标注模型，以我们的模型输出概率最大的标签作为草稿标签 $y_i^*=\text{argmax}\left(\boldsymbol{p}_i\right)$，然后向量 $\boldsymbol{p}_i$ 的不确定性可以通过计算概率向量的熵得到：

$$u_i=H(\boldsymbol{p}_i)=-\sum_{c=1}^{C}p_c\lg p_c$$

其中，c 表示总类别数量。在第二阶段，基于草稿标签及原始序列，使用双流自注意模型对草稿标签进行修正。该模型的两个流的注意力权重分别计算如下：

$$\begin{aligned}\boldsymbol{A}_{i,j}^{x2x}&=\boldsymbol{E}_{x_i}^{\mathrm{T}}\boldsymbol{W}_{qx}^{\mathrm{T}}\boldsymbol{W}_{kx}\boldsymbol{E}_{x_j}+\boldsymbol{E}_{x_i}^{\mathrm{T}}\boldsymbol{W}_{qx}^{\mathrm{T}}\boldsymbol{W}_{kR}\boldsymbol{R}_{i-j}+\boldsymbol{u}_x^{\mathrm{T}}\boldsymbol{W}_{kx}\boldsymbol{E}_{x_j}+\boldsymbol{v}_x^{\mathrm{T}}\boldsymbol{W}_{kR}\boldsymbol{R}_{i-j}\\\boldsymbol{A}_{i,m}^{x2l}&=\boldsymbol{E}_{x_i}^{\mathrm{T}}\boldsymbol{W}_{ql}^{\mathrm{T}}\boldsymbol{W}_{kl}\boldsymbol{E}_{y_m^*}+\boldsymbol{E}_{x_i}^{\mathrm{T}}\boldsymbol{W}_{ql}^{\mathrm{T}}\boldsymbol{W}_{kR}\boldsymbol{R}_{i-m}+\boldsymbol{u}_l^{\mathrm{T}}\boldsymbol{W}_{kl}\boldsymbol{E}_{y_m^*}+\boldsymbol{v}_l^{\mathrm{T}}\boldsymbol{W}_{kR}\boldsymbol{R}_{i-m}\end{aligned}$$

其中，第一项表示第 i 个词对第 j 个词的注意力权重，$\boldsymbol{E}$ 为词嵌入矩阵，$\boldsymbol{R}$ 为位置编码矩阵；第二项表示第 i 个词对第 m 个词的草稿标签的注意力权重。一层中的一个注意力头的具体计算过程表示如下：

$$\begin{aligned}\boldsymbol{V}_x&=\boldsymbol{E}_x\boldsymbol{W}_x,\boldsymbol{a}_x=\text{Softmax}\left(\boldsymbol{A}^{x2x}\right)\boldsymbol{V}_x\\\boldsymbol{V}_l&=\boldsymbol{E}_{y^*}\boldsymbol{W}_l,\boldsymbol{a}_l=\text{Softmax}\left(\boldsymbol{A}^{x2l}\right)\boldsymbol{V}_l\\\boldsymbol{o}_x&=\text{LayerNorm}\left(\text{Linear}\left(\boldsymbol{a}_x\right)+\boldsymbol{E}_x\right)\end{aligned}$$

$$\boldsymbol{o}_l = \text{LayerNorm}\left(\text{Linear}\left(\boldsymbol{a}_l\right) + \boldsymbol{E}_{y^*}\right)$$
$$\boldsymbol{H}_x = \text{FeedForward}\left(\boldsymbol{o}_x\right)$$
$$\boldsymbol{H}_l = \text{FeedForward}\left(\boldsymbol{o}_l\right)$$

在最终的标签整合阶段，通过一个预设的不确定性阈值，只对草稿标签中不确定性值大于该阈值的标签进行修正，并保持其余草稿标签不变，以此来减轻第一阶段不正确的草稿标签对正确的草稿标签产生的负面影响。具体来说，当训练结束时，可以从变分 LSTM 模块上获得以下草稿标签 $Y^* = \left\{y_1^*, y_2^*, \cdots, y_n^*\right\}$ 和其对应的不确定性 $U = \left\{u_1, u_2, \cdots, u_n\right\}$，以及从双流自注意力模块上获得的修正后的标签 $\hat{Y} = \left\{\hat{y}_1, \hat{y}_2, \cdots, \hat{y}_n\right\}$。为了避免正确的标签被错误地修正，设置了一个不确定性阈值 $\varGamma$ 来区分哪些标签应该被修正。当 $u_i > \varGamma$ 时，使用修正后的标签；当 $u_i < \varGamma$ 时，使用草稿标签。举例来说，假设 $u_1 > \varGamma, u_2 > \varGamma, u_n > \varGamma$，则最终的解码标签为 $\left\{\hat{y}_1, y_2^*, \cdots, \hat{y}_n\right\}$。

对比基线模型的实验结果，发现提出的模型取得了更好的预测准确率（表 5），而且具有明显的速度优势（表 6）。BiLSTM-UANet 在 CoNLL2003，OntoNotes 和 WSJ 数据集上分别能每秒处理 1630，1262 和 1192 个句子，这比 BiLSTM-CRF 模型在 3 个数据集上的处理速度分别快了 13.7%，32.8%和 48.8%。还可以看到，对于句子平均长度越长的数据集，我们的模型的优势会更加明显。另外，由于我们的模型通过并行多次采样同样

表 5　3 个序列标注数据集上的实验结果

模型	CoNLL2003	Ontonotes	WSJ
Chiu 等人（2016 年）	90.91	86.28	—
Strubell 等人（2017 年）	90.54	86.84	—
Liu 等人（2018 年）	91.24	—	97.53
Chen 等人（2019 年）	91.44	87.67	—
BiLSTM-CRF[41]	91.21	86.99	97.51
BiLSTM-Softmax[52]	90.77	83.76	97.51
BiLSTM-Seq2seq[49]	91.22	—	97.59
Rel-Transformer[53]	90.70	87.45	97.49
BiLSTM-LAN[23]	90.77	88.16	97.58
BiLSTM-UANet($M = 8$)	91.60	88.39	97.62

表 6 不同模型推断速度的比较

	CoNLL2003	Ontonotes	WSJ
Average Sentence Length	13	18	24
BiLSTM-CRF[41]	1433	950	801
BiLSTM-LAN[23]	949	773	943
BiLSTM-Seq2seq[49]	1084	842	751
BiLSTM-UANet($M=1$)	1630	1262	1192
BiLSTM-UANet($M=8$)	1474	1129	1044

的输入形成一个批次的数据来计算不确定性，所以当采样次数 $M=8$ 时，我们的模型最终的推断速度也没有受到较大影响。

3.4 融合实体知识的预训练

为深度融合实体知识和预训练语言模型，提出上下文化的语言和知识的联合预训练模型 CoLAKE[54]，以更好地融入结构化知识。之前的工作一般仅将所需的实体表示向量融入预训练语言模型中，未考虑实体在知识图谱中的上下文。CoLAKE 在融入对应实体知识时，不仅考虑文本中提及的实体，还考虑其在知识图谱中的邻接三元组，将其以知识子图的形式与文本融合，得到一种图形式的数据结构作为模型输入。

具体地，首先设计了一种新型的数据结构，将文本与三元组形式的知识整合在一张异质图中，称为单词-知识图谱（word-knowledge graph）。该图谱的构建方式如图 4 所示，

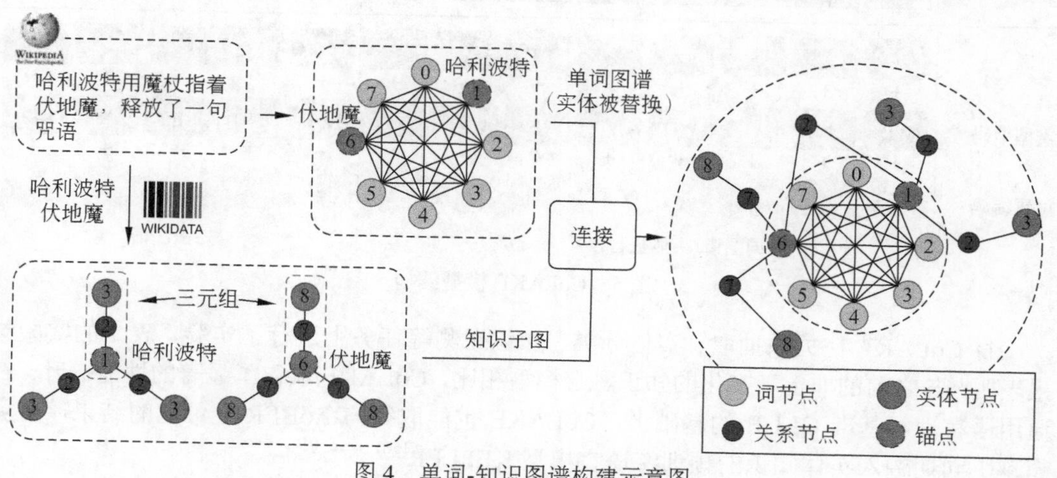

图 4 单词-知识图谱构建示意图

对于一个句子，首先通过实体链接（entity linking）将其中提及的实体链接到外部知识图谱中，抓取这些实体的邻接三元组得到知识子图，接着将这些知识子图以文本中提及的实体为锚点连接到词图上组成单词-知识图谱。

为使 Transformer 编码器[16]适配这种输入数据结构，对 Transformer 编码器进行了如下三点修改：

（1）将绝对位置编码改为软位置编码（soft position embedding）[55]，给插入的实体和关系分配合适的位置索引；

（2）添加节点类型编码，包含单词、实体、关系三种类型，增强模型表达能力；

（3）使用掩码注意力机制根据图的邻接矩阵控制信息流。

修改后的编码器如图 5 所示。最后，在输入的图结构上进行掩码语言模型（MLM）[36]训练，学习上下文化的文本、实体和关系的统一语义表示。

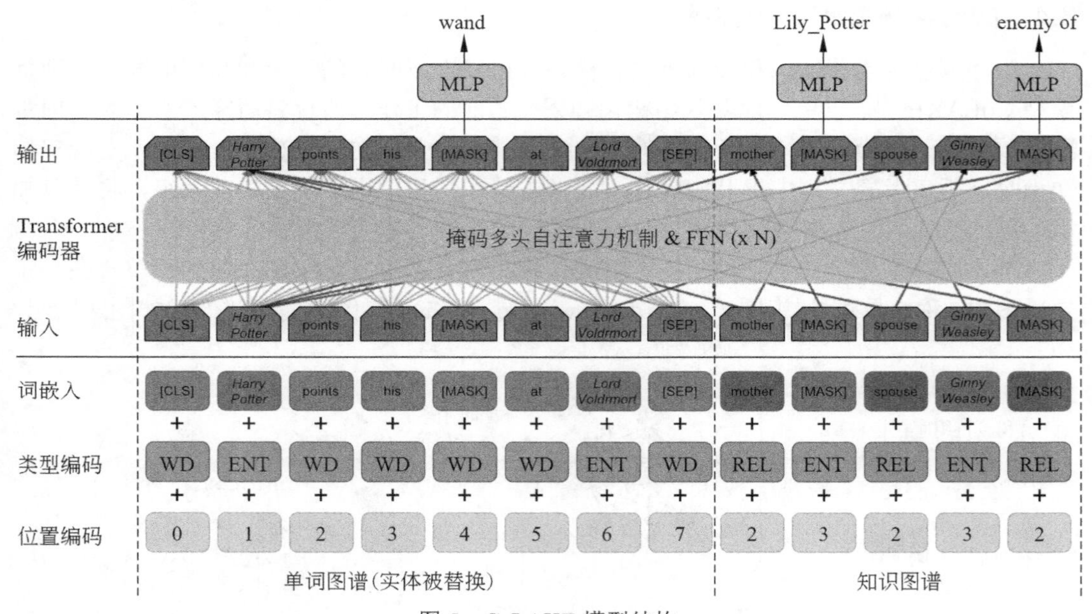

图 5　CoLAKE 模型结构

将 CoLAKE 在关系抽取、实体分类、知识探测等任务上进行了实验，表 7 的实验结果表明，与之前的非上下文化的知识融合模型相比，CoLAKE 取得了显著的性能提升。在通用语言理解基准 GLUE 的基础上，CoLAKE 也能取得与 RoBERTa 可比的结果，显著超越了同期融入实体知识的预训练语言模型 KEPLER[56]。

表 7 实体分类和关系抽取任务上的实验结果

模型	Open Entity			FewRel		
	P	R	F	P	R	F
Bert[36]	76.4	71.0	73.6	85.0	85.1	84.9
RoBERTa[37]	77.4	73.6	75.4	85.4	85.4	85.3
ERNIE[39]	78.4	72.9	75.6	88.5	88.4	88.3
KnowBERT[40]	78.6	73.7	76.1	—	—	—
KEPLER[56]	77.8	74.6	76.2	—	—	—
E-BERT[57]	—	—	—	88.6	88.5	88.5
CoLAKE	77.0	75.7	76.4	90.6	90.6	90.5

4 总结和展望

基于深度学习的命名实体识别已经取得了极大的成功，但仍然存在一些待解决的问题。本文对基于深度学习的 NER 的背景及现有工作进行了简单梳理，并着重介绍了针对其存在的四个挑战性问题开展的研究工作。具体而言，针对未登录词问题，设计了基于 Student-Teacher 的学习范式来训练未登录词表示预测网络，有效提高了未登录词的识别率；针对语义依赖问题和中文命名实体识别的边界歧义问题，设计了基于 Transformer 的 TENER 及 FLAT 模型，全方位优化了现有编码器模型的效率并对长距离依赖及中文 NER 中的词信息编码进行建模；针对标签解码问题，设计了基于不确定度量的两阶段的解码，提升了解码效率的同时优化了标签依赖的建模；最后，设计了一种深度融合实体知识和预训练语言模型的预训练方式，通过更好地融合实体知识提升了预训练语言模型在命名实体识别任务上的表现。

以上研究仅仅是一些初步的尝试。随着深度命名实体识别的发展与应用，实际任务中涉及的许多问题仍有待解决。首先，命名实体的识别涉及对世界知识的有效利用与建模。语言作为对世界知识的一种描述，在描述实体时仍存在大量的信息损失，单纯靠语言知识、数据驱动的方法在很多情况下可能无法解决特定语境下的实体识别问题。因此，如何对世界知识这一符号系统进行有效的表示并融入到深度学习模型中是一个值得思考的问题；其次，随着互联网的发展，社交媒体上的文本包含大量有价值的信息，然而现有模型在获取短文本和非规范文本的语义表示上仍有不足。同时，图片、视频、音频等多模态信息往往伴随着文本信息一起出现，这些多模态信息可以成为建模文本语义的有效补充；最后，嵌套实体、非连续实体在医学、法律等领域广泛存在，序列标注作为命名实体识别的常用框架，在嵌套实体和非连续实体的标注上存在很大困难，设计一套统一的标签解码方案来有效解决也是未来值得探索的问题。

参考文献

[1] UGAWA A, TAMURA A, NINOMIYA T, et al. Neural machine translation incorporating named entity[C]//Proceedings of the 27th International Conference on Computational Linguistics. 2018: 3240-3250.

[2] HUANG Z, XU W, YU K. Bidirectional lstm-crf models for sequence tagging[Z]. arXiv preprintarXiv:1508.01991，2015.

[3] ZHOU J T, ZHANG H, JIN D, et al. Dual adversarial neural transfer for low-resource named entity recognition[C]//Proceedings of the 57th Annual Meeting of the Association for Computational Linguistics. 2019: 3461-3471.

[4] YAN H, DENG B, LI X, et al. Tener: Adapting transformer encoder for named entity recognition[Z]. arXiv preprint arXiv:1911.04474, 2019.

[5] PENNINGTON J, SOCHER R, MANNING C D. Glove: Global vectors for word representation[C]// Proceedings of the 2014 conference on empirical methods in natural language processing (EMNLP). 2014:1532-1543.

[6] MIKOLOV T, CHEN K, CORRADO G, et al. Efficient estimation of word representations in vector space[Z]. arXiv preprint arXiv:1301.3781, 2013.

[7] BOJANOWSKI P , GRAVE E , JOULIN A, et al. Enriching word vectors with subword information[J]. Transactions of the Association for Computational Linguistics, 2017, 5: 135-146.

[8] SANTOS C N, GUIMARAES V. Boosting named entity recognition with neural character embeddings[Z]. arXiv preprint arXiv:1505.05008, 2015.

[9] LING W, LUÍS T, MARUJO L, et al. Finding function in form: Compositional character models for open vocabulary word representation[Z]. arXiv preprint arXiv:1508.02096, 2015.

[10] YANG Z, SALAKHUTDINOV R, COHEN W. Multi-task cross-lingual sequence tagging from scratch[Z]. arXiv preprint arXiv:1603.06270, 2016.

[11] CHIU J P C, NICHOLS E. Named entity recognition with bidirectional LSTM-CNNs[J]. Transactions of the Association for Computational Linguistics, 2016，4: 357-370.

[12] LAMPLE G, BALLESTEROS M, SUBRAMANIAN S, et al. Neural architectures for named entity recognition[Z]. arXiv preprint arXiv:1603.01360, 2016.

[13] HUANG Z, XU W, YU K. Bidirectional LSTM-CRF models for sequence tagging[Z]. arXiv preprint arXiv:1508.01991, 2015.

[14] STRUBELL E, VERGA P, BELANGER D, et al. Fast and accurate entity recognition with iterated dilated convolutions[J]. arXiv preprint arXiv:1702.02098, 2017.

[15] CHEN H, LIN Z, DING G, et al. GRN: Gated relation network to enhance convolutional neural network for named entity recognition[C]//Proceedings of the AAAI Conference on Artificial Intelligence. 2019, 33(1): 6236-6243.

[16] VASWANI A, SHAZEER N, PARMAR N, et al. Attention is all you need[J]. arXiv preprint arXiv:1706.03762, 2017.

[17] GUO Q, QIU X, LIU P, et al. Star-transformer[Z]. arXiv preprint arXiv:1902.09113, 2019.

[18] Liu W, Xu T, Xu Q, et al. An encoding strategy based word-character LSTM for Chinese NER[C]//Proceedings of the 2019 Conference of the North American Chapter of the Association for Computational Linguistics: Human Language Technologies. 2019: 2379-2389.

[19] PENG M, MA R, ZHANG Q, et al. Simplify the usage of lexicon in Chinese NER[Z]. arXiv preprint arXiv:1908.05969, 2019.

[20] ZHANG Y, YANG J. Chinese NER using lattice LSTM[Z]. arXiv preprint arXiv:1805.02023, 2018.

[21] GUI T, MA R, ZHANG Q, et al. CNN-based Chinese NER with lexicon rethinking[J]. IJCAI, 2019: 4982-4988.

[22] RATINOV L, ROTH D. Design challenges and misconceptions in named entity recognition[C]//Proceedings of the 13th Conference on Computational Natural Language Learning (CoNLL-2009). 2009: 147- 155.

[23] CUI L, ZHANG Y. Hierarchically-refined label attention network for sequence labeling[Z]. arXiv preprint arXiv:1908.08676, 2019.

[24] KRISHNAN V, MANNING C D. An effective two-stage model for exploiting non-local dependencies in named entity recognition[C]//Proceedings of the 21st International Conference on Computational Linguistics and 44th Annual Meeting of the Association for Computational Linguistics. 2006: 1121-1128.

[25] WANG Y, LI Y, ZHU Z, et al. SC-NER: A sequence-to-sequence model with sentence classification for named entity recognition[C]//Proceedings of Pacific-Asia Conference on Knowledge Discovery and Data Mining. 2019: 198-209.

[26] STRAKOVÁ J, STRAKA M, HAJIČ J. Neural architectures for nested NER through linearization[Z]. arXiv preprint arXiv:1908.06926, 2019.

[27] GUI T, YE J, ZHANG Q, et al. Leveraging document-level label consistency for named entity recognition[C]//Proceedings of the 29th International Joint Conference on Artificial Intelligence. IJCAI, 2020: 3976-3982.

[28] LI X, FENG J, MENG Y, et al. A unified mrc framework for named entity recognition[Z]. arXiv preprint arXiv:1910.11476, 2019.

[29] YU J, BOHNET B, POESIO M. Named entity recognition as dependency parsing[Z]. arXiv preprint arXiv:2005.07150, 2020.

[30] JUE W, SHOU L, CHEN K, et al. Pyramid: A layered model for nested named entity recognition[C]//

Proceedings of the 58th Annual Meeting of the Association for Computational Linguistics. 2020: 5918-5928.
[31] XU M, JIANG H, WATCHARAWITTAYAKUL S. A local detection approach for named entity recognition and mention detection[C]//Proceedings of the 55th Annual Meeting of the Association for Computational Linguistics. 2017: 1237-1247.
[32] LUAN Y, WADDEN D, HE L, et al. A general framework for information extraction using dynamic span graphs[Z]. arXiv preprint arXiv:1904.03296, 2019.
[33] PETERS M E, NEUMANN M, IYYER M, et al. Deep contextualized word representations[Z]. arXiv preprint arXiv:1802.05365, 2018.
[34] AKBIK A, BLYTHE D, VOLLGRAF R. Contextual string embeddings for sequence labeling[C]// Proceedings of the 27th international conference on computational linguistics. 2018：1638-1649.
[35] AKBIK A, BERGMANN T, VOLLGRAF R. Pooled contextualized embeddings for named entity recognition[C]//Proceedings of the 2019 Conference of the North American Chapter of the Association for Computational Linguistics: Human Language Technologies. 2019: 724-728.
[36] DEVLIN J, CHANG M W, LEE K, et al. Bert: Pre-training of deep bidirectional transformers for language understanding[Z]. arXiv preprint arXiv:1810.04805, 2018.
[37] LIU Y, OTT M, GOYAL N, et al. Roberta: A robustly optimized bertpretrainingapproach[Z]. arXiv preprint arXiv:1907.11692, 2019.
[38] BAEVSKI A, EDUNOV S, LIU Y, et al. Cloze-driven pretraining of self-attention networks[Z]. arXiv preprint arXiv:1903.07785, 2019.
[39] ZHANG Z, HAN X, LIU Z, et al. ERNIE: Enhanced language representation with informative entities[Z]. arXiv preprint arXiv:1905.07129, 2019.
[40] PETERS M E, NEUMANN M, LOGAN IV R L, et al. Knowledge enhanced contextual word representations[Z]. arXiv preprint arXiv:1909.04164, 2019.
[41] MA X, HOVY E. End-to-end sequence labeling via bi-directional lstm-cnns-crf[Z]. arXiv preprint arXiv:1603.01354, 2016.
[42] MADHYASTHA P S, BANSAL M, GIMPEL K, et al. Mapping unseen words to task-trained embedding spaces[Z]. arXiv preprint arXiv:1510.02387, 2015.
[43] FIELDS C R. Conditional random fields: Probabilistic models for segmenting and labeling sequence data[C]//Proceedings of ICML, 2001.
[44] KHODAK M, SAUNSHI N, LIANG Y, et al. A la carte embedding: Cheap but effective induction of semantic feature vectors[Z]. arXiv preprint arXiv:1805.05388, 2018.
[45] LAZARIDOU A, MARELLI M, BARONI M. Multimodal word meaning induction from minimal exposure to natural text[J]. Cognitive science, 2017，41: 677-705,.
[46] SCHICK T, SCHÜTZE H. Learning semantic representations for novel words: Leveraging both form and

context[C]//Proceedings of the AAAI Conference on Artificial Intelligence. 2019，33(01): 6965-6973.
[47] GUI T, ZOU Y, ZHANG Q, et al. A lexicon-based graph neural network for chinesener[C]//Proceedings of the 2019 Conference on Empirical Methods in Natural Language Processing and the 9th International Joint Conference on Natural Language Processing (EMNLP-IJCNLP). 2019：1039-1049.
[48] VASWANI A, BISK Y, SAGAE K, et al. Supertagging with lstms[C]//Proceedings of the 2016 Conference of the North American Chapter of the Association for Computational Linguistics: Human Language Technologies. 2016: 232-237.
[49] ZHANG Y, CHEN H, ZHAO Y, et al. Learning tag dependencies for sequence tagging[J]. IJCAI, 2018: 4581-4587.
[50] BENGIO S, VINYALS O, JAITLY N, et al. Scheduled sampling for sequence prediction with recurrent neural networks[Z]. arXiv preprint arXiv:1506.03099, 2015.
[51] GAL Y, GHAHRAMANI Z. Dropout as a bayesian approximation: Representing model uncertainty in deep learning. international conference on machine learning[J]. PMLR, 2017: 1050-1059.
[52] YANG J, LIANG S L, ZHANG Y. Design challenges and misconceptions in neuralsequence labeling[C]// Proceedings of the 27th Inter-national Conference on Computational Linguistics. 2018: 3879-3889.
[53] DAI Z H, YANG Z L, YANG Y M, et al. Transformer-XL: Attentive language modelsbeyond a fixed-length context[C]//Proceedings of the 57th Annual Meeting of the Associationfor Computational Linguistics. 2019: 2978-2988.
[54] SUN T, SHAO Y, QIU X, et al. CoLAKE: Contextualized language and knowledge embedding[Z]. arXiv preprint arXiv:2010.00309, 2020.
[55] LIU W, ZHOU P, ZHAO Z, et al. K-bert: Enabling language representation with knowledge graph[C]// Proceedings of the AAAI Conference on Artificial Intelligence. 2020, 34(3): 2901-2908.
[56] WANG X, GAO T, ZHU Z, et al. KEPLER: A unified model for knowledge embedding and pre-trained language representation[Z]. arXiv preprint arXiv:1911.06136, 2019.
[57] NINA P, ULLI W, HINRICH S. BERT is not a knowledge base (yet): Factual knowledge vs. name-based reasoning in unsupervised QA[J]. CoRR, abs/1911.03681, 2019.

从 Transformer 到 BERT：自然语言表示学习的新进展

邱锡鹏

（复旦大学计算机学院，上海 200433）

1 引言

随着深度学习的发展，各类神经网络模型开始被广泛用于解决自然语言处理（natural language processing，NLP）任务，比如卷积神经网络（convolutional neural networks, CNN）、循环神经网络（recurrent neural network，RNN）、图神经网络（graph neural networks，GNN）和注意力机制（attention mechanism）等。在传统的非神经网络 NLP 模型中，模型性能通常过于依赖手工设计或选择文本特征，因此训练一个高性能的 NLP 模型通常开发周期较长。而神经网络模型的优势是可以大幅缓解特征工程问题，通过使用在特定 NLP 任务中学习的低维稠密向量（分布式表示）隐式地表示文本的句法和语义特征。因此，神经网络方法简化了开发各类 NLP 系统的难度。

尽管神经网络模型在 NLP 任务中已取得较好的效果，但其相对于非神经网络模型的优势并没有像在计算机视觉领域中那么明显。该现象的主要原因可归结于当前 NLP 任务的数据集相对较小（除机器翻译任务）。深度神经网络模型通常包含大量参数，因此在较小规模的训练集中易过拟合，且泛化性较差。因此，用于 NLP 任务的早期网络模型相对较浅，且通常仅由一至三层网络组成。

近期有大量工作表明，通过海量无标注语料来预训练神经网络模型可以学习到有益于下游 NLP 任务的通用语言表示，并可避免从零训练新模型。预训练模型一直被视为一种训练深度神经网络模型的高效策略。早在 2006 年，深度学习的突破便来自逐层无监督预训练+微调方式。在计算机视觉领域，已出现在大型 ImageNet 数据集上训练的预训练模型并进一步在不同任务的小数据集上微调的范式。该方法比随机初始化参数模型更有效，因为模型能学到可被用于不同视觉任务上的通用图像特征。在 NLP 领域，无论是浅层词嵌入还是深层网络模型，大规模语料下的预训练语言模型均被证实对下游 NLP 任务有益。随着算力的快速发展、以 Transformer 为代表的深度模型的不断涌现及训练技巧的逐步提升，预训练模型（pre-trained models，PTMs）[1]也由浅至深。自然语言处理中预训练模型的发展可以分为两个阶段。第一代预训练模型着力于学习词嵌入（也称为词向量），而用于学习词向量的模型本身在下游任务中已不需要使用。因此基于计算效率考量，这类模型通常比较简单，如 CBOW，Skip-Gram 和 GloVe 等。尽管这些预训练词嵌入可捕获词语语义，但是它们是静态的，无法表示与上下文相关的信息，比如一词多义、句法结构、语义角色和共指等。第二代预训练模型聚焦于学习上下文相关的词嵌入，如上下文向量（context vector，CoVe）[2]，ELMo（embedding from language model）[3]，OpenAI GPT（generative pre-training）[4]和 BERT（bidirectional encoder representation from transformer）[5]等。该类预训练模型会直接在下游任务中使用来表示文本的上下文特征。

2 背景介绍

2.1 语言表示学习

如 Bengio 所述[6]，一个好的表示（特征）应该能够表达与任务无关的通用先验，且可能对待解决的下游任务提供帮助。就语言而言，好的表示应捕获隐藏于文本中的隐式语言规则和常识，如词汇意义、句法结构、语义角色甚至语用等。使用低维稠密向量来表示文本意义是分布式表示的核心思想。向量的每个维度无特定意义，联合起来却表示一个具体的概念。图 1 展示了 NLP 中的通用神经网络架构，其中包含两种词嵌入：非上下文嵌入和上下文词嵌入，而两者的区别在于词嵌入是否根据出现的上下文发生动态变化。

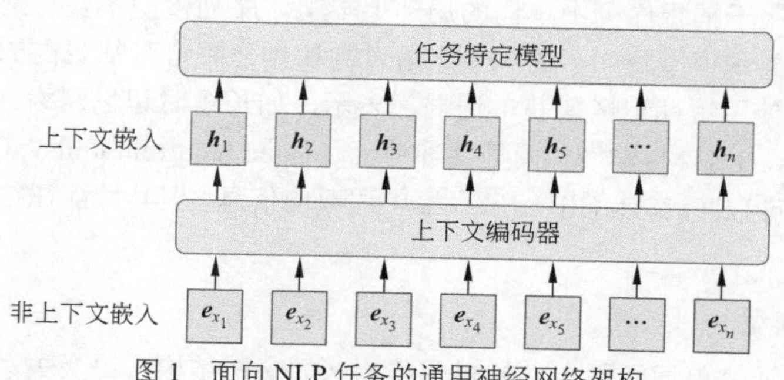

图 1　面向 NLP 任务的通用神经网络架构

（1）非上下文嵌入：语言表示的第一步是将离散的语言符号映射至一个分布式嵌入空间中，并将得到的嵌入与模型参数在任务数据中共同训练。然而，该类嵌入存在两个主要缺陷。第一个是该类词嵌入在任意上下文中保持相同，故无法建模多义词。第二个是未登录词问题。为缓解这些问题，在各类 NLP 任务中广泛运用了字符级别或子词级别的表示，如字符级卷积神经网络（CharCNN）[7]、FastText[8]和字节对编码（byte pair encoder，BPE）[9]等。

（2）上下文嵌入：为处理多义词问题并考虑词语的上下文依赖性质，需区分词语在不同上下文中的语义。通常需要依赖一个好的上下文编码器。

2.2 上下文编码器

大多数上下文编码器可被分为两大类：序列模型和非序列模型，如图 2 所示。

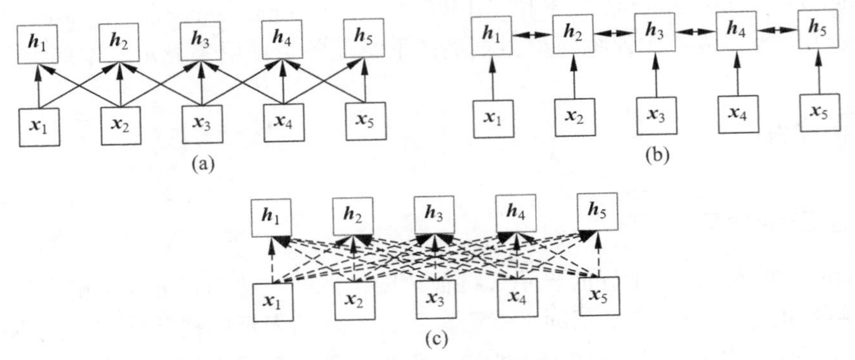

图 2　神经上下文编码器

（a）卷积模型；（b）循环模型；（c）全连接自注意力模型

2.2.1 序列模型

序列模型常用于捕捉序列中单词的局部上下文。序列模型主要有两种模型：卷积模型和循环模型。卷积模型将句子的词嵌入通过卷积操作聚合其邻域内的局部信息来捕获词义。循环模型使用短期记忆捕获单词局部表示，如长短时记忆网络（long short term memory networks，LSTMs）[10]和门控循环单元（gated recurrent units，GRUs）[11]。在实践中，常使用双向 LSTMs 或 GRUs 来聚集单词双向信息，但其性能往往受限于长距离依赖问题。

2.2.2 非序列模型

非序列模型通过单词间预定义的树图结构（如句法结构或语义关系等）学习上下文

表示。部分流行的非序列模型包括递归网络、Tree-LSTM 模型[12]和图神经网络（graph neural network，GNN）[13]等。尽管这些图结构可以刻画语言的先验知识，然而如何构建合适的图结构仍具有挑战性。此外，这些图结构通常依赖于专家知识或句法解析器等外部 NLP 工具。

在实际中，使用全连接图对每两个词之间的关系进行建模，从而使模型自己学到词之间的依赖关系是一种更加直接的方式。Transformer[1]便是全连接自注意力模型中的一个典型成功样例。Transformer 通过自注意力机制动态计算连接权重，从而隐式对词间连接进行建模，并通过位置编码、层归一化、残差连接和逐位前馈网络层等组件来构建一个非常深的网络。

2.2.3 分析

使用序列模型学习词语的上下文表示具有局部性，因而难以捕获词与词之间的长距离交互。但由于序列模型通常易于训练并有较好的效果，因此也在各类 NLP 任务中大量使用。

Transformer 作为一个全连接自注意力模型的具体实现，由于可直接对序列中任意两词之间的依赖关系进行建模，从而更适用于对语言的长距离交互进行建模。然而，由于其结构较复杂且模型约束较小，Transformer 通常需要大量训练语料，且易在中小规模数据集上过拟合。当有大量数据进行训练时，Transformer 会超过其他模型，因此 Transformer 已成为当前预训练模型的主流架构。

2.3 预训练重要性

随着深度学习的发展，模型参数显著增长，从而需要越来越大的数据集用于充分训练模型参数并预防过拟合。然而，因大部分 NLP 任务的标注成本极为高昂，尤其是句法和语义相关任务，构建大规模标注数据集尤为困难。

相比较而言，大规模无标注数据集相对易于构建。为更好地利用海量无标签文本数据，可以首先从这些数据中学到较好的文本表示，然后再将其用于其他任务。许多研究已表明，在大规模无标注语料中训练的预训练语言模型得到的表示可以使许多 NLP 任务获得显著的性能提升。预训练的优势可总结为以下几点：

（1）在海量文本中通过预训练可以学习到一种通用语言表示，并有助于完成下游任务。

（2）预训练可提供更好的模型初始化，从而具有更好的泛化性并在下游任务上更快收敛。

（3）预训练可被看作是在小数据集上避免过拟合的一种正则化方法。

3 预训练模型概述

本节将重点介绍预训练任务，并给出预训练模型的分类方法。

3.1 预训练任务

预训练任务对于学习语言的通用表示至关重要。本节将预训练任务分为三类：监督学习、无监督学习及自监督学习。

（1）监督学习（supervised learning，SL）通过学习一个函数，根据输入-输出对组成的训练数据将输入映射至输出。

（2）无监督学习（unsupervised learning，UL）从无标记数据中寻找一些内在知识，如簇、密度、潜在表示等。

（3）自监督学习（self-supervised learning，SSL）介于监督学习和无监督学习之间，其学习范式与监督学习相同，而训练数据标签自动生成。自监督学习的关键思想是通过输入的一部分信息来预测其他部分信息。例如，掩码语言模型（masked language model，MLM）[5]是一种自我监督的任务，就是将句子中的某些词删掉，并通过剩下的其他词来预测这些被删掉的词。

下面介绍一些现有预训练模型中被广泛使用的预训练任务。

3.1.1 语言模型

NLP 中最常见的无监督任务为概率语言模型（language model，LM），是一个经典的概率密度估计问题。在实践中语言模型通常特指自回归语言模型或单向语言模型。单向语言模型的缺点在于每个词的表示仅对包括自身的上文编码，而更好的文本上下文表示应从两个方向对上下文信息进行编码。一种改进策略为双向语言模型[3]，即由一个前向语言模型和一个反向语言模型组成。

3.1.2 掩码语言模型

掩码语言模型（masked language model，MLM）首先将输入句子的一些词替换为[MASK]，然后训练模型通过剩余词来预测被替换的词。然而，在实际使用时句子中并不会出现[MASK]，导致这种预训练方法在预训练阶段和微调阶段之间存在不一致。为缓解该现象，Devlin 等人[5]在每个序列中替换 15%的词块，其中每个词以 80%的概率使用[MASK]替换，10%的概率替换为随机词，10%的概率保持不变。

通常情况下，MLM 可以作为分类任务处理，也可以使用编码器-解码器架构处理，即在编码器中输入掩码序列，接着解码器以自回归形式预测出被掩码的词。同时，也存在部分强化掩码语言模型，如 RoBERTa[14]的动态掩码。UniLM[15]使用单向、双向和序列到序列预测的语言模型任务拓展掩码预测任务。

3.1.3 乱序语言模型

为缓解 MLM 中因[MASK]导致的预训练和微调阶段的差异问题，有研究者提出使用乱序语言模型（permuted language modeling，PLM）[16]替代 MLM。PLM 是一种基于输入序列随机排列的语言建模任务。该任务从所有可能的排列中随机抽取一种排列，接着选择其中某些词为目标，训练模型依据剩下的词块和位置信息预测这些词。该置换方式并不影响序列的自然顺序，仅定义了词块预测的顺序。

3.1.4 降噪自编码器

降噪自编码器（denoised auto-encoder，DAE）是将部分损坏的序列作为输入并恢复原始的序列。就语言来说，常使用如标准 Transformer 等序列至序列模型重构原始输入文本。常见的损坏文本方式有词屏蔽、词删除、文本填充、句序打乱和文档内容置换等。

3.1.5 对比学习

对比学习（contrastive learning，CTL）[17]假设一些观察到的成对文本在语义上比随机抽样的文本更相似。CTL 背后的理念是“通过比较来学习”。与语言模型相比，CTL 通常具有较少的计算复杂度，因此是预训练模型理想的训练方式。

Deep InfoMax（DIM）[18]最早在图像领域被提出，它通过最大化图像表示与图像局部区域之间的互信息来改进表示质量。替换标签检测（replaced token detection，RTD）与噪声对比估计（noise-contrastive estimation，NCE）相同，但前者根据其周围的上下文来预测某个词块是否被替换。ELECTRA[19]使用生成器替换序列中的部分词，并在预训练后抛弃生成器，而判别器用于判断每个词是否被替换。在下游任务上使用微调的判别器作为最终的模型。下一句预测任务（next sentence prediction，NSP）[5]训练模型区分输入的两个句子是否是训练预料中的两个连续片段。后续也有工作质疑 NSP 任务的有效性[14]。句序预测任务（sentence order prediction，SOP）[20]将同一个文档中的两个连续片段作为正例，将两个连续片段调换顺序作为反例。

3.2 预训练模型分类

为阐明 NLP 领域中现有预训练模型的关系，本文建立了预训练模型分类法，从四个不同的角度对现有预训练模型进行分类。图 3 展示了四种分类及对应的部分代表性预训练模型。

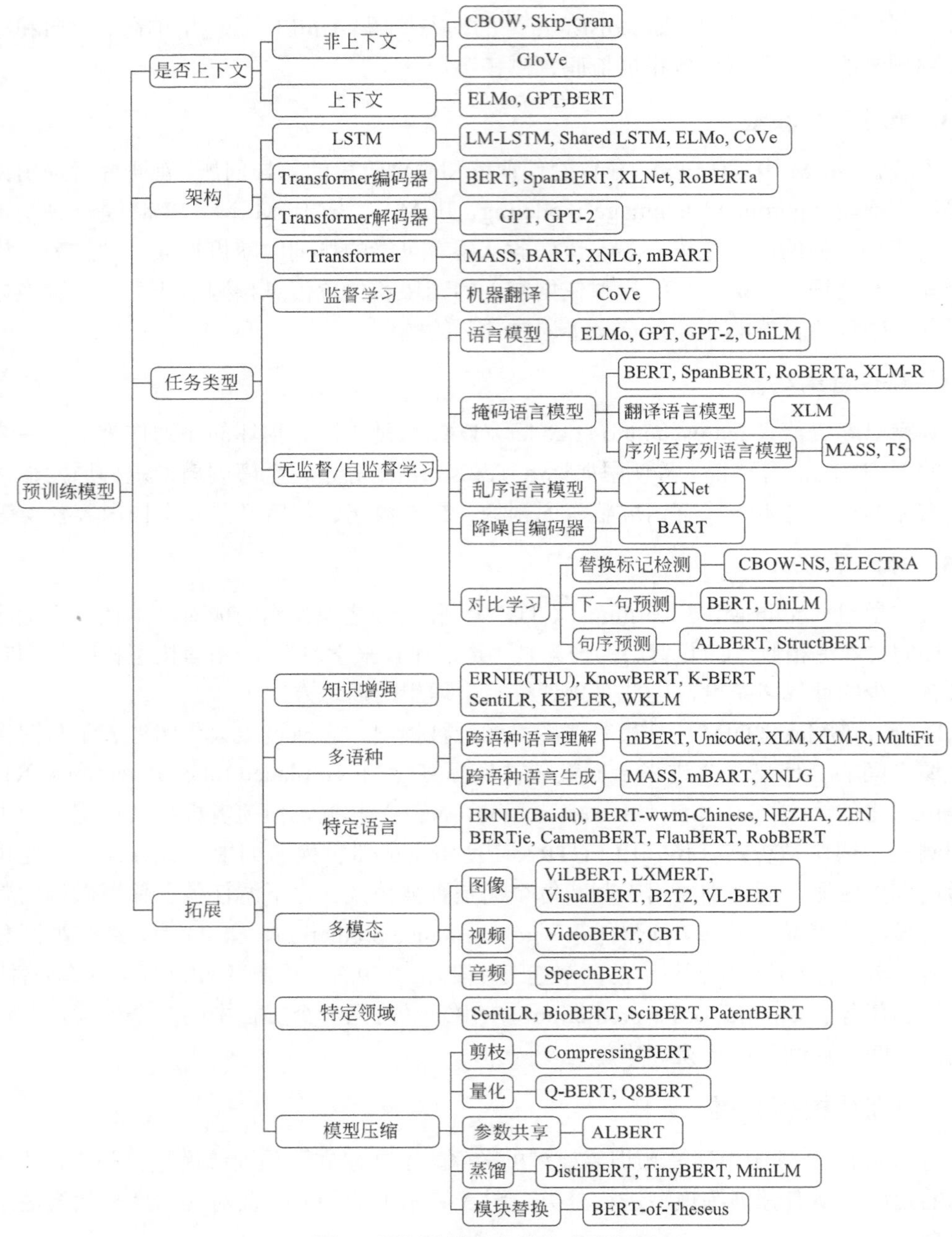

图 3　预训练模型分类与典型例子

（1）表示类型：根据用于下游任务的表示，可将预训练模型分为非上下文预训练模型和上下文预训练模型。

（2）架构：预训练模型使用的骨干网络，包括 LSTM、Transformer 编码器、Transformer 解码器和完整的 Transformer 架构。

（3）预训练任务类型：预训练模型使用的预训练任务类型。

（4）拓展：为各种场景设计的预训练模型，包括知识增强预训练模型、多语言或特定语言的预训练模型、多模态预训练模型、特定领域的预训练模型和预训练模型的压缩等。

4 预训练模型拓展

4.1 知识增强型预训练模型

预训练模型通常从通用大规模文本语料中学习通用语言表示，但缺乏特定的领域知识。对于下游任务，将外部知识库中的领域知识整合至预训练模型中是有效的。外部知识包括语言学、语义学、通识、事实和领域特定知识等。一方面，可在预训练时将外部知识注入模型。LIBERT[21]通过额外的语言约束任务整合语言知识。ERNIE（THU）[22]将知识图谱（knowledge graph，KG）上预训练的实体嵌入并与文本中提及的相应实体相结合，从而增强文本表示。KnowBERT[23]使用端到端范式联合训练 BERT 和实体链接模型来整合实体的表示。K-BERT[24]将从 KG 中提取的相关三元组显式地注入到句子中，得到 BERT 的扩展树形输入。另一方面，可将外部知识整合至预训练过的模型中，而不必从零开始对其进行再训练。例如，K-BERT 允许在对下游任务进行微调时注入事实性知识。

4.2 多语言预训练模型

在许多跨语言 NLP 任务中，跨语言共享的多语言文本表示起着重要作用。早期的研究大多集中在多语言词嵌入的学习上，在单一语义空间中表示多种语言的文本。然而，这些方法通常需要语言之间的弱对齐。mBERT（multilingual Bert）[5]使用 MLM 在 104 种语言的维基百科语料中进行了预训练。每个训练样本都是一个单语言文档，没有专门设计跨语言目标，也没有任何跨语言的数据。即便如此，mBERT 在跨语言泛化方面仍然表现出色。跨域预训练语言模型（cross-lingual language model，XLM）[25]通过翻译语言模型提升 mBERT 的性能，即在拼接的平行双语句子对中执行 MLM 任务。在跨语种语言生成中，屏蔽序列到序列预训练（masked sequence to sequence pre-training，MASS）[26]利用单语序列到序列 MLM 在多种语言上对模型进行预训练，并在无监督的机器翻译上得到了显著改进。

4.3 多模态预训练模型

由于预训练模型在许多任务上取得了显著成功，部分研究者已开始关注跨模态预训练模型。这些模型中的绝大多数为通用视觉和语言特征编码设计，并在大量跨模态数据语料库上进行预训练，如包含语音的视频或带有字幕的图像，并结合扩展预训练任务来充分利用多模态特征。通常，在多模态预训练中广泛使用了基于视觉的 MLM、掩码视觉特征建模、视觉-语言匹配等任务，如 VideoBERT[27]、VisualBERT[28]和 ViLBERT[29]等。

4.4 预训练模型压缩

通常情况下，由于预训练模型至少包含数亿参数，因此在实际应用中部署在在线服务和资源受限的设备上较为困难。模型压缩是一种潜在的减少模型尺寸和提高计算效率的方法。当前主要存在四种压缩预训练模型的方法：

（1）模型剪枝：移除影响较小的参数，如剪去 Transformer 中的部分层或自注意力头，对下游任务影响较小。

（2）权重量化：将参数从高精度压缩为低精度，但该操作通常需要特定的硬件设备支持。

（3）参数共享：使用跨层参数共享来减少预训练模型的参数。尽管模型参数量减少，但其训练和推理时间不会减少。

（4）知识蒸馏：使用蒸馏方法将大模型的知识蒸馏至相对较小的模型。部分研究工作甚至尝试将 Transformer 架构的参数蒸馏至循环神经网络或卷积神经网络架构中。

5 展望和总结

虽然预训练模型已在多种 NLP 任务中展现出强大的性能，但由于语言的复杂性，挑战仍然存在。本节对预训练模型的未来研究趋势的展望与总结如下。

（1）预训练模型的上界：当前的预训练模型还未达到其上界，大多数预训练模型仍然可通过更多的训练步长和更多语料得到进一步提升。NLP 中最先进的模型仍然能通过更深的堆砌获得性能的增长。研究人员仍然尝试使用预训练模型学习语言内在的通识知识甚至世界知识。然而该类预训练模型通常需要更深的架构、更大的语料和更具挑战的预训练任务，因而进一步提高了训练代价。同时，从工程实现角度来说，训练超大型预训练模型也需要更复杂和更高效的训练技巧，如分布式训练、混合精度和梯度累积等。因此，更实际的方向或许是设计更为高效的模型架构、自监督预训练任务、优化器及现有软硬件条件下的训练技巧。

（2）预训练模型的架构：Transformer 的一个主要弊端是其计算复杂度，而当前预训练模型因 GPU 显存限制，通常难以处理长度大于 512 的序列，突破这类限制须从改进 Transformer 的架构入手。因此，使用神经架构搜索等技术自动构建深层架构，以使模型具备捕获更长距离上下文依赖的能力，或许是一种解决方案。

（3）任务导向的预训练模型和模型压缩：在实际使用中，不同难度的下游任务往往需要具有不同能力的预训练模型。同时，尽管大型预训练模型的性能通常更佳，但如何在低资源和低延迟场景下部署该类大型预训练模型仍然是一个实际面临的问题。因此，可以为下游任务精心设计特定的架构和预训练任务，或从现有预训练模型中提取出与任务相关的知识。相比于从零开始训练任务导向型的预训练模型，还可考虑使用知识蒸馏等技术，将现成的大型通用预训练模型知识蒸馏至小模型中。

（4）超越微调的知识迁移：微调是当前将预训练模型的知识迁移至下游任务中的主要方法，但该方法存在参数低效问题，即每个下游任务有其相应的微调参数。一种改进方案是保持预训练模型的原始参数，并为特定任务添加小的微调模块，从而节省参数。实际上，预训练模型也可以作为外部知识，从而更灵活地将预训练模型中的知识迁移到下游任务。

（5）预训练模型的可解释性和可靠性：尽管预训练模型性能惊艳，但其深度非线性架构使决策过程处于高度非透明状态。当前较多的预训练模型解释性工作依赖于注意力机制，而注意力机制在可解释性上的合理性仍备受争议。此外，由于预训练模型在对抗攻击下表现出的脆弱性，其可靠性也随着预训练模型的广泛部署而备受学者关注。对抗攻击研究有助于探究预训练模型的弱点，对抗防御研究则有助于提升预训练模型的鲁棒性。

本文从背景、模型架构、拓展等方面对 NLP 中的预训练模型进行了系统的归纳与总结，并从四个方面提出了一种新的分类方法，最后对预训练模型的研究趋势和方向进行了展望。关于预训练模型方面的详细综述可以进一步阅读文献[1]。

参考文献

[1] QIU X, SUN T, XU Y, et al. Huang, Pre-trained models for natural language processing: A survey[J]. Science China Technological Sciences, 2020, 63(10).

[2] MCCANN B, BRADBURY J, XIONG C, et al. Learned in translation: Contextualized word vectors[C]// Proceedings of NeruIPS, 2017.

[3] PETERS M E, NEUMANN M, IYYER M, et al. Deep contextualized word representations[C]// Proceedings of NAACL-HLT, 2018.

[4] RADFORD A, NARASIMHAN K, SALIMANS T, et al. Improving language understanding by generative pre-training [J]. 2018.

[5] DEVLIN J, CHANG M W, LEE K, et al. Bert: Pre-training of deep bidirectional transformers for language understanding[C]// Proceedings of NAACL-HLT, 2019.

[6] BENGIO Y, COURVILLE A, VINCENT P. Representation learning: A review and new perspectives[J]. IEEE transactions on pattern analysis and machine intelligence, 2013, 35(8): 1798-1828.

[7] KIM Y, JERNITE Y, SONTAG D, et al. Character-aware neural language models[C]//Proceedings of the AAAI conference on artificial intelligence. 2016, 30.

[8] BOJANOWSKI P, GRAVE E, JOULIN A, et al. Enriching word vectors with subword information[C]// Transactions of the Association for Computational Linguistics. 2017, 6:135-146.

[9] SENNRICH R, HADDOW B, BIRCH A. Neural machine translation of rare words with subword units[C]//ACL, 2016.

[10] HOCHREITER S, SCHMIDHUBER J. Long short-term memory[J]. Neural computation, 1997, 9(8): 1735-8170.

[11] CHUNG J, GULCEHRE C, CHO K, et al. Empirical evaluation of gated recurrent neural networks on sequence modeling[Z]. arXiv:1412.3555.

[12] TAI K S, SOCHER R, MANNING C D. Improved semantic representations from tree-structured long short-term memory networks[C]//ACL, 2015.

[13] KIPF T N, WELLING M. Semi-supervised classification with graph convolutional networks[C]// ICLR, 2017.

[14] LIU Y, OTT M, GOYAL N, et al. RoBERTa: A robustly optimized bert pretraining approach[Z]. arXiv: 1907.11692.

[15] DONG L, YANG N, WANG W, et al. Unified language model pre-training for natural language understanding and generation[C]// NeurIPS. 2019: 13042-13054.

[16] YANG Z, DAI Z, YANG Y, et al. XLNet: Generalized autoregressive pretraining for language understanding[C]// NeurIPS. 2019: 5754-5764.

[17] SAUNSHI N, PLEVRAKIS O, ARORA S, et al. A theoretical analysis of contrastive unsupervised representation learning[C]//International Conference on Machine Learning. 2019: 5628-5637.

[18] HJELM R D, FEDOROV A, LAVOIE-MARCHILDON S, et al. Learning deep representations by mutual information estimation and maximization[C]// ICLR 2019.

[19] CLARK K, LUONG M T, LE Q V, et al. ELECTRA: Pre-training text encoders as discriminators rather than generators[C]//ICLR, 2020.

[20] LAN Z, CHEN M, GOODMAN S, et al. ALBERT: A lite BERT for self-supervised learning of language representations[C]//ICLR, 2020.

[21] LAUSCHER A, VULIC I, PONTI E M, et al. Informing unsupervised pretraining with external linguistic knowledge [Z]. arXiv preprint arXiv:1909.02339, 2019.

[22] ZHANG Z, HAN X, LIU Z, et al. ERNIE: Enhanced language representation with informative entities[C]//ACL 2019.

[23] PETERS M E, NEUMANN M, LOGAN IV RL, et al. Knowledge enhanced contextual word representations[C]// EMNLP-IJCNLP, 2019.

[24] LIU W, ZHOU P, ZHAO Z, et al. K-Bert: Enabling language representation with knowledge graph[C]// Proceedings of the AAAI Conference on Artificial Intelligence. 2020, 34: 2901-2908.

[25] LAMPLE G, CONNEAU A. Cross-lingual language model pretraining[C]//NeurIPS. 2019: 7057-7067, 2019.

[26] SONG K, TAN X, QIN T, et al. MASS: Masked sequence to sequence pre-training for language generation[C]//Proceedings of Machine Learning Research. 2019: 5926-5936.

[27] SUN C, MYERS A, VONDRICK C, et al. VideoBERT: A joint model for video and language representation learning[C]//Proceedings of the IEEE/CVF International Conference on Computer Vision. 2019: 7464-7473.

[28] LI L H, YATSKAR M, YIN D, et al. VisualBERT: A simple and performant baseline for vision and language[Z]. arXiv preprint arXiv:1908.03557.

[29] LU J, BATRA D, PARIKH D, et al. Pretraining task-agnostic visiolinguistic representations for vision-and-language tasks[C]// Proceedings of NeurIPS. 2019: 13-23.

基于机器学习的脑解码方法研究

张道强 [1,2]　黄　硕 [1,2]　Muhammad Yousefnezhad[1,2]

([1]南京航空航天大学计算机科学与技术学院，南京 211106
[2]模式分析与机器智能工业和信息化部重点实验室，南京 211106)

1　引言

神经科学是关于神经系统的科学研究。在传统的研究中，人们通常认为神经科学是生物学的一个分支学科。但实际上，神经科学是一门跨学科的科学，它与包括认知科学、工程学、计算机科学、语言学、化学、数学、医学（包括神经学）和遗传学在内的多个学科相互交叉。神经科学的主要挑战是破解神经编码，这需要人们深刻地理解感知、记忆、思想和知识在大脑激活模式中的表示方式，在此基础上才能提出信息编码为神经活动的算法及从测量到的神经活动中提取信息的算法。

目前，随着技术的不断发展，对于神经活动的测量已有多种方式，包括事件相关光信号（event related optical signal, EROS）、正电子发射断层扫描（positron emission tomography, PET）、单光子发射计算机断层扫描（single photon emission computed tomography, SPECT）、近红外光谱（near infrared spectroscopy, NIRS）、脑磁图（magnetoencephalography, MEG）、脑皮层电图（electrocorticography, ECoG）、脑电图（electroencephalography, EEG），以及功能磁共振成像（functional magnetic resonance Imaging, fMRI）[1-7]。其中，功能磁共振成像技术通过使用血氧水平依赖（blood-oxygen-level-dependent, BOLD）对比作为神经激活的代用指标来测量神经活动，主要思想是通过对神经活动的测量来阐明认知过程。事实上，功能磁共振成像使我们能够了解人类大脑的某个区域内究竟蕴含了哪些信息，以及这些信息是如何被编码的[1]。功能磁共振成像技术之所以能够从众多技术中脱颖而出，主要有以下两个方面的原因：一方面，功能磁共振成像是一种无创成像技术[2,5]。另一方面，与其他非侵入式脑成像技术相比，它还具有超高的时空分辨率，且没有任何已知的副作用[1]。

神经活动可以用包括图结构、连续信号和基于成分的表示在内的多种不同形式表示。在 fMRI 数据集中，神经活动通常以体素（voxel，脑影像中的体积元素）的形式表示。人类大脑解码的核心概念是具有高维和大数据特征的表征空间[4]。以一个被试者的功能磁共振影像数据为例，假设该被试的神经响应信号包括感兴趣区域（region of interest，ROI）中的 5 个时间点和 2000 个体素，那么这个响应要由 10000 维空间中的向量来定义[1]。也就是说，所有被试者的表征空间可以由矩阵进行数值化表示，通常在 fMRI 试验中，每个时间点对被试者施加一个刺激（比如展示一张图片），也就是说每个时间点产生一个样本，因此，在表征空间矩阵中，每列表示人脑特征（即表示某一体素的所有时间点的激活强度），而每行表示与单个刺激有关的响应向量。图 1 给出了一个表征空间的例子，其中三个视觉刺激被观察和描绘成 3 个三维的向量。使用表征空间概念的主要优势在于可以在不同的测量模式中推广机器学习方法的应用[1-2,5]。

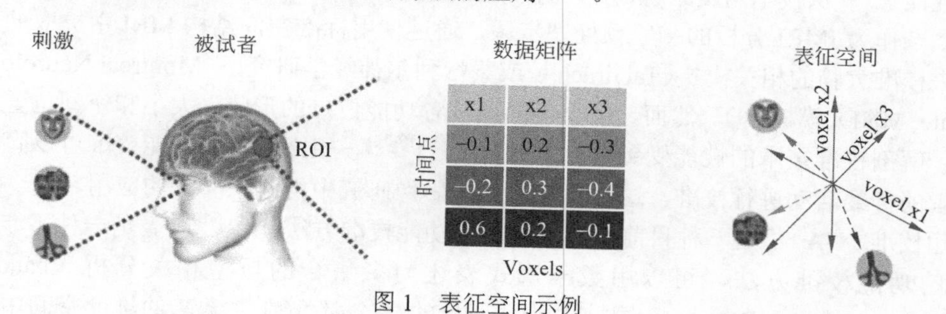

图 1　表征空间示例

多变量模式（multivariate pattern，MVP）分类是神经科学和计算机科学的结合，主要通过应用分类方法提取和解码大脑模式[8]。事实上，它可以预测与不同认知状态相关的神经活动模式[9-10]，也可以定义决策面来区分不同的刺激，以便解码大脑并理解其工作原理[1,11]。MVP 分类也可以帮助我们了解大脑如何存储和处理不同的刺激。此外，MVP 分类法使用机器学习算法对响应模式进行分类，将每个神经响应与一个实验条件联系起来。模式分类需要在神经表征空间中定义扇区，同一扇区中所有响应向量表示相同类别的信息，如刺激类别[1]、伴随刺激[12]或认知状态[9-10]。

针对上述挑战进行了一系列的研究，并取得了一些进展。针对脑影像的功能校准，提出了一系列的方法。针对非线性转换问题，提出了深度超校准方法，扩展了校准过程中响应空间的转换模式，显著提高了多被试的神经解码效果。为了有效地利用数据中的标签信息，提出了一种有监督的超校准方法，该方法有效地利用相关刺激的标签，最大化同一类别的相关性，最小化不同类别的相关性。为了解决多站点小样本数据的问题，提出了一种共享空间迁移学习方法，将不同站点的数据映射到共享空间中，并在新生成的空间中进行校准。该方法有效地扩展了单个站点的数据量，显著提高了网络的训练效果和多数据集模型的泛化性能。此外，不平衡 AdaBoost 二分类方法被提出，作为一种新的类别不平衡学习技术，该方法适用于一对一的分类任务。我们还开发了深度表征相似性学习，这是 RSA 的深度扩展，适用于分析具有大量被试者的 fMRI 数据集和高维数据中各种认知任务之间的相似性。

2　多被试者神经影像的功能校准

作为功能磁共振成像研究的一个基本挑战，MVP 模型必须在被试者之间进行泛化和验证[1,4,6,13-16]。然而，多被试者的 fMRI 数据集中的神经元活动必须进行校准以提高最终结果的性能[6,14]。从技术角度来说，有两种不同的校准技术，即解剖校准和功能校准[6,14-16]。解剖校准作为 fMRI 分析的一般预处理步骤，通过使用标准空间结构 MRI 中提取的解剖特征来校准大脑的相关模式（Talairach[17]或蒙特利尔神经学研究所（Montreal Neurological Institute, MNI）模板[18]）。然而，由于不同被试者功能位点的形状、大小和空间位置各有不同，解剖校准技术的性能受到了一定的限制[16,19-20]。相比之下，功能校准可以直接对被试者的神经活动进行校准，这在功能磁共振影像研究中得到了广泛的应用。

超校准（HA）[6]是一种目前被广泛使用的功能校准方法[1,4,6,13-16,21]，是一种不涉及解剖学的功能校准方法，可以用数学形式表述为多集合的典型相关分析（canonical correlation analysis, CCA）问题[4,13-14,21]。原始的 HA 不能在非常高的维度空间中被使用[15,22]。为了将 HA 扩展到现实问题中，Xu 等人[13]开发了正则化超校准（regularized hyperalignment, RHA）方法，利用 EM 算法迭代寻找正则化最优参数。Lorbert 等人[4]阐述了 HA 方法在 fMRI 响应的线性表示方面的局限性。他们还提出了核超校准（kernel hyperalignment, KHA）方法作为功能校准的一种非线性替代，在嵌入空间中解决 HA 限制。虽然 KHA 可以解决非线性和高维问题，但其性能受到所采用的固定核函数的限制。Sui 等人[23-24]针对多模态数据使用了多模态 CCA 和 ICA 方法，以识别跨模态关联的唯一和共享方差。进一步，Chen 等人[14]开发了奇异值分解超校准（singular value decomposition hyperalignment, SVDHA），首先通过奇异值分解进行降维，然后在降维空间通过 HA 方法对功能响应进行校准。在另一项研究中，Chen 等人[15]引入了共享响应模型（shared response model, SRM），它在技术上等同于概率 CCA。Guntupalli 等人[25]基于原始 HA 开发了一个共享表征空间的线性模型，该模型可以通过响应调优的基本功能追踪不同神经响应之间的细微差别，而这些功能在被试者和模型中都是通用的。事实上，这种方法实际上是拟 CCA 模型的集合，适合大脑图像的 patch 块。作为另一种非线性的超校准方法，Chen 等人[26]开发了用于全脑功能校准的卷积自编码器（convolutional autoencoder, CAE）。实际上，该方法将 SRM 重新定义为多视图的自编码器，然后使用标准的“探照灯”（search light, SL）进行分析，以提高生成的分类模型（认知解码模型）的稳定性和鲁棒性。由于 CAE 同时采用 SRM 和 SL，其时间复杂度很高。Turek 等人[27]提出了一种半监督的超校准方法，可以同时进行校准和响应模式分析。实际上，该方法也使用了 SRM 进行校准，然后使用多项逻辑回归进行分类。

脑影像相关描述：设 S 为被试者的数量，V 为体素的数量（将其视为一个一维向量），T 为以重复时间（time of repetitions, TRs）为单位的时间点的数量。第 ℓ 个被试者预处理后的脑影像（神经响应）定义为 $\boldsymbol{X}^{(\ell)} \in \mathbb{R}^{T\times V}$，$\ell = 1,2,\cdots,S$。这里，我们对神经响应矩阵的每一列进行了标准化，即 $\boldsymbol{X}^{(\ell)} \sim \mathcal{N}(0,1)$，$\ell = 1,2,\cdots,S$。提出如下的目标函数作为可扩展的功能校准：

$$\arg\min_{\boldsymbol{R}^{(\ell)},G} \sum_{\ell=1}^{S} \| \boldsymbol{X}^{(\ell)}\boldsymbol{R}^{(\ell)} - G \|_F^2 \qquad (1)$$
$$\text{s.t. } \boldsymbol{G}^{\mathrm{T}}\boldsymbol{G} = \boldsymbol{I}$$

其中，$\boldsymbol{G} \in \mathbb{R}^{T\times V}$ 是共享空间；$\boldsymbol{R}^{(\ell)} \in \mathbb{R}^{V\times V}$ 是第 ℓ 个被试者的映射矩阵，可以将属于第 ℓ 个被试者的原始体素空间转换到共享空间。这里的目标函数与经典的功能校准技术主要有两个区别。首先，经典的功能校准方法中的共享空间是所有特征在映射后简单地取平均值，即 $\boldsymbol{G} = {}^{1}\!/_{S}\sum_{\ell=1}^{S}\boldsymbol{X}^{(\ell)}\boldsymbol{R}^{(\ell)}$。这个简单的平均值无法有效地控制共享空间中的噪声。此外，它的计算效率也不高。第二个区别是约束条件。这里，假设我们的共享空间必须使 $\boldsymbol{G}^{\mathrm{T}}\boldsymbol{G} = \boldsymbol{I}$。在经典的校准方法中，它们只是分别施加每个被试协方差矩阵，即 $\left(\boldsymbol{X}^{(\ell)}\boldsymbol{R}^{(\ell)}\right)^{\mathrm{T}}\boldsymbol{X}^{(\ell)}\boldsymbol{R}^{(\ell)} = \boldsymbol{I}$。虽然这个假设可以控制被试层面上的过拟合，但无法保证仍然能够施加到最终的共享空间（作为所有映射特征的总和）。因此，提出的目标函数不会遇到在经典的功能校准技术中提到的问题。

2.1 有监督的超校准

图 2 解释了如何生成有监督的和无监督的共享空间。如图 2(b)所示，有监督的共享空间 $\boldsymbol{W} \in \mathbb{R}^{L\times V}$ 是一个向量空间，其中每一类刺激在所有被试者中都有一个独特的神经特征。这些特征是通过考虑所有类内和类间的关系而生成的。如图 2(c)所示，无监督的共

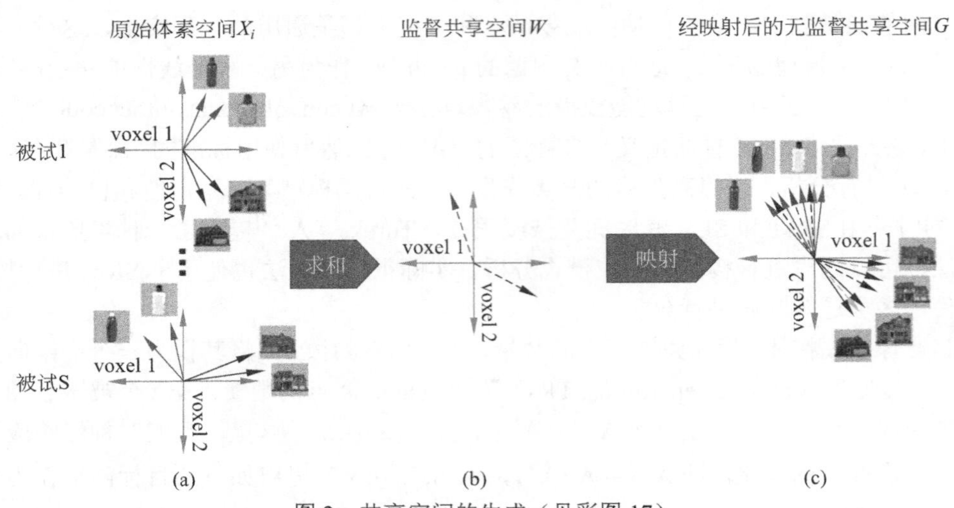

图 2 共享空间的生成（见彩图 17）
（a）原始的神经活动空间；（b）监督共享空间；（c）无监督共享空间

享空间$\boldsymbol{G} \in \mathbb{R}^{T \times V}$是一个向量空间，其中第$i$行是属于第$i$个刺激（时间点）的共享神经活动。实际中，监督空间中的每个神经特征都代表一个抽象概念，如房子或瓶子，而在无监督共享空间中的映射神经活动属于一个独特的刺激，如白色的房子或灰色的瓶子。例如，在图 2 中，为每个被试者提供了两组刺激（用红色和绿色箭头表示）。首先，有监督的超校准（supervised hyperalignment, SHA）为每个刺激类别（图中蓝色（瓶子）和棕色（房子）所示）在所有被试者中生成一个有监督的共享神经签名。最后，计算无监督的共享神经活动（例如，紫色箭头代表瓶子，黄色箭头代表房子），然后将原始的神经活动映射到这个空间。

属于第ℓ个被试者的神经活动可以表示为$\boldsymbol{X}^{(\ell)} \in \mathbb{R}^{T \times V}$，$\ell = 1, 2, \cdots, S$。类别标签则表示为$\boldsymbol{Y}^{(\ell)} = \left\{ y_{mn}^{(\ell)} \right\}, \boldsymbol{Y}^{(\ell)} \in \{0,1\}^{L \times T}, m = 1:L, n = 1:T, L > 1$。这里，$L$是类别（刺激类别）的数量。为了将监督信息引入 HA 问题，定义监督项如式（2）所示：

$$\boldsymbol{K}^{(\ell)} \in \mathbb{R}^{L \times T} = \boldsymbol{Y}^{(\ell)} \boldsymbol{H} \tag{2}$$

其中，归一化矩阵$\boldsymbol{H} \in \mathbb{R}^{T \times T}$表示为

$$\boldsymbol{H} = \boldsymbol{I}_T - \gamma \boldsymbol{1}_T \tag{3}$$

其中，$\boldsymbol{1}_T \in \{1\}^{T \times T}$表示大小为$T$的矩阵；$\gamma$用于平衡类内样本和类间样本。

根据式（2），SHA 的目标函数定义为

$$\max_{\boldsymbol{R}^{(i)},\boldsymbol{R}^{(j)}}\left\{\frac{2}{S-1}\sum_{i=1}^{S}\sum_{j=i+1}^{S}\operatorname{tr}\left((\boldsymbol{K}^{(i)}\boldsymbol{X}^{(i)}\boldsymbol{R}^{(i)})^{\mathrm{T}}(\boldsymbol{K}^{(j)}\boldsymbol{X}^{(j)}\boldsymbol{R}^{(j)})\right)+\epsilon\sum_{\ell=1}^{S}\|\boldsymbol{R}^{(\ell)}\|_F^2\right\} \tag{4}$$

$$\text{s.t.}\ (\boldsymbol{R}^{(\ell)})^{\mathrm{T}}\left(\left(\boldsymbol{K}^{(\ell)}\boldsymbol{X}^{(\ell)}\right)^{\mathrm{T}}\boldsymbol{K}^{(\ell)}\boldsymbol{X}^{(\ell)}+\epsilon\boldsymbol{I}_V\right)\boldsymbol{R}^{(\ell)}=\boldsymbol{I}_V$$

其中，$\ell=1,2,\cdots,S;\boldsymbol{R}^{(\ell)}\in\mathbb{R}^{V\times V}$ 为映射矩阵；ϵ 为正则化参数。

考虑式(4)中 SHA 的目标函数，可将 SHA 的优化表示为[28]

$$\boldsymbol{U}\boldsymbol{W}=\boldsymbol{W}\boldsymbol{\Lambda} \tag{5}$$

其中，Λ 和 $\boldsymbol{W}$ 分别表示矩阵 $\boldsymbol{U}$ 的特征值和特征向量。我们可以对 $\boldsymbol{U}=\tilde{\boldsymbol{U}}\tilde{\boldsymbol{U}}^{\mathrm{T}}$ 进行 Cholesky 分解[28]，其中 $\tilde{\boldsymbol{U}}\in\mathbb{R}^{L\times\mu}$，$\mu=L\times S$，记为

$$\tilde{\boldsymbol{U}}=\left[\boldsymbol{I}_T-\boldsymbol{A}^{(1)}\boldsymbol{D}^{(1)},\cdots,\boldsymbol{I}_T-\boldsymbol{A}^{(S)}\boldsymbol{D}^{(S)}\right] \tag{6}$$

其中，$\boldsymbol{W}$ 可视作 $\tilde{\boldsymbol{U}}=\boldsymbol{W}\tilde{\Sigma}\tilde{\boldsymbol{B}}$ 的左奇异向量。由于 $\tilde{\boldsymbol{U}}$ 可能过大，无法适应内存，因此使用增量 PCA[29]计算左奇异向量 $\boldsymbol{W}$。为了同时映射训练集和测试集中的特征，通过使用监督共享空间，在体素空间上定义一个无监督的模板，如下式所示：

$$\boldsymbol{G}=\frac{1}{S}\left(\sum_{\ell=1}^{S}\boldsymbol{W}^{\mathrm{T}}\boldsymbol{K}^{(\ell)}\right)^{\mathrm{T}} \tag{7}$$

其中，$\boldsymbol{G}\in\mathbb{R}^{T\times V}$ 为无监督模板。值得注意的是，$\boldsymbol{W}$ 和 $\boldsymbol{G}$ 都是两个不同层次的神经活动共享的表征空间，即 $\boldsymbol{W}$ 在类别层次上定义，$\boldsymbol{G}$ 在体素层次上表示。我们提出的算法具体如下：

算法 1 有监督的超校准(SHA)

输入：训练集 $\boldsymbol{X}^{(i)},i=1:S$，测试集 $\bar{\boldsymbol{X}}^{(j)},j=1:\bar{S}$，训练集类别标签 $\boldsymbol{Y}^{(i)}$，正则化系数 $\epsilon=10^{-4}$

1. 初始化 $\boldsymbol{K}^{(\ell)},\boldsymbol{H}$
2. 根据公式（6）计算 $\tilde{\boldsymbol{U}}$
3. 在 $\tilde{\boldsymbol{U}}$ 上使用 SVD[28]分解 $\boldsymbol{W}$
4. 使用 $\boldsymbol{W}$ 和式(7)计算共享空间 $\boldsymbol{G}$
5. 生成训练数据 $\boldsymbol{Z}^{(\ell)},\ell=1:S$
6. 生成测试数据 $\bar{\boldsymbol{Z}}^{(\ell)},\ell=1:\bar{S}$
7. 通过 $\boldsymbol{Z}^{(\ell)},\boldsymbol{Y}^{(\ell)}(\ell=1:S)$ 训练分类器
8. 使用 $\bar{\boldsymbol{Z}}^{(\ell)},\ell=1:\bar{S}$ 对分类器进行评估

输出：分类性能（分类精度（ACC），曲线下面积（area under the curve, AUC））

在实验中，使用来自 Open Neuro（https://openneuro.org/）任务态 fMRI 公开数据平台上的数据集。选择其中与认知任务相关的数据集，包括 DS005，DS105，DS107，DS116，DS117 和 CMU。图 3(a)说明了二分类分析的准确性，图 3(b)给出了多类数据集的分类精度。如图所示，未经功能校准处理的数据集与校准数据相比，分类精度有限。由于 SHA 利用监督信息来校准神经活动，因此它的性能优于其他方法。在多类数据集上，监督信息对功能校准性能的影响更为突出。

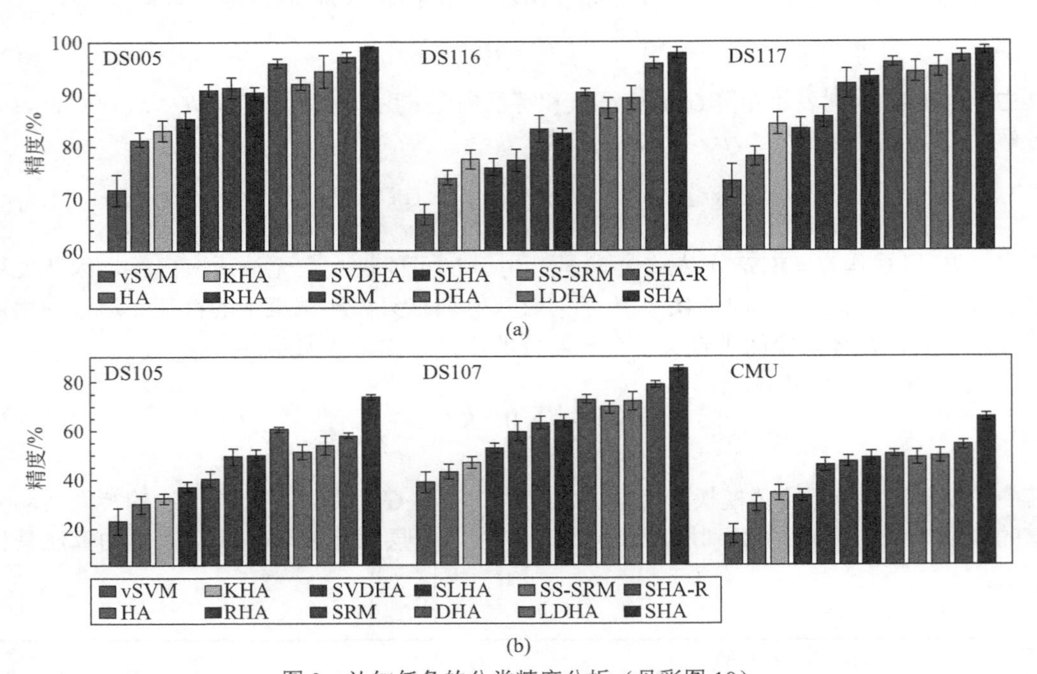

图 3 认知任务的分类精度分析（见彩图 18）

（a）不同 HA 方法在校准后二分类任务中的精度；（b）不同 HA 方法在校准后多分类任务中的精度

2.2 深度超校准方法

深度超校准（deep hyper alignment, DHA）主要采用深度网络，即非线性转换的多层堆叠层，作为参数化的核函数，并使用秩为 m 的奇异值分解和随机梯度下降（stochastic gradient descent, SGD）进行优化。因此，DHA 在大数据集上的运行时间更短。而且，当 DHA 计算一个新被试的功能校准时，无须使用前期的训练样本。此外，DHA 不受有限的固定表征空间的限制，因为 DHA 的内核是一个多层神经网络，可以分别为每个被试者实现任何非线性函数[22,29-30]，将大脑神经响应转换到一个公共空间。DHA 的目标函数如下：

$$\min_{G,R^{(i)},\theta^{(i)}} \sum_{i=1}^{S} \| G - f_i(X^{(i)};\theta^{(i)})R^{(i)} \|_F^2 \tag{8}$$

$$\begin{cases} G = \dfrac{1}{S}\sum_{j=1}^{S} f_j(X^{(j)};\theta^{(j)})R^{(j)} \\ \text{s.t.} \left(R^{(\ell)}\right)^{\mathrm{T}} \left(f_\ell(X^{(\ell)};\theta^{(\ell)})\right)^{\mathrm{T}} f_\ell(X^{(\ell)};\theta^{(\ell)}) + \epsilon \boldsymbol{I} = \boldsymbol{I} \end{cases} \tag{9}$$

其中，$\theta^{(\ell)} = \{W_m^{(\ell)}, b_m^{(\ell)}, m = 2:C\}$ 表示属于第 ℓ 个被试者的第 ℓ 个深度网络中的所有参数；$R^{(\ell)} \in \mathbb{R}^{V_{\text{new}} \times V_{\text{new}}}$ 是第 ℓ 个被试者的 DHA 的解；$V_{\text{new}} \leqslant V$ 表示转换后的特征数量；正则化参数 ϵ 是一个值很小的常数，如10^{-8}。深度多层核函数 $f_\ell(X^{(\ell)};\theta^{(\ell)}) \in \mathbb{R}^{T \times V_{\text{new}}}$ 表示如下:

$$f_\ell(X^{(\ell)};\theta^{(\ell)}) = \text{mat}\left(\boldsymbol{h}_C^{(\ell)}, T, V_{\text{new}}\right) \tag{10}$$

其中，T 为时间点的数量；$C \geqslant 3$ 为深度网络层数；$\text{mat}\left(\boldsymbol{x}, m, n\right): \mathbb{R}^{mn} \to \mathbb{R}^{m \times n}$ 为重塑（矩阵化）函数；$\boldsymbol{h}_C^{(\ell)} \in \mathbb{R}^{TV_{\text{new}}}$ 为以下多层深度网络的输出层:

$$\boldsymbol{h}_m^{(\ell)} = g\left(W_m^{(\ell)} \boldsymbol{h}_{m-1}^{(\ell)} + b_m^{(\ell)}\right) \tag{11}$$

其中，$\boldsymbol{h}_1^{(\ell)} = \text{vec}(\boldsymbol{X}^{(\ell)})$；$m = 2:C$；$g: \mathbb{R} \to \mathbb{R}$ 是应用分量的非线性函数；$\text{vec}: \mathbb{R}^{m \times n} \to \mathbb{R}^{mn}$ 表示向量函数，因此 $\boldsymbol{h}_1^{(\ell)} = \text{vec}(\boldsymbol{X}^{(\ell)}) \in \mathbb{R}^{TV}$。值得注意的是，我们认为 vec() 和 mat() 函数都是线性转换，其中对于任意矩阵 $\boldsymbol{X}$，$\boldsymbol{X} \in \mathbb{R}^{m \times n} = \text{mat}(\text{vec}(\boldsymbol{X}), m, n)$。通过考虑第 m 中间层的 $U^{(\mathrm{m})}$ 单元，$f_\ell(X^{(\ell)};\theta^{(\ell)})$ 的不同层的参数由以下属性定义：对于输出层，$\boldsymbol{W}_C^{(\ell)} \in \mathbb{R}^{TV_{\text{new}} \times U^{(C-1)}}$，$\boldsymbol{b}_C^{(\ell)} \in \mathbb{R}^{TV_{\text{new}}}$；对于第一个中间层，$\boldsymbol{W}_2^{(\ell)} \in \mathbb{R}^{U^{(2)} \times TV}$，$\boldsymbol{b}_2^{(\ell)} \in \mathbb{R}^{U^{(2)}}$；对于第 m 个中间层($3 \leqslant m = C-1$)，$\boldsymbol{W}_m^{(\ell)} \in \mathbb{R}^{U^{(m)} \times U^{(m-1)}}$，$\boldsymbol{b}_m^{(\ell)} \in \mathbb{R}^{U^{(m)}}$，$\boldsymbol{h}_m^{(\ell)} \in \mathbb{R}^{U^{(m)}}$。

与以往的 HA 问题研究一样[4,13-15]，DHA 的解并不是唯一的。如果一个 DHA 模板 G 是为一个特定的 HA 问题计算的，那么 $\boldsymbol{Q}$G 是该特定 HA 问题的另一个解，其中 $\boldsymbol{Q} \in \mathbb{R}^{V_{\text{new}} \times V_{\text{new}}}$ 可以是任何正交矩阵。因此，如果为特定数据集训练两个独立的模板 G_1，G_2，则可以通过计算 $\| G_2 - \boldsymbol{Q}G_1 \|$ 将解相互映射，其中 $\boldsymbol{Q}$ 可以作为第一个解的功能校准系数，以便将其结果与第二个解进行比较。实际上，G_1 和 G_2 在同一条等高线上处于不同位置[2,9,22]。

使用秩为 m 的奇异值分解（SVD）和随机梯度下降（SGD)来对 DHA 的目标函数进行优化。该方法通过两步迭代求解 DHA 目标函数式（8）和式（9）的最优解。通过考虑固定的网络参数($\theta^{(\ell)}$)，首先通过深度网络对少量神经响应进行校准。然后，采用反向传播算法更新网络参数。求解 DHA 目标函数的主要挑战是不能寻求关联对象到两个以上随机变量的自然扩展。因此，功能校准被堆叠在 $S \times S$ 矩阵中，并最大化该矩阵的特定矩阵范数[29]。

同样，选择 Open Neuro 数据平台中的数据集对所提出的方法进行实验验证，另外四种用于功能校准的经典非监督 HA 方法被用作对比方法。不同方法的分类精度见表 1。可以看出，不使用 HA 方法的分类性能明显较低，接近随机抽样。与其他无监督方法相比，DHA 在绝大多数数据集上获得了最好的分类精度，尤其在二分类问题上，DHA 具有更好的表现。

表 1　不同 HA 方法的分类精度（最大值 ± 标准差）

数据集	ν-SVM	RHA	SRM	CAE	DHA
R105	18.21 ± 2.32	24.90 ± 0.04	35.64 ± 0.05	38.06 ± 0.06	**43.76 ± 0.23**
R107	27.71 ± 3.86	**52.41 ± 0.63**	40.51 ± 3.32	45.08 ± 0.34	45.87 ± 0.01
R232	28.12 ± 1.87	31.39 ± 1.20	37.42 ± 0.82	42.75 ± 0.93	**45.36 ± 0.07**
W001	26.45 ± 0.31	35.74 ± 0.21	35.00 ± 0.64	39.57 ± 0.33	**42.14 ± 0.14**
W002D	63.81 ± 2.09	66.72 ± 1.12	73.02 ± 0.60	76.22 ± 0.08	**80.04 ± 0.00**
W002P	65.21 ± 1.51	70.55 ± 0.98	71.41 ± 0.71	79.48 ± 0.06	**82.56 ± 0.01**
W005	33.59 ± 1.33	42.05 ± 0.24	50.32 ± 0.79	53.24 ± 0.15	**60.32 ± 0.05**
W011D	42.70 ± 0.90	62.88 ± 1.70	60.60 ± 0.57	65.07 ± 0.13	**70.43 ± 0.05**
W105	16.81 ± 1.77	24.65 ± 0.62	30.06 ± 0.19	35.49 ± 0.26	**40.97 ± 0.07**
W107	30.69 ± 2.04	47.42 ± 0.94	49.52 ± 0.95	57.04 ± 0.27	**69.32 ± 0.01**
W231	30.62 ± 1.20	58.55 ± 0.54	62.21 ± 0.88	63.42 ± 0.95	**67.15 ± 0.05**
W232	26.79 ± 0.52	40.21 ± 0.73	47.66 ± 0.29	50.37 ± 0.30	**55.17 ± 0.06**

使用 Open Neuro 数据库中的三个基于 ROI 的数据集（R105，R107 和 R203）对所提出的方法和其他 HA 方法的运行时间进行比较。图 4 展示了不同方法在上述三个数据集上的运行时间。为了形象地表现 DHA 与其他方法的区别，以 DHA 的运行时间为基础（单位“1”），其他方法的运行时间是基于 DHA 的比率。如图 4 所示，CAE 的运行时间最长，因为它同时使用了 SRM 和 SearchLight（SL）的改进版本来实现功能校准。此外，LDA 的时间复杂度也很高，因为它必须针对每个类别分别进行矩阵分解。通过综合考虑 DHA 方法的性能，它在无监督方法中所需的运行时间是可以接受的。此外，由于 LDHA 使用了有监督的公共空间表示，并且不需要对每个类别分别进行矩阵分解，因此它比 LDA 具有更短的运行时间。值得注意的是，全脑数据集上的实验结果也呈现出同样的趋势。

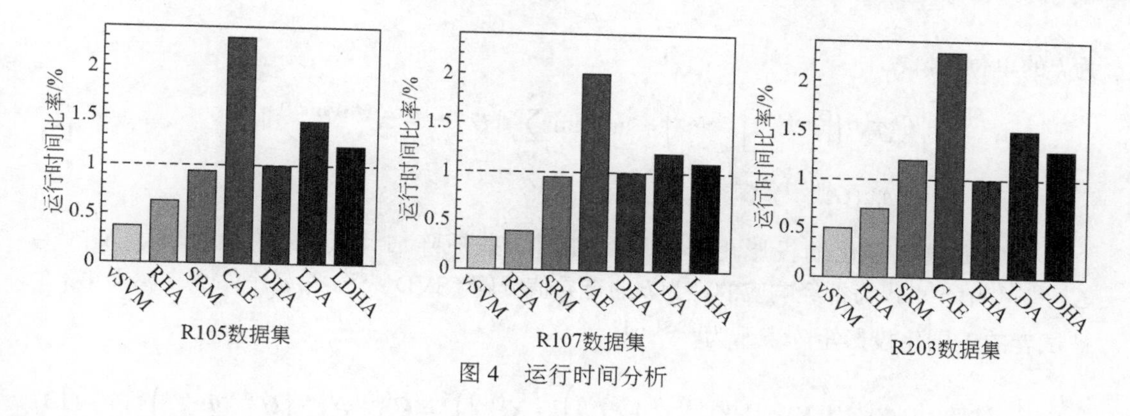

图 4 运行时间分析

3 多站点功能影像的共享空间迁移学习

在第 2 节中，我们所提出的方法主要针对的是单站点的 fMRI 数据。尽管直接对单站点 fMRI 数据进行分析是一个相对简单的过程，但这种方法不能直接被应用到具有不同实验范式的多站点数据集[30-31]。另一个问题是实验过程产生的批次效应（batch effects）[32]。它具体是指一组可能影响收集到的 fMRI 数据在每个站点的分布的外部因素，例如，环境噪声、fMRI 机器供应商使用的标准等。为了解决这些问题，最近的研究[30,33-35]表明，通过利用同质认知任务的现有领域知识，迁移学习（transfer learning, TL）可以显著提高多站点 fMRI 数据集分类模型的性能。

我们提出了共享空间迁移学习（shared space transfer learning, SSTL）作为一种新的迁移学习方法，它可以通过多个站点的 fMRI 数据集生成鲁棒的、广义的、准确的分类模型，然后在每个站点上被有效地使用。SSTL 通过使用一个分层的两步过程来学习一个共享的特征空间。首先，为每个站点的所有被试者提取一组公共特征，然后使用 TL 将这些特定于站点的特征映射到一个独立于站点的共享空间。此外，SSTL 使用了一种可扩展的优化算法，该算法对具有大量被试者的高维 fMRI 数据集有效。优化过程采用单次迭代多视图方法提取每个站点的公共特征，然后将这些公共特征映射到站点独立的共享空间中。

首先，开发了一种无监督的多视图方法，可以从每个站点分别提取特定站点的公共特征。设 k 为公共特征空间中的特征数。计算映射矩阵 $\boldsymbol{R}^{(d,s)} \in \mathbb{R}^{V\times k}, k \leqslant V$，将每个被试者的神经响应通过转换旋转到共同特征空间 $\boldsymbol{G}^{(d,S_d)} \in \mathbb{R}^{T_d\times k} = \left\{ g_{tv}^{(d,S_d)} | t=1,2,\cdots,T_d, v=1,2,\cdots,k \right\}$。使用以下目标函数提取第 d 个位置的映射矩阵和公共特征空间，其中 $\boldsymbol{I}_k \in \{0,1\}^{k\times k}$ 是大小

为 k 的单位矩阵:

$$\mathcal{J}_C^{(d)}\left(\left[\boldsymbol{X}^{(d,s)}\right]_{s=1,\cdots,S_d}\right)=\underset{\boldsymbol{R}^{(d,s)},\boldsymbol{G}^{(d,S_d)}}{\operatorname{argmin}}\sum_{s=1}^{S_d}\|\boldsymbol{G}^{(d,S_d)}-\boldsymbol{X}^{(d,s)}\boldsymbol{R}^{(d,s)}\|_F^2 \quad (12)$$
$$\text{s.t.}(\boldsymbol{G}^{(d,S_d)})^T\boldsymbol{G}^{(d,S_d)}=I_k$$

首先提出正则化的投影矩阵，然后使用这些矩阵估计式(12)的最优解。$\boldsymbol{X}^{(d,s)}=\boldsymbol{U}^{(d,s)}\Sigma^{(d,s)}(\boldsymbol{V}^{(d,s)})^{\mathrm{T}}$ 为神经响应的秩为 k 的奇异值分解（SVD）[36]。第 d 个站点中属于第 s 个被试者的正则化投影矩阵表示为[22,31,28]

$$\boldsymbol{P}^{(d,s)}=\boldsymbol{X}^{(d,s)}\left(\boldsymbol{X}^{(d,s)}\left(\boldsymbol{X}^{(d,s)}\right)^{\mathrm{T}}+\epsilon\boldsymbol{I}_{T_d}\right)^{-1}\left(\boldsymbol{X}^{(d,s)}\right)^{\mathrm{T}}=\boldsymbol{U}^{(d,s)}\boldsymbol{\Phi}^{(d,s)}\left(\boldsymbol{U}^{(d,s)}\boldsymbol{\Phi}^{(d,s)}\right)^{\mathrm{T}} \quad (13)$$

$$\boldsymbol{\Phi}^{(d,s)}(\boldsymbol{\Phi}^{(d,s)})^{\mathrm{T}}=\Sigma^{(d,s)}\left(\Sigma^{(d,s)}\left(\Sigma^{(d,s)}\right)^{\mathrm{T}}+\epsilon\boldsymbol{I}_{T_d}\right)^{-1}\left(\Sigma^{(d,s)}\right)^{\mathrm{T}} \quad (14)$$

我们的研究目的是通过使用特定站点的共同特征来创建一个多站点 fMRI 分析的迁移学习模型，但不是通过直接迁移原始神经响应[37]，也不是通过基于一组被试者出现在每对站点的全局共享空间[30]。我们将站点划分为一组训练站点和一组测试站点。基于训练站点的共同特征，构建了一个全局共享空间。然后，利用这个全局空间来映射测试站点数据的神经响应。设 $\tilde{D}$ 为训练站点个数，$\tilde{T}=\sum_{d=1}^{\tilde{D}}T_d$，$\hat{D}$ 为测试站点个数，即 $D=\tilde{D}+\hat{D}$。通过每个站点特定的公共空间 $\boldsymbol{G}^{(d,S_d)}$，$d=1,2,\cdots,\tilde{D}$ 来生成一个全局共享空间 $\boldsymbol{G}$。

我们希望找到一个全局的共享空间，其转换后的共同特征具有最小分布失配。设 $\boldsymbol{g}_{t.}\in\mathbb{R}^k$ 为矩阵 $\boldsymbol{G}$ 的第 t 行，这里要找到一对编码/解码转换函数。编码转换函数 $\boldsymbol{q}_{t.}=\mathcal{J}_1(\boldsymbol{g}_{t.};\theta_1)$ 将共同特征（来自特定位置）映射到全局共享空间中，其中 $\boldsymbol{q}_{t.}$ 是共享空间中第 t 次转换后的共同特征。解码函数 $\overline{\boldsymbol{g}}_{t.}=\mathcal{J}_2(\boldsymbol{q}_{t.};\theta_2)$ 从共享空间中重构出特定站点的共同特征。一般来说，可以使用以下目标函数来找到这些编码/解码转换:

$$\mathcal{J}_G(\boldsymbol{G})=\underset{\theta_1,\theta_2}{\operatorname{argmin}}\sum_{t=1}^{\tilde{T}}\|\boldsymbol{g}_{t.}-\overline{\boldsymbol{g}}_{t.}\|_F^2+\Omega(\theta_1,\theta_2) \quad (15)$$

其中，Ω 是参数 θ_1,θ_2 上的正则化函数。机器学习解决问题有几种标准方法（式(15)）。例如，可以使用正则化的自动编码器来寻找这些转换，其中 $\mathcal{J}_1,\mathcal{J}_2$ 是对称多层感知器（multilayer perceptron, MLP）[22]。然而，复杂的模型需要大量的样本对模型性能进行提升，而大多数多站点 fMRI 数据集没有足够多的样本数量[37]。这里，提出了线性 Karhunen-Loeve

转换（KLT）[38]来学习全局共享空间：

$$\tilde{\mathcal{J}}_G(\boldsymbol{G}) = \underset{\boldsymbol{W}}{\operatorname{argmin}} \| \boldsymbol{G} - \boldsymbol{G}\boldsymbol{W}\boldsymbol{W}^{\mathrm{T}} \|_F^2 \\ \text{s.t.}\ \boldsymbol{W}^{\mathrm{T}}\boldsymbol{W} = I_k \tag{16}$$

其中，$\boldsymbol{W} \in \mathbb{R}^{k \times k}$ 表示转换矩阵；$\boldsymbol{Q} = \mathcal{J}_1(\boldsymbol{G};\boldsymbol{W}) = \boldsymbol{G}\boldsymbol{W}$；$\bar{\boldsymbol{G}} = \mathcal{J}_2(\boldsymbol{G};\boldsymbol{W}^{\mathrm{T}}) = \boldsymbol{Q}\boldsymbol{W}^{\mathrm{T}}$。

在实验中，使用了来自五个站点（A, B, C, G, H）的数据集进行多站点 fMRI 分析，这些站点中，有部分重叠的被试者（不止出现在一个站点中）。我们分别进行了两个层次的分析，即双站点分析和多站点分析。在双站点分析中，首先从一个站点学习 TL 模型来预测另一个站点的神经响应，例如，a $\rightleftharpoons$ B。图 5 第一行展示了双站点分析的实验结果。当分析过程中有两个以上站点时，我们还对 TL 模型的性能进行基准测试，如(A, B) $\rightleftharpoons$ C。图 5 下面一行显示了这些多站点分析的准确性。如图 5 所示，原始神经响应在 MNI 空间表现不佳，可能是因为不同站点的分布不匹配。虽然单视图方法（MIDA 和 SIDeR）表现得更好，但它们在多站点分析中表现不佳。我们看到，SRM，MDDL，MDMS 和 SSTL 中的多视图技术使它们能够生成比单一视图方法更精确的 TL 模型。最后，SSTL 提供了最精确的 TL 模型，首先使用多视图方法生成特定站点的公共特征，然后使用这些公共特征（而不是有噪声的原始神经响应）将数据传输到全局共享空间，从而获得更好的性能

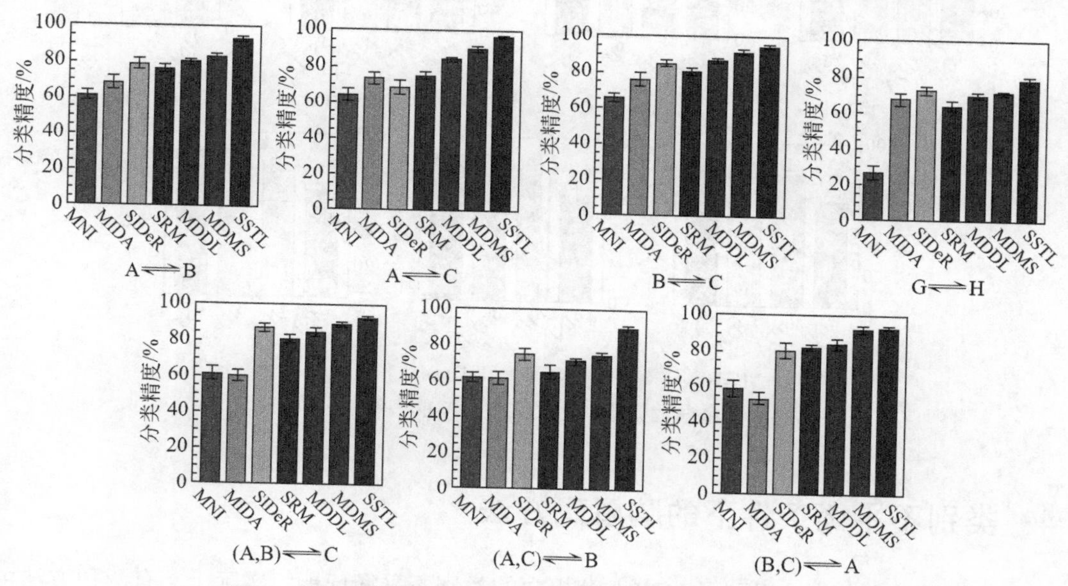

图 5　有重叠数据集（存在公共被试）的多站点分类分析

（注：图 5 的 7 个图中的每一个都是比较 SSTL 和 χ_1，对于 6 个不同的 $\chi_1 \in$ {MNI, MIDA, SIDeR, SRM, MDDL, MDMS}，总共是 $7 \times 6 = 42$ 组对照。双侧 t 检验发现 42 例患者的 $p < 0.05$）。

类似地，进行了更通用的多站点 fMRI 分析，其中每组数据集中没有相同的被试者出现，即数据集之间无重叠。结果如图 6 所示。可以看出，相比其他方法，SSTL 能够在无被试者重叠的数据集中取得更稳定、具有泛化性的分类性能。

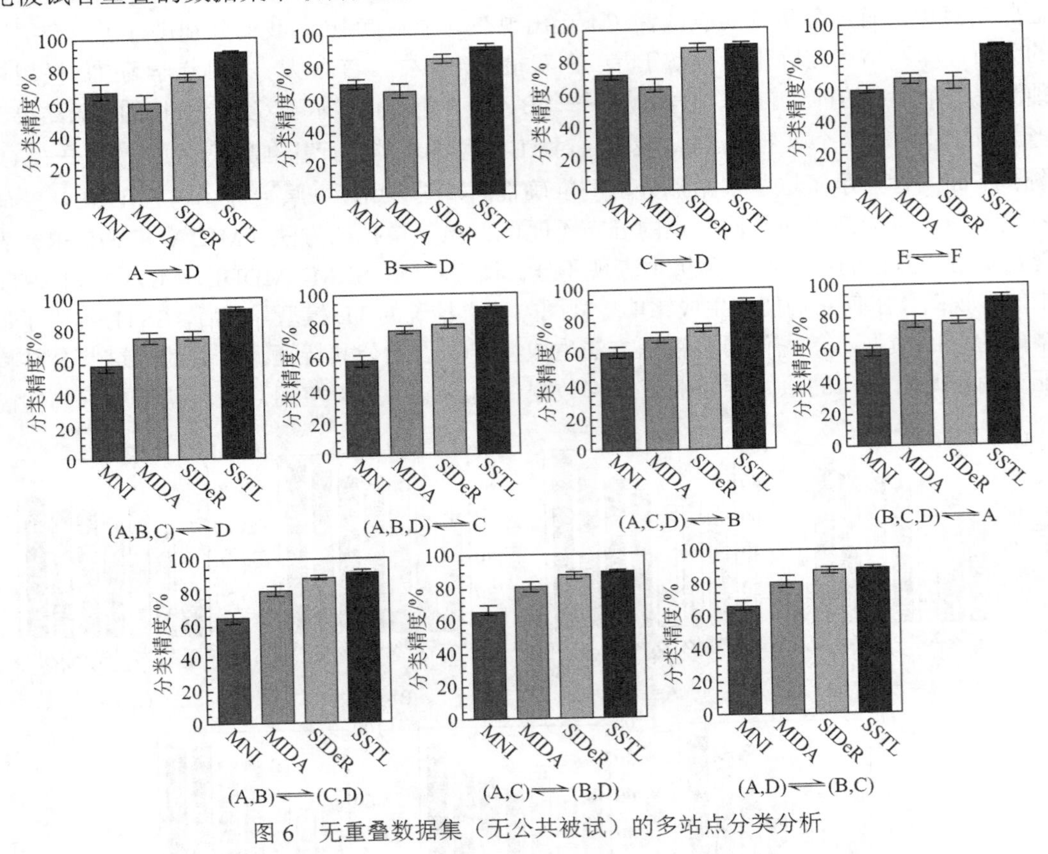

图 6　无重叠数据集（无公共被试）的多站点分类分析

4　类别不平衡条件下的脑解码方法

在前面的章节中，提到了 MVPA 分析中的类别不平衡问题。实际上，有两种方法可以解决这一问题，即设计一个不平衡分类器或将不平衡问题转化为类别平衡的分类模型的集成。先前的研究表明，不平衡分类器的性能可能不稳定，特别是当我们的数据集较为稀疏或存在噪声的时候[39-41]。由于 fMRI 数据集大多包含噪声和稀疏性，我们选择了集成方法。从技术上讲，集成学习也包含两种不同的解决方案，即 bagging 或 boosting。bagging 同时生成所有的分类器，并将它们组合成最终的模型；boosting 则逐步地创建每个分类器，通过跟踪之前迭代的错误来提高每次迭代的性能。需要注意的是，集成学习可以用于平衡和不平衡问题。事实上，两者主要的区别在于抽样的策略。在平衡问题中，抽样方法被应用于整个数据集，而在不平衡问题中，样本量更大的类别被抽样[41]。

如图 7 所示，为了显著提高模型最终在 fMRI 分析中的性能，提出了 AdaBoost 算法的一个新分支，称为不平衡的 AdaBoost 二分类（imbalance AdaBoost binary classification, IABC）。该算法首先将一个不平衡的 MVPA 问题转化为一系列的平衡问题。然后，迭代地将决策树[41]应用于每个平衡分类问题。最后，使用 AdaBoost 生成最终的模型。在该方法中，根据之前迭代的错误（失败预测）生成最终组合的每个分类器（树）的权值，逐步提高最终模型的性能。

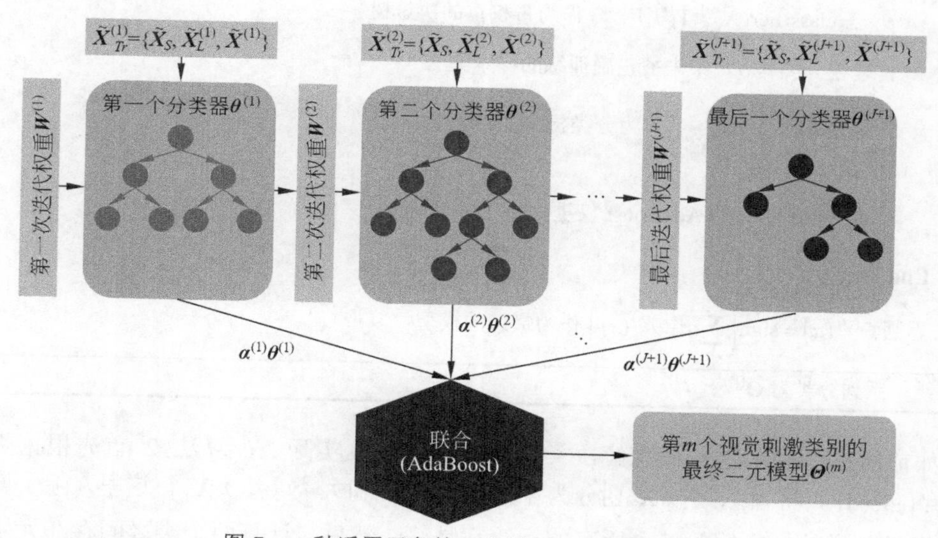

图 7　一种适用于鲁棒二分类的 AdaBoost 算法

为了应用二分类问题，本文将提取的特征 $\boldsymbol{X}$ 随机划分为训练集 $\tilde{\boldsymbol{X}}$ 和测试集 $\hat{\boldsymbol{X}}$ 。算法 2 是 AdaBoost 算法的一个新分支，使用训练集 $\tilde{\boldsymbol{X}}$ 进行二分类训练。然后，利用 $\hat{\boldsymbol{X}}$ 来估计最终模型的性能。如前所述，fMRI 分析的二分类主要是不平衡，特别是使用一对一的策略。因此，这些二进制类中的一个类的样本数比其他类要少。

算法 2 不平衡 AdaBoost 二分类

输入： 训练集 $\tilde{\boldsymbol{X}}$，类别标签 $\boldsymbol{Y}^{(m)}$

1. 根据类别标签 $\boldsymbol{Y}^{(m)}$ 划分训练集 $\tilde{\boldsymbol{X}}=\{\tilde{\boldsymbol{X}}_S,\tilde{\boldsymbol{X}}_L\}$
2. 计算平衡子集的数量 $J=\operatorname{int}(\tilde{\boldsymbol{X}}_S/\tilde{\boldsymbol{X}}_L)$
3. 随机采样：$\tilde{\boldsymbol{X}}_L=\left\{\tilde{\boldsymbol{X}}_L^{(1)},\tilde{\boldsymbol{X}}_L^{(2)},\cdots,\tilde{\boldsymbol{X}}_L^{(n)},\cdots,\tilde{\boldsymbol{X}}_L^{(J)}\right\}$
4. 初始化 $\bar{\boldsymbol{X}}^{(1)}=\bar{\boldsymbol{Y}}^{(1)}=\varnothing$
5. **For** $(n=1,2,\cdots,J)$：
6. $\tilde{\boldsymbol{X}}_{T_r}^{(n)}=\{\tilde{\boldsymbol{X}}_S,\tilde{\boldsymbol{X}}_L^{(n)},\bar{\boldsymbol{X}}^{(n)}\}$ 作为此次迭代的训练集
7. $\tilde{\boldsymbol{Y}}_{T_r}^{(n)}=\{\tilde{\boldsymbol{Y}}_S,\tilde{\boldsymbol{Y}}_L^{(n)},\bar{\boldsymbol{Y}}^{(n)}\}$ 作为此次迭代的类别标签
8. $\boldsymbol{W}^{(n)}=\begin{cases}1, & \tilde{\boldsymbol{X}}_S \text{ or } \bar{\boldsymbol{X}}^{(n)}\\ 1-\left|\operatorname{corr}\left(\tilde{\boldsymbol{X}}_S,\tilde{\mathbf{Y}}_L^{(n)}\right)\right|, & \tilde{\boldsymbol{X}}_L^{(n)}\end{cases}$
9. $\theta^{(n)}=\operatorname{classifier}(\tilde{\boldsymbol{X}}_{T_r}^{(n)},\tilde{\boldsymbol{Y}}_{T_r}^{(n)},\boldsymbol{W}^{(n)})$ 作为带权重的决策树
10. 构建在分类器 $\theta^{(n)}$ 中未被正确训练的样本集 $\bar{\boldsymbol{X}}^{(n+1)}$
11. 计算 $\epsilon^{(n)}=\left|\bar{\boldsymbol{X}}^{(n+1)}\right|\Big/\left|\tilde{\boldsymbol{X}}_{T_r}^{(n)}\right|$ 作为分类损失
12. 计算分类器 $\theta^{(n)}$ 的 AdaBoost 权重，$\alpha^{(n)}=\dfrac{1}{2}\ln\left(\dfrac{1-\epsilon^{(n)}}{\epsilon^{(n)}}\right)$
13. **End For**
14. 返回 $\Theta^{(m)}(x)=\operatorname{sign}\left(\sum_{n=1}^{J+1}\alpha^{(n)}\theta^{(n)}(x)\right)$ 作为最终的模型

输出： 一系列分类器 $\boldsymbol{\Theta}^{(m)}$.

如前所述，利用这一概念解决类别不平衡问题。实际上，算法 2 首先根据类标签 $\boldsymbol{Y}^{(m)}\in\{+1,-1\}$，将训练数据 $\tilde{\boldsymbol{X}}$ 划分为小类（组）$\tilde{\boldsymbol{X}}_S$ 和大类（组）$\tilde{\boldsymbol{X}}_L$。这里，除了属于第 m 类视觉刺激的样本标签外，所有的标签都是–1。然后，计算两个类之间存在元素的尺度 J。我们必须注意 int() 定义了 floor 函数。下一步，将这个大类（组）$\tilde{\boldsymbol{X}}_L$ 随机划分为 J 个部分。实际上，J 是不平衡数据集生成的平衡子集的数量。因此，集成迭代的次数为 J。在每个平衡子集中，训练数据 $\tilde{\boldsymbol{X}}_{T_r}^{(n)}$ 是由小类 $\tilde{\boldsymbol{X}}_S$ 的所有样本、大类 $\tilde{\boldsymbol{X}}_L$ 中划分的 1 折样本及在前一次迭代中被预测错误的样本 $\bar{\boldsymbol{X}}^{(n)}$ 组成的。然后，利用训练样本间的皮尔逊相关（Pearson correlation）计算最终组合 $\boldsymbol{W}^{(n)}\in[0,1]$ 的训练权重，数值越大，学习灵敏度越高。

实际上，这些权重总是将小类的样本和先前迭代的误分类样本最大化。此外，其他样本的权重是大类和小类之间关联的一个尺度。因此，这些权重会根据以前迭代的性能在每次迭代中更新。作为每次迭代的最后一步，该方法生成分类模型 $\boldsymbol{\theta}^{(n)}$ 和最终组合的权重 $\boldsymbol{\alpha}^{(n)}$。我们采用一种简单的加权决策树[41]作为分类器 classifier() 的加权分类算法。最后，通过将 AdaBoost 方法应用到生成的平衡分类器，创建了最终的模型。

在二分类问题中，通过实验给出了二值 ℓ_1 正则化支持向量机（support vector machine, SVM）的性能。实际上，这种方法在文献[9]中被用于分类来自人脑的不同类别的刺激。作为文献[9]中介绍的用于解码大脑模式的正则化方法，我们比较了 Elastic Net 的性能。Elastic Net 的参数在文献[9]中被认为是最优的。我们还同时分析了文献[42]中提出的基于图的方法的性能。进一步，将我们提出的方法与多层感知机（MLP）的性能进行了比较，多层感知机是为了解码大脑模式而引入的[43]。我们使用文献[43]中提出的相同的网络参数作为 MLP 网络的优化解，即两个单元大小相同的隐含层。其中，T 和 V 分别为数据集中的时间点和体素的数量。此外，除了 R107 使用了“留 4 被试”策略外，所有的评估都采用了“留 1 被试”法进行交叉验证。例如，在 R105 中选取 5 个被试者的数据，在每次迭代中训练分类器，然后使用另外 1 个被试者的脑活动模式来对模型进行测试。需要注意的是，对于所有对比方法，每次迭代都使用了相同的训练集和测试集。针对不同数据集的实验结果见表 2。可以看出，与其他方法相比，我们所提出的方法在绝大多数数据集上获得了最佳的分类性能，因为它提供了更好的集成方法对 fMRI 数据集进行分析。

表 2　不同分类方法的分类精度（最大值 ± 标准差）

数据集	ℓ_1-SVM	Elastic Net	MLP	Graph-based	IABC
R105	16.72 ± 1.89	17.90 ± 0.21	**40.86 ± 0.11**	30.27 ± 0.41	39.32 ± 0.45
R107	26.39 ± 2.60	26.75 ± 0.97	37.54 ± 0.56	34.71 ± 0.08	**39.02 ± 0.17**
R232	30.65 ± 0.79	31.68 ± 0.69	40.40 ± 0.61	36.32 ± 0.31	**54.59 ± 0.02**
W001	25.47 ± 0.36	24.42 ± 0.14	33.73 ± 0.84	31.02 ± 0.67	**36.37 ± 0.59**
W002D	61.69 ± 0.86	62.91 ± 0.79	64.42 ± 0.38	64.81 ± 0.32	**65.24 ± 0.60**
W002P	64.98 ± 0.15	69.55 ± 0.61	72.59 ± 0.80	60.85 ± 0.70	**75.85 ± 0.76**
W005	32.80 ± 0.93	36.13 ± 0.73	43.09 ± 0.06	37.40 ± 0.94	**49.96 ± 0.36**
W011D	41.26 ± 0.13	45.80 ± 0.41	65.53 ± 0.52	70.67 ± 0.82	**71.78 ± 0.09**
W105	17.72 ± 0.10	30.18 ± 0.39	34.12 ± 0.41	32.06 ± 0.17	**37.72 ± 0.93**
W107	32.24 ± 1.61	39.37 ± 0.63	48.83 ± 0.52	50.72 ± 0.05	**54.49 ± 0.28**
W231	27.66 ± 0.92	30.09 ± 0.21	60.48 ± 0.17	**64.43 ± 0.07**	62.62 ± 0.38
W232	29.97 ± 0.46	24.88 ± 0.88	30.35 ± 0.26	**39.71 ± 0.91**	30.07 ± 0.07

利用基于ROI的数据集分析了所提出的IABC方法和其他分类方法的运行时间。图8展示了上述方法的运行时间，其中其他算法的运行时间是基于IABC进行缩放的。也就是说，所提出的IABC方法的运行时间被视为单位“1”。如图8所示，MLP耗费了最久的运行时间，尤其是当体素或时间点维度很高的时候。由于基于图的方法必须跨被试者将体素空间转换为图形空间，该方法的运行时间也较长，特别是对于具有较多被试者的数据集，如R107数据集。对于所提出的IABC方法，它通过将大量的不平衡样本划分为一组小的平衡样本，然后创建分类（认知）模型，从而生成合适的运行时间。值得注意的是，全脑数据集的运行时间也有同样的趋势。

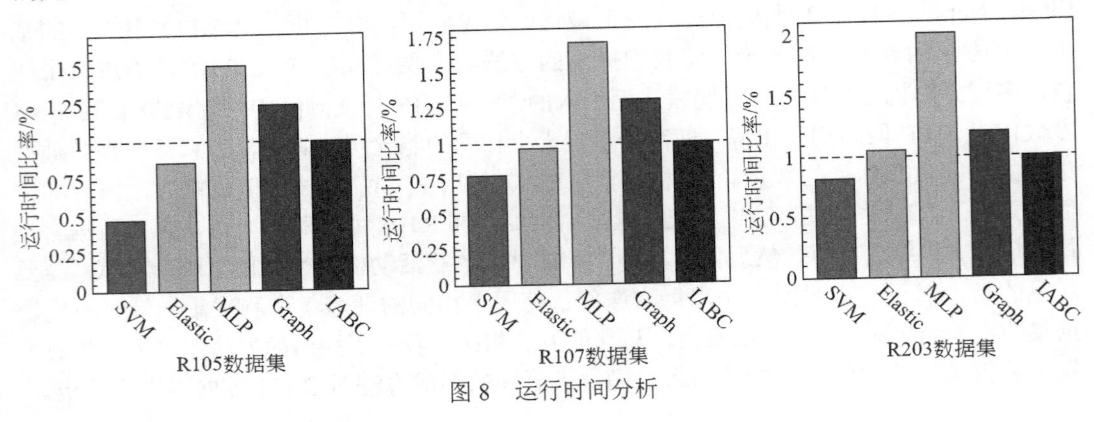

图8　运行时间分析

5　深度表征相似性学习

作为fMRI分析的基本方法之一，表征相似性分析（representational similarity analysis, RSA）[7,44-45]评估不同认知任务之间的相似性（或距离）[46]。实际上，经典的RSA[7,45]可以从数学上表述为一个多集合回归问题，即神经活动矩阵与设计矩阵之间的线性映射模型[3,7]。经典的RSA采用基本的线性方法，例如，普通最小二乘（ordinary least squares, OLS）[7]或一般线性模型（general linear model, GLM）[45]。

最近的研究表明，这些方法不能在大型真实世界数据集上提供准确的性能，例如，具有广泛兴趣区域（ROI）的数据集或全脑功能磁共振成像数据[3,22,46-47]。这些经典的方法存在几个问题[21,48]。首先，经典的RSA方法可能无法提供准确的相似性分析，特别是当神经活动的协方差具有较低的信噪比（signal-to-noise ratio, SNR）时[3,21]。换句话说，经典的RSA计算协方差矩阵的逆没有正则化项[7,21,44-45]，而大多数fMRI数据集可能不是满秩的，即体素的数量大于时间点[1,22,46]。下一个问题是，为了应用相似性分析，这些技术需要同时存储所有的数据点。因此，它们对于大型数据集的计算效率不高[49-50]。

在收集基于任务的 fMRI 数据集时，我们生成两个元素，即功能磁共振影像和设计矩阵。其中，功能磁共振影像的定义和表示与上文相同。此外，为每个被试者定义一个设计矩阵，描述与每个认知任务相关的事件。$\boldsymbol{D}^{(\ell)}=\left\{d_{ik}^{(\ell)}\right\}\in\mathbb{R}^{T\times P},d_{ik}\in\mathbb{R},1\leqslant i\leqslant T,1\leqslant k\leqslant P$ 定义了属于第 ℓ 个被试者的设计矩阵。这里，P 表示不同类别刺激的数量。此外，$\boldsymbol{d}_{.k}^{(\ell)}\in\mathbb{R}^{T},1\leqslant k\leqslant P$ 作为设计矩阵的第 k 列，是第 k 类（$\boldsymbol{o}_{.k}^{(\ell)}\in\mathbb{R}^{T}$）的起始点与 Ξ 作为血流动力学响应函数（HRF）信号的卷积（$\boldsymbol{d}_{.k}^{(\ell)}=\boldsymbol{o}_{.k}^{(\ell)}*\Xi$）[3,39]。

如图 9 所示，DRSL 通过转换函数将神经活动映射到一个信息丰富的空间，即 $\boldsymbol{x}\in\mathbb{R}^{V_{\text{org}}}\to f(\boldsymbol{x})\in\mathbb{R}^{V}$，其中 $V_{\text{org}}\geqslant V$ 表示在线性嵌入空间中映射的特征数量。虽然 f 可以是任何受限制的固定转换函数（如高斯或多项式），但本文在 $\boldsymbol{x}$ 中使用了多层叠加的非线性转换函数。

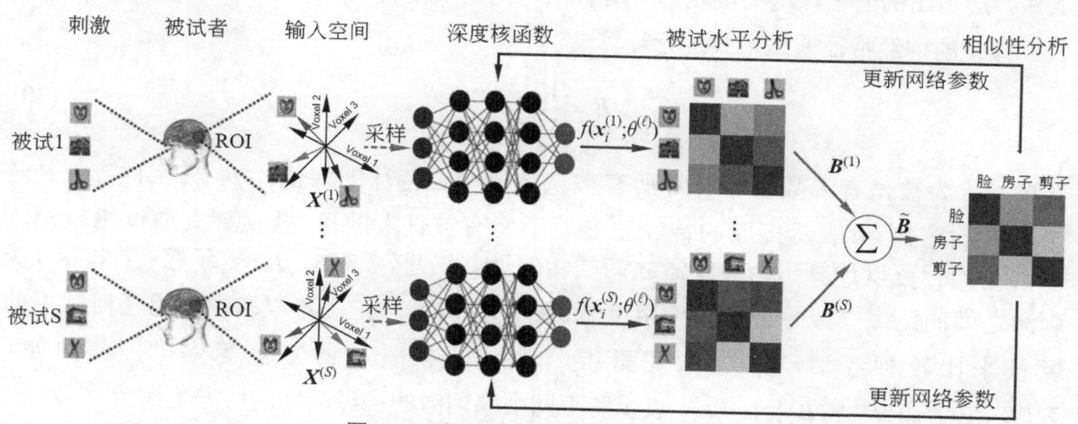

图 9　深度表征相似性学习模型框架

利用所提出的转换函数，在被试层次内，DSRL 的目标函数可以表示为

$$J_{R}^{(k,\ell)}=\sum_{i\in\Psi^{(k,\ell)}}\|f\left(\boldsymbol{x}_{i.}^{(\ell)};\theta^{(\ell)}\right)-\boldsymbol{d}_{i.}^{(\ell)}\boldsymbol{B}^{(\ell)}\|_{2}^{2}+r\left(\boldsymbol{B}^{(\ell)}\right) \tag{17}$$

通过解决以下优化问题可以获得最优解:

$$\min_{\boldsymbol{B}^{(\ell)},\theta^{(\ell)}} J_{R}^{(k,\ell)} \tag{18}$$

其中，$\boldsymbol{B}^{(\ell)}=\left\{\beta_{kj}^{(\ell)}\right\}\in\mathbb{R}^{P\times V},\beta_{kj}^{(\ell)}\in\mathbb{R},1\leqslant k\leqslant P,1\leqslant j\leqslant V$ 表示估计回归矩阵，包括抽取的第 ℓ 个被试者的神经特征。此外，$\Psi^{(k,\ell)}$ 是一组与第 k 次迭代和第 ℓ 个被试者有关的随机选

择的时间点。这个集合（$\left|\Psi^{(k,\ell)}\right|<T$）的大小等于 batch 的大小。此外，$\boldsymbol{x}_{i.}^{(\ell)}\in\mathbb{R}^{1\times V_{\text{org}}}$ 表示属于第 ℓ 个被试者和第 i 个时间点的神经活动 $\boldsymbol{X}^{(\ell)}$ 的所有体素，$\boldsymbol{d}_{i.}^{(\ell)}\in\mathbb{R}^{1\times P}$ 则是属于第 ℓ 个被试者的设计矩阵 $\boldsymbol{D}^{(\ell)}$ 中的第 ℓ 行。

与深度转换函数的其他应用不同[22,51-52]，我们考虑的是所有被试者的深度网络层的固定结构，包括 $f(\boldsymbol{x}_{i.}^{(\ell)},\theta^{(\ell)})$，而不是 $f_\ell(\boldsymbol{x}_{i.}^{(\ell)},\theta^{(\ell)})$。这种表示法可以提高生成结果的稳定性，也可以减少必须为每个 RSA 问题估计的参数数量。因此，与传统的 RSA 方法相比，我们只需要对每个问题估计额外的网络参数 $(\theta^{(\ell)},1\leqslant\ell\leqslant S)$。

DRSL 正则化项定义如下:

$$r(\boldsymbol{B})=\sum_{j=1}^{V}\sum_{k=1}^{P}\alpha\left|\beta_{kj}\right|+10\alpha\left(\beta_{kj}\right)^{2} \tag{19}$$

其中，α 是比例因子。这里，我们对所有归一化数据使用 $\alpha=10$，即 $\boldsymbol{X}^{(\ell)}\sim\mathcal{N}(0,1)$。

最终的神经特征通过下式进行计算:

$$\tilde{\boldsymbol{B}}=\frac{1}{S}\sum_{\ell=1}^{S}\boldsymbol{B}^{(\ell)} \tag{20}$$

其中，S 为被试者人数；$\tilde{\boldsymbol{B}}$ 计算所有被试者的神经特征的均值；$\tilde{\beta}_{i.}\in\mathbb{R}^{V},1\leqslant i\leqslant P$ 是矩阵 $\tilde{\boldsymbol{B}}$ 的第 i 行，表示提取的神经签名属于所有被试者的第 i 类刺激。通过比较矩阵 $\tilde{\boldsymbol{B}}$ 的不同行可以计算出两类刺激之间的相似性。这里，$\mathcal{D}(\tilde{\beta}_{i.},\tilde{\beta}_{j.})$ 表示第 i 类和第 j 类刺激之间的距离（或相似性）使用度量 $\mathcal{D}()$。我们可以在机器学习中使用任何常用的度量来比较神经活动，如欧几里得、余弦、相关、协方差等[21,46-48]。例如，$\mathbb{E}[\left(\tilde{\boldsymbol{B}}-\mathbb{E}(\tilde{\boldsymbol{B}})\right)\left(\tilde{\boldsymbol{B}}-\mathbb{E}(\tilde{\boldsymbol{B}})\right)^{\mathrm{T}}]\in\mathbb{R}^{P\times P}$ 表示跨不同类别刺激的神经活动的协方差比较[3]。

我们比较了 DRSL 与现有 6 种相似度方法的性能。类间相关性最大值见表 3。传统 RSA 算法生成的神经特征具有高度的相关性。LASSO 和 RSL 通过应用正则化方法提供了更好的特征。在这里，RSL 通过使用一个定制的正则化项可以处理稀疏性和噪声，与常规的 ℓ_1 范数相比，RSL 优于 LASSO。接下来，BRSA 和 PCM 通过估计 fMRI 数据集上的高斯分布来生成神经特征。尽管 PCM 算法在 R232 中的性能优于其他算法，但高斯先验假设并不能提供比该方法更好的性能。如前所述，GRSA 不能提供稳定的分析。表 3 比较了 GRSA 与其他技术产生的神经特征的方差。最后，与其他方法相比，DRSL 具有更好的性能。实际上，它通过使用深度神经网络提供了更好的特征表示，即将神经活动映射到线性空间。

表 3　类间相关性分析（最大值 ± 标准差）

方法	R105	R107	R232	W005	W203	W231
RSA[27]	0.947 ± 0.042	0.922 ± 0.053	0.927 ± 0.033	0.891 ± 0.035	0.951 ± 0.045	0.888 ± 0.113
LASSO	0.751 ± 0.242	0.715 ± 0.147	0.900 ± 0.026	0.823 ± 0.036	0.851 ± 0.077	0.761 ± 0.114
RSL[33]	0.821 ± 0.120	0.531 ± 0.123	0.631 ± 0.193	0.699 ± 0.076	0.572 ± 0.155	0.421 ± 0.273
BRSA[3]	0.389 ± 0.010	0.458 ± 0.076	0.871 ± 0.100	0.519 ± 0.045	0.383 ± 0.051	0.521 ± 0.082
PCM[52]	0.401 ± 0.064	0.357 ± 0.041	**0.490 ± 0.057**	0.591 ± 0.083	0.405 ± 0.075	0.487 ± 0.107
GRSA[53]	0.584 ± 0.142	0.426 ± 0.099	0.754 ± 0.107	0.392 ± 0.069	0.411 ± 0.094	0.267 ± 0.243
LRSL	0.451 ± 0.081	0.142 ± 0.092	0.555 ± 0.112	0.361 ± 0.021	0.451 ± 0.012	0.231 ± 0.063
DRSL	**0.372 ± 0.016**	**0.135 ± 0.000**	0.496 ± 0.093	**0.139 ± 0.019**	**0.270 ± 0.039**	**0.126 ± 0.002**

不同对比方法的分类精度见表 4。在分类实验分析中，RSA（无正则化项）的性能类似于随机抽样。LASSO 和 RSL 的结果表明正则化对线性模型的性能有积极的影响。通过对数据流形拟合高斯分布，与仅通过正则化函数控制结果质量的方法相比，BRSA 和 PCM 提供了更好的性能。通过比较线性 LRSL 和提出的 DRSL，可以了解非线性转换函数对分析性能的积极影响。在大多数 fMRI 数据集中，DRSL 有两个重要的特点：①灵活的非线性核；②有效的正则化方法。

表 4　不同对比方法的分类精度（平均值 ± 标准差）

方法	R105	R107	R232	W005	W203	W231
RSA[27]	18.69 ± 2.37	33.57 ± 3.06	31.78 ± 2.71	27.10 ± 1.02	24.40 ± 2.13	18.63 ± 3.37
LASSO[30]	47.63 ± 1.04	38.18 ± 1.27	43.66 ± 1.80	31.03 ± 0.71	35.64 ± 0.81	29.43 ± 0.96
RSL[33]	51.29 ± 1.13	37.97 ± 2.08	50.30 ± 1.27	46.17 ± 1.36	60.06 ± 2.41	**68.25 ± 1.07**
BRSA[3]	60.39 ± 0.69	65.24 ± 0.58	61.06 ± 0.19	66.58 ± 0.73	76.90 ± 0.59	50.26 ± 1.00
PCM[52]	63.91 ± 0.92	69.11 ± 0.73	59.74 ± 0.88	68.24 ± 0.93	72.58 ± 0.33	49.51 ± 0.98
GRSA[53]	58.36 ± 1.03	66.05 ± 1.76	65.14 ± 1.01	61.43 ± 0.92	83.15 ± 0.24	52.76 ± 1.61
LRSL	59.73 ± 2.81	72.31 ± 0.04	63.27 ± 1.30	60.00 ± 0.72	75.68 ± 0.80	51.68 ± 1.42
DRSL	**78.13 ± 0.19**	**84.26 ± 0.64**	**70.31 ± 0.55**	**81.37 ± 0.11**	**91.40 ± 0.24**	65.74 ± 1.00

6　easyfMRI——人脑解码和可视化工具箱

除了上述理论和实验研究，还开发了一款免费、开源的人脑解码和可视化工具箱——easyfMRI（https://easyfmri.learningbymachine.com/）。easyfMRI 是基于脑影像数据结构（brain imaging data structure, BIDS）文件结构设计的，支持 Matlab 数据文件，能够基于设计矩阵的数据集自动标记标签，并同时支持有监督信息和无监督信息的脑影像分析。

如图 10 所示，easyfMRI 采用先进的机器学习技术和高性能计算来分析基于任务的 fMRI 数据集。它为应用特征分析、超校准（hyperalignment, HA）、多体素模式分析（multi-voxel pattern analysis, MVPA）、表征相似性分析（representational similarity analysis, RSA）等提供了一个友好的基于图形用户界面的环境。此外，easyfMRI 是与 FSL（用于预处理步骤）、SciKit-Learn（用于模型分析）、PyTorch（用于深度学习方法）及 AFNI 和 SUMA（用于三维可视化）集成的。

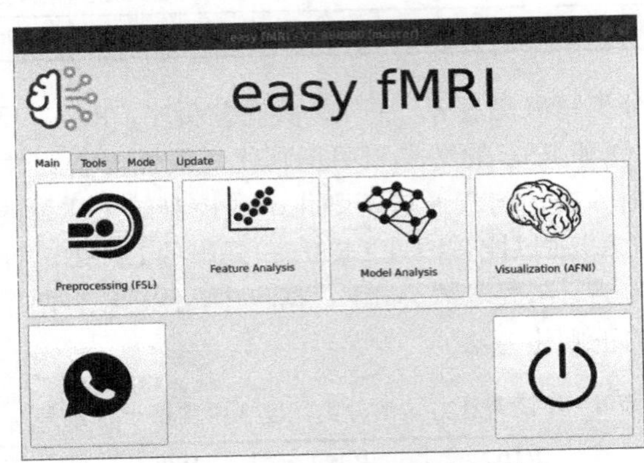

图 10　easyfMRI 工具箱主界面

在预处理步骤中，easyfMRI 为 FSL 生成预处理脚本，尤其可以同时为所有文件生成预处理脚本，支持多个预处理进程的并行计算，并且可编程为不同格式的事件文件（event files）；在 AFNI 和 SUMA 中可视化输出，将分析模型转换到三维空间上，支持标准可视化空间；软件采用深度学习方法，基于 GPU 运行基本的方法；拥有用于更进一步分析的编程控制台窗口，可以编辑 Matlab / Ezdata 的文件。

7　总结与展望

21 世纪最大的挑战之一是理解人类大脑如何工作。作为跨学科的研究领域，计算神经科学可以通过使用数学、物理、心理学、精神病学和机器学习等学科的不同概念来破解神经编码模式。本文着重针对我们在基于机器学习的脑解码方面的进展进行介绍。对于多被试者脑影像的功能校准，提出了一种深度超校准方法，扩展了校准过程中响应空间的转换模式，显著提高了多被试的神经解码效果。为了有效地利用数据中的标签信息，提出了一种有监督的功能校准方法，该方法有效地利用样本标签，最大化同一类别的相关性，最小化不同类别的相关性。为了解决多站点小样本数据的问题，提出了一种共享空间迁移学习方法，将不同站点的数据映射到共享空间中，并在新生成的空间进行脑解码相关研究。该方法有效地扩展了单个站点的数据量，显著提高了网络的训练效果和多数据集模型的泛化性能。此外，不平衡 AdaBoost 二分类（IABC）被提出作为一种新的二元不平衡分类学习技术，它非常适合于一对一的分类分析。我们还开发了深度表征相似学习（DRSL）框架，这是 RSA 的深度扩展，适用于分析具有大量被试者的 fMRI 数据集和高维数据中各种认知任务之间的相似性。

未来，在基于机器学习的脑解码研究中，有几个问题值得深入关注。首先，现有研究大多基于单模态数据进行，而多模态数据融合往往能够提供更为充分、互补的信息。例如，将功能磁共振影像与结构磁共振影像（sMRI）、脑电图（EEG）等模态数据结合，挖掘更加丰富的响应特征。其次，现有的方法大多没有很好地利用整个大脑的全局信息。在未来的研究中，计划在了解全脑数据的内在信息的基础上，开发基于信息的模型，对小区域的数据信息进行平滑处理。这样能够使全脑数据中的有效区域信息更加清晰，为后续的特征选择和表征相似性分析提供更好的输入信息。最后，如何以认知机制为先验知识增加复杂模型的可解释性是一个重要的研究方向。

参考文献

[1] HAXBYJ V, CONNOLLY A C, GUNTUPALLI J S. Decoding neural representational spacesusing multivariate pattern analysis[J]. Annual Review of Neuroscience, 2014, 37: 435-456.

[2] CHEN P H. Multi-view Representation learning with applications to functional neuroimaging data[D].Princeton: Princeton University, 2017.

[3] CAI MB, SCHUCK WN, PILLOW W J, et al. A Bayesian method for reducing biasin neural representationalsimilarity analysis[C]//Proceedings of Advances in Neural Information Processing Systems. 2016, 29: 4951-4959.

[4] LORBERT A, RAMADGEP J. Kernel hyperalignment[C]//Proceedings of Advances in Neural Information Processing Systems. 2012, 25: 1790-1798.

[5] LORBERT A. Alignment and supervised learning with functional neuroimaging data[D]. Princeton: Princeton University, 2012.

[6] HAXBY J V, GUNTUPALLI J S, CONNOLLY A C, et al. A common, high-dimensional model of the representationalspace in human ventral temporal cortex[J]. Neuron, 2011, 72(2): 404-416.

[7] KRIEGESKORTE N, GOEBEL R, BANDETTINI P. Information-based functional brain mapping[C]//Proceedings of the National academy of Sciences of the United States of America. 2006, 103(10): 3863-3868.

[8] ANDERSON M, OATES T. A critique of multi-voxel pattern analysis[C]//Proceedings of the Annual Meeting of the Cognitive Science Society. 2010, 32(32).

[9] MOHR H, WOLFENSTELLER U, FRIMMEL S, et al.Sparse regularization techniques provide novel insights into outcome integration processes[J]. NeuroImage, 2015, 104: 163-176.

[10] MCMENAMIN B W, DEASON R G, STEELE V R, et al. Separability of abstractcategory and specificexemplar visual object subsystems: Evidence from fMRI pattern analysis[J]. Brain and Cognition, 2015, 93: 54-63.

[11] HANSON S J, MATSUKA T, HAXBY J V. Combinatorial codes in ventral temporal lobe for object recognition: Haxby (2001) revisited: is there a "face" area?[J]. NeuroImage, 2004, 23(1): 156-166.

[12] YOUSEFNEZHAD M, ZHANG D Q. Multi-Region Neural Representation: A novel model for decoding visual stimuli in human brains[C]//Proceedings of the 2017 SIAM International Conference on Data Mining. Society for Industrial and Applied Mathematics. 2017: 54-62.

[13] XU H, LORBERT A, RAMADGE P J, et al. Regularized hyperalignment of multi-set fMRI data[C]// Proceedings of 2012 IEEE Statistical Signal Processing Workshop (SSP). IEEE, 2012: 229-232.

[14] CHEN P H, GUNTUPALLI J S, HAXBY J V, et al. Joint SVD-Hyperalignment for multi-subject fMRI data alignment[C]//Proceedings of 2014 IEEE International Workshop on Machine Learning for Signal Processing (MLSP). IEEE, 2014: 1-6.

[15] CHEN P H, CHEN J, YESHURUN Y, et al. A reduced-dimension fMRI shared response model[C]// Proceedings of Advances in Neural Information Processing Systems. 2015, 28: 460-468.

[16] YOUSEFNEZHAD M, ZHANG D Q. Local discriminant hyperalignment for multi-subject fMRI data alignment[C]//Proceedings of the AAAI Conference on Artificial Intelligence. 2017, 31(1): 59-65.

[17] TALAIRACH J AND TOURNOUX P. Co-planar stereotaxic atlas of the human brain. 3-Dimensional proportional system: An approach to cerebral imaging[J]. Thieme Medical Publishers, Inc., Georg ThiemeVerlag, Stuttgart, New York, 1988.

[18] MAZZIOTTA J, TOGA A, EVANS A, et al. A probabilistic atlas and reference system for the human brain: International Consortium for Brain Mapping (ICBM)[J]. Philosophical Transactions of the Royal Society of London. Series B: Biological Sciences, 2001, 356(1412): 1293-1322.

[19] WATSON J D G, MYERS R, FRACKOWIAK R S J, et al. Area V5 of the human brain: evidence from a combined study using positron emission tomography and magnetic resonance imaging[J]. Cerebral Cortex, 1993, 3(2): 79-94.

[20] RADEMACHER J, CAVINESS JR V S, STEINMETZ H, et al. Topographical variation of the human primary cortices: implications for neuroimaging, brain mapping, and neurobiology[J]. Cerebral Cortex, 1993, 3(4): 313-329.

[21] DIEDRICHSEN J, KRIEGESKORTE N. Representational models: A common framework for understanding encoding, pattern-component, and representational-similarity analysis[J]. PLoS computational biology, 2017, 13(4): e1005508.

[22] YOUSEFNEZHAD M, ZHANG D Q. Deep Hyperalignment[C]//Proceedings of Advances in Neural Information Processing Systems. 2017, 1603-1611.

[23] SUI J, PEARLSON G, CAPRIHAN A, et al. Discriminating schizophrenia and bipolar disorder by fusing fMRI and DTI in a multimodal CCA+ joint ICA model[J]. NeuroImage, 2011, 57(3): 839-855.

[24] SUI J, HE H, PEARLSON G D, et al. Three-way (N-way) fusion of brain imaging data based on mCCA+ jICA and its application to discriminating schizophrenia[J]. NeuroImage, 2013, 66: 119-132.

[25] GUNTUPALLI J S, HANKE M, HALCHENKO Y O, et al. A model of representational spaces in human cortex[J]. Cerebral cortex, 2016, 26(6): 2919-2934.

[26] CHEN P H, ZHU X, ZHANG H J, et al. A convolutional autoencoder for multi-subject fMRI data aggregation[C]//Proceedings of 29th Workshop of Representation Learning in Artificial and Biological Neural Networks, 2016.

[27] TUREK J S, WILLKE T L, CHEN P H, et al. A semi-supervised method for multi-subject fMRI functional alignment[C]//Proceedings of 2017 IEEE International Conference on Acoustics, Speech and Signal Processing (ICASSP). IEEE, 2017: 1098-1102.

[28] RASTOGI P, VAN DURME B, ARORA R. Multiview LSA: Representation learning via generalized CCA[C]//Proceedings of the 2015 conference of the North American chapter of the Association for Computational Linguistics: human language technologies. 2015: 556-566.

[29] BRAND M. Incremental singular value decomposition of uncertain data with missing values[C]//Proceedings of European Conference on Computer Vision. Springer. Berlin, Heidelberg, 2002: 707-720.

[30] ZHANG H J, CHEN P H, RAMADGE P. Transfer learning on fMRI datasets[C]//Proceedings of International Conference on Artificial Intelligence and Statistics. PMLR, 2018: 595-603.

[31] YOUSEFNEZHAD M, SELVITELLA A, HAN L X, et al. Supervised hyperalignment for multi-subject fMRI data alignment[J]. IEEE Transactions on Cognitive and Developmental Systems, 2020.

[32] VEGA R, GREINER R. Finding effective ways to (Machine) learn fMRI-based classifiers from multi-site data[M]//Understanding and Interpreting Machine Learning in Medical Image Computing Applications. Springer, 2018: 32-39.

[33] WANG M L, ZHANG D Q, HUANG J S,et al. Identifying autism spectrum disorder with multi-site fMRI via low-rank domain adaptation[J]. IEEE transactions on medical imaging, 2019, 39(3): 644-655.

[34] GAO Y F, ZHOU B, ZHOU Y J,et al. Transfer learning-based behavioural task decoding from brain activity[C]//Proceedings of The International Conference on Healthcare Science and Engineering. Springer, Singapore, 2018: 71-81.

[35] THOMAS A W, MÜLLER K R, SAMEK W. Deep transfer learning for whole-brain FMRI analyses[M]//OR 2.0 Context-Aware Operating Theaters and Machine Learning in Clinical Neuroimaging. Springer, 2019: 59-67.

[36] ARORA R, LIVESCU K. Kernel CCA for multi-view learning of acoustic features using articulatory measurements[C]//Proceedings of Symposium on Machine Learning in Speech and Language Processing. 2012, 34-37.

[37] ZHOU S, LI W W, COX C,et al. Side information dependence as a regularizer for analyzing human brain conditions across cognitive experiments[C]//Proceedings of the AAAI Conference on Artificial Intelligence. 2020, 34(4): 6957-6964.

[38] SAPATNEKAR S S. Overcoming variations in nanometer-scaletechnologies[J]. IEEE Journal on Emerging and Selected Topics in Circuits and Systems, 2011, 1(1): 5-18.

[39] YOUSEFNEZHAD M, ZHANG D Q. Anatomical pattern analysis for decoding visual stimuliin human brains[J]. Cognitive Computation, 2018, 10(2): 284-295.

[40] YOUSEFNEZHAD M, ZHANG D Q.Decoding visual stimuli in human brain by using Anatomical Pattern Analysis on fMRI images[C]//Proceedings of International Conference on Brain Inspired Cognitive Systems. Springer, 2016: 47-57.

[41] LIU X-Y, WU J-X, ZHOU Z-H. Exploratory undersampling for class-imbalance learning[J]. IEEE Transactions on Systems, Man, and Cybernetics, Part B (Cybernetics), 2008, 39(2): 539-550.

[42] OSHER D E, SAXE R R, KOLDEWYN K, et al. Structural connectivity fingerprints predict cortical selectivity for multiple visual categories across cortex[J]. Cerebral cortex, 2016, 26(4): 1668-1683.

[43] ANDERSON M AND OATES T. A critique of multi-voxel pattern analysis[C]//Proceedings of the Annual Meeting of the Cognitive Science Society. 2010, 32(32).

[44] KRIEGESKORTE N, MUR M, BANDETTINI P A. Representational similarity analysis-connecting the branches of systems neuroscience[J]. Frontiers in systems neuroscience, 2008, 2: 4.

[45] CONNOLLY A C, GUNTUPALLI J S, GORS J, et al. The representation of biological classes in the human brain[J]. Journal of Neuroscience, 2012, 32(8): 2608-2618.

[46] KHALIGH-RAZAVI S M, KRIEGESKORTE N. Deep supervised, but not unsupervised, models may explain IT cortical representation[J]. PLoS computational biology, 2014, 10(11): e1003915.

[47] KHALIGH-RAZAVI S M, HENRIKSSON L, KAY K, et al. Fixed versus mixed RSA: Explaining visual representations by fixed and mixed feature sets from shallow and deep computational models[J]. Journal of Mathematical Psychology, 2017, 76: 184-197.

[48] WALTHER A, NILI H, EJAZ N, et al. Reliability of dissimilarity measures for multi-voxel pattern analysis[J]. NeuroImage, 2016, 137: 188-200.

[49] OSWAL U, COX C, LAMBON-RALPH M, et al. Representational similarity learning with application to brain networks[C]//Proceedings of International Conference on Machine Learning. PMLR, 2016: 1041-1049.

[50] FIGUEIREDO M, NOWAK R. Ordered weighted ℓ_1 regularized regression with strongly correlated covariates: Theoretical aspects[C]//Proceedings of Artificial Intelligence and Statistics. PMLR, 2016: 930-938.

[51] ANDREW G, ARORA R, BILMES J, et al. Deep canonical correlation analysis[C]//Proceedings of International conference on machine learning. PMLR, 2013: 1247-1255.

[52] BENTON A, KHAYRALLAH H, GUJRAL B, et al. Deep generalized canonical correlation analysis[EB/OL]. (2017-06-15) [2018-12-01]. http://arXiv Preprint arXiv:1702.02519.

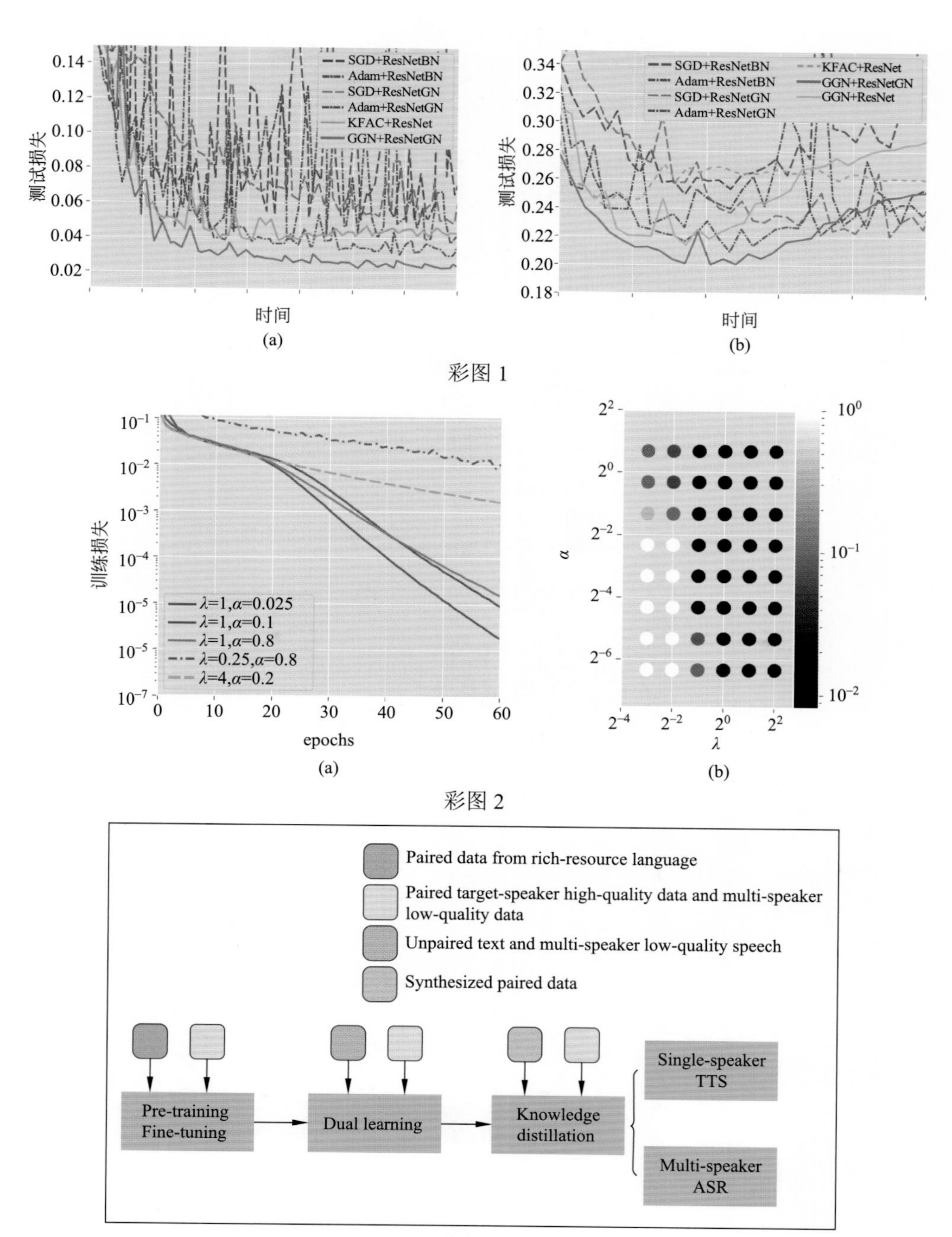

SGD+ResNetBN
Adam+ResNetBN
SGD+ResNetGN
Adam+ResNetGN
KFAC+ResNet
GGN+ResNetGN
GGN+ResNet
测试损失
时间
(a)
(b)
训练损失
epochs
λ=1,α=0.025
λ=1,α=0.1
λ=1,α=0.8
λ=0.25,α=0.8
λ=4,α=0.2
α
λ
Paired data from rich-resource language
Paired target-speaker high-quality data and multi-speaker low-quality data
Unpaired text and multi-speaker low-quality speech
Synthesized paired data
Pre-training
Fine-tuning
Dual learning
Knowledge distillation
Single-speaker TTS
Multi-speaker ASR

彩图 1

彩图 2

彩图 3

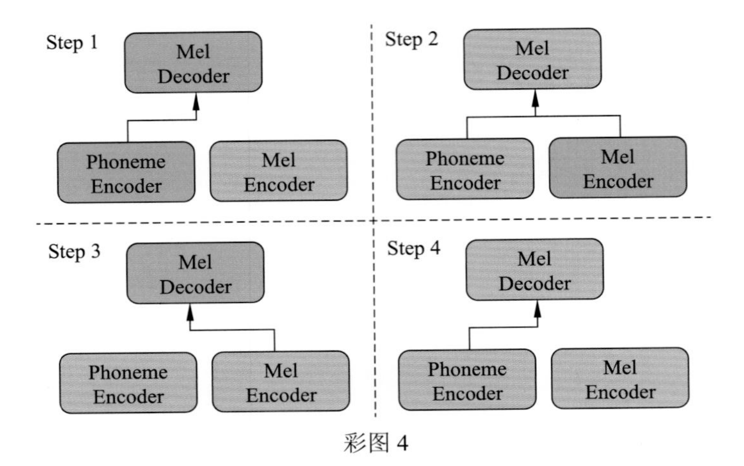
Step 1
Mel Decoder
Phoneme Encoder
Mel Encoder
Step 2
Mel Decoder
Phoneme Encoder
Mel Encoder
Step 3
Mel Decoder
Phoneme Encoder
Mel Encoder
Step 4
Mel Decoder
Phoneme Encoder
Mel Encoder

彩图 4

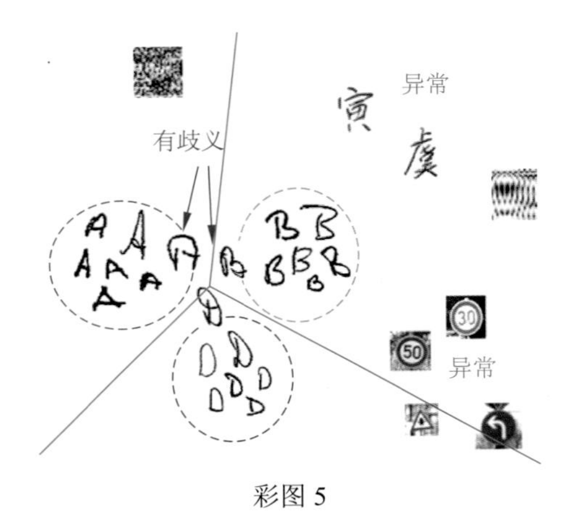
异常
有歧义
异常

彩图 5

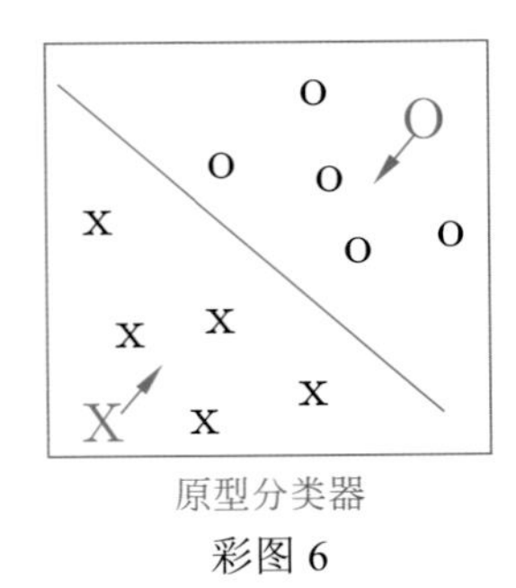
原型分类器

彩图 6

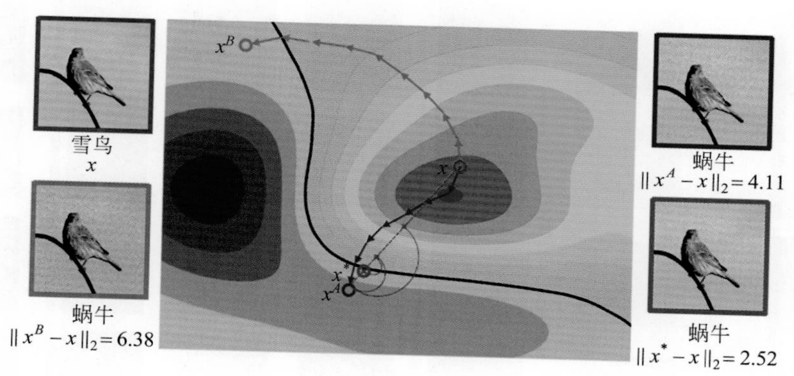

雪鸟
x
x^B
x
x^*
x^A
蜗牛
$\|x^A - x\|_2 = 4.11$
蜗牛
$\|x^B - x\|_2 = 6.38$
蜗牛
$\|x^* - x\|_2 = 2.52$

彩图 7

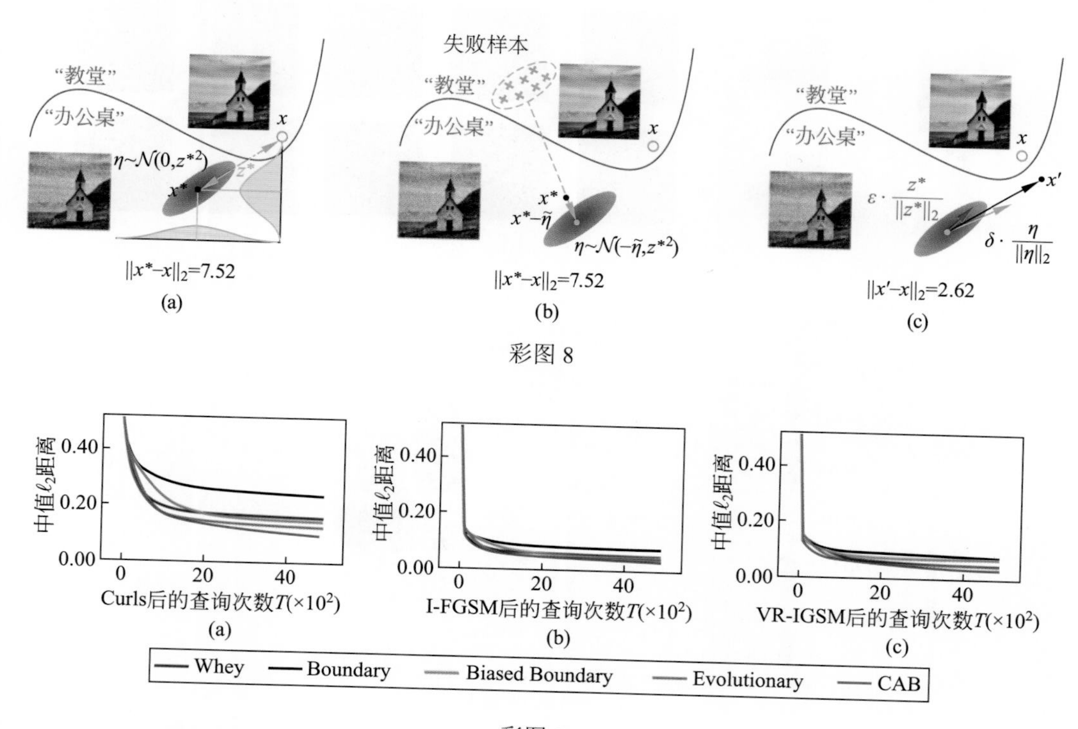

"教堂"
"办公桌"
x
$\eta \sim \mathcal{N}(0, z^{*2})$
x^*
$\|x^* - x\|_2 = 7.52$
(a)
失败样本
"教堂"
"办公桌"
x
x^*
$x^* - \tilde{\eta}$
$\eta \sim \mathcal{N}(-\tilde{\eta}, z^{*2})$
$\|x^* - x\|_2 = 7.52$
(b)
"教堂"
"办公桌"
x
x'
$\varepsilon \cdot \frac{z^*}{\|z^*\|_2}$
$\delta \cdot \frac{\eta}{\|\eta\|_2}$
$\|x' - x\|_2 = 2.62$
(c)
彩图 8
中值ℓ_2距离
0.40
0.20
0.00
0
20
40
Curls后的查询次数$T(\times 10^2)$
(a)
中值ℓ_2距离
0.40
0.20
0.00
0
20
40
I-FGSM后的查询次数$T(\times 10^2)$
(b)
中值ℓ_2距离
0.40
0.20
0.00
0
20
40
VR-IGSM后的查询次数$T(\times 10^2)$
(c)
Whey
Boundary
Biased Boundary
Evolutionary
CAB

彩图 9

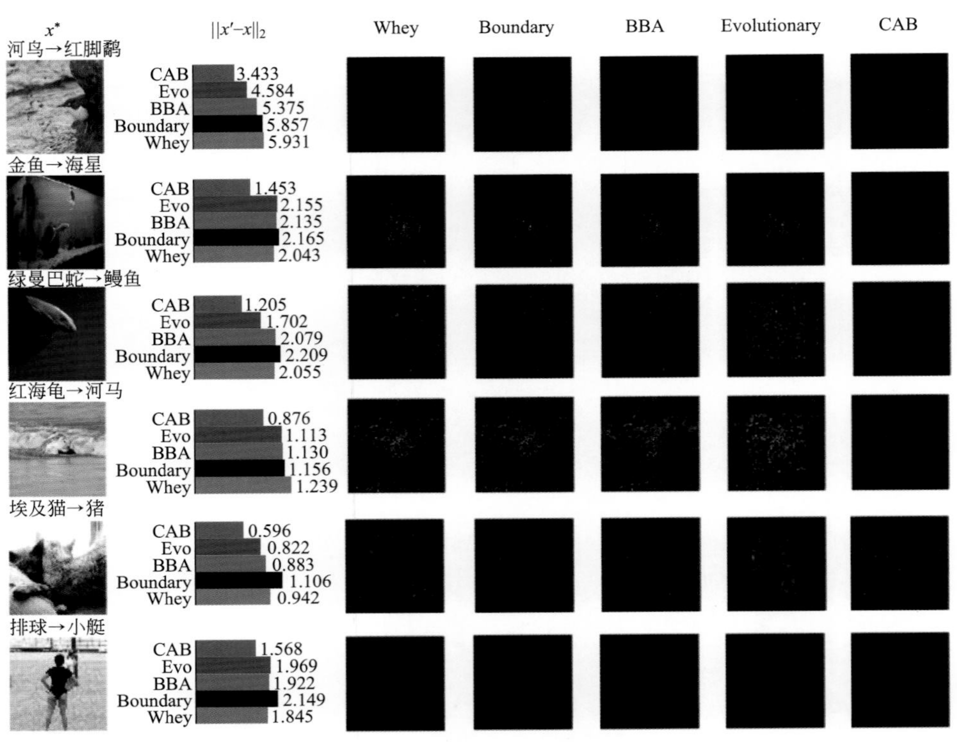
x*
‖x′−x‖₂
Whey
Boundary
BBA
Evolutionary
CAB
河鸟→红脚鹬
CAB 3.433
Evo 4.584
BBA 5.375
Boundary 5.857
Whey 5.931
金鱼→海星
CAB 1.453
Evo 2.155
BBA 2.135
Boundary 2.165
Whey 2.043
绿曼巴蛇→鳗鱼
CAB 1.205
Evo 1.702
BBA 2.079
Boundary 2.209
Whey 2.055
红海龟→河马
CAB 0.876
Evo 1.113
BBA 1.130
Boundary 1.156
Whey 1.239
埃及猫→猪
CAB 0.596
Evo 0.822
BBA 0.883
Boundary 1.106
Whey 0.942
排球→小艇
CAB 1.568
Evo 1.969
BBA 1.922
Boundary 2.149
Whey 1.845

彩图 10

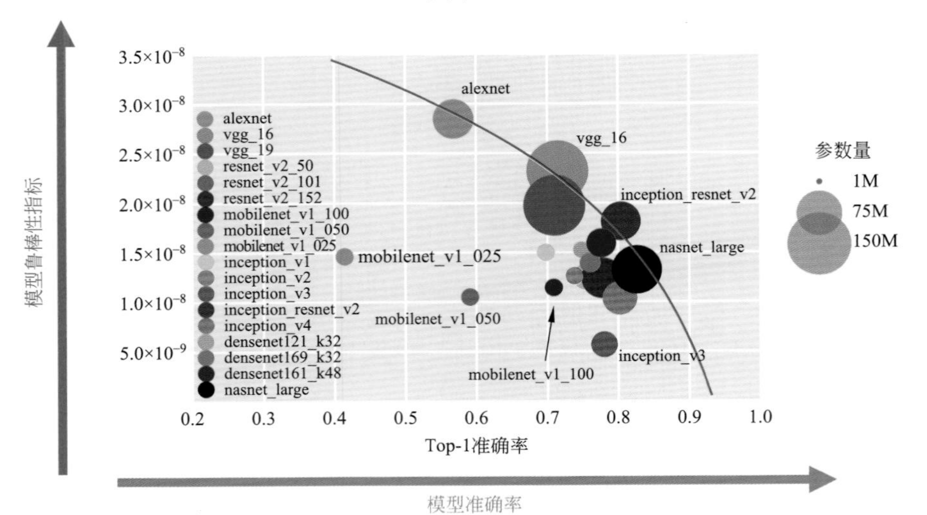
模型鲁棒性指标
3.5×10⁻⁸
3.0×10⁻⁸
2.5×10⁻⁸
2.0×10⁻⁸
1.5×10⁻⁸
1.0×10⁻⁸
5.0×10⁻⁹
alexnet
vgg_16
vgg_19
resnet_v2_50
resnet_v2_101
resnet_v2_152
mobilenet_v1_100
mobilenet_v1_050
mobilenet_v1_025
inception_v1
inception_v2
inception_v3
inception_resnet_v2
inception_v4
densenet121_k32
densenet169_k32
densenet161_k48
nasnet_large
alexnet
vgg_16
inception_resnet_v2
nasnet_large
mobilenet_v1_025
mobilenet_v1_050
mobilenet_v1_100
inception_v3
参数量
1M
75M
150M
0.2 0.3 0.4 0.5 0.6 0.7 0.8 0.9 1.0
Top-1准确率
模型准确率

彩图 11

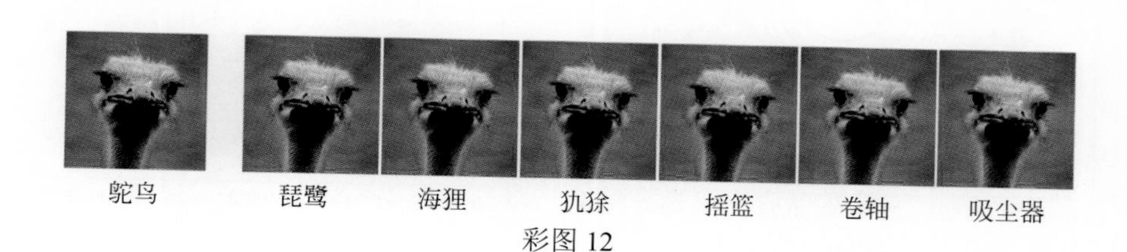

彩图 12

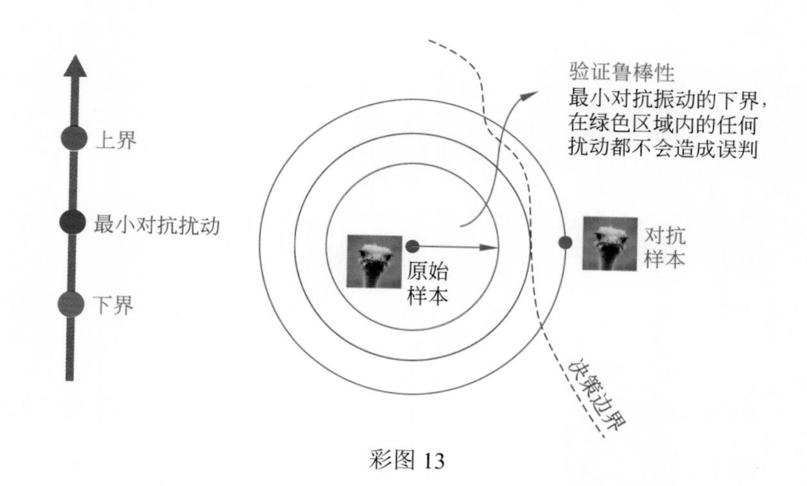

彩图 13

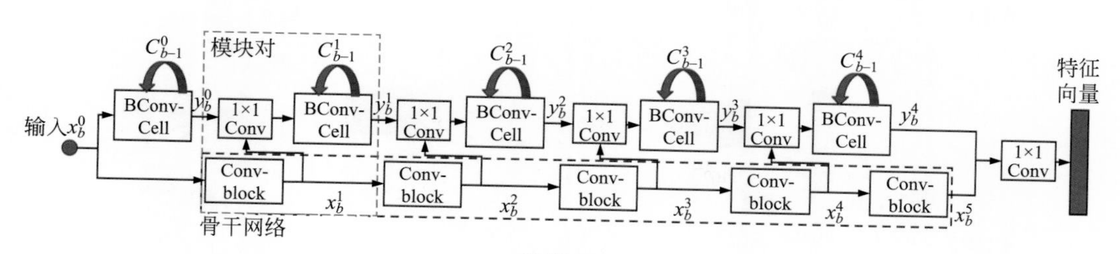

彩图 14

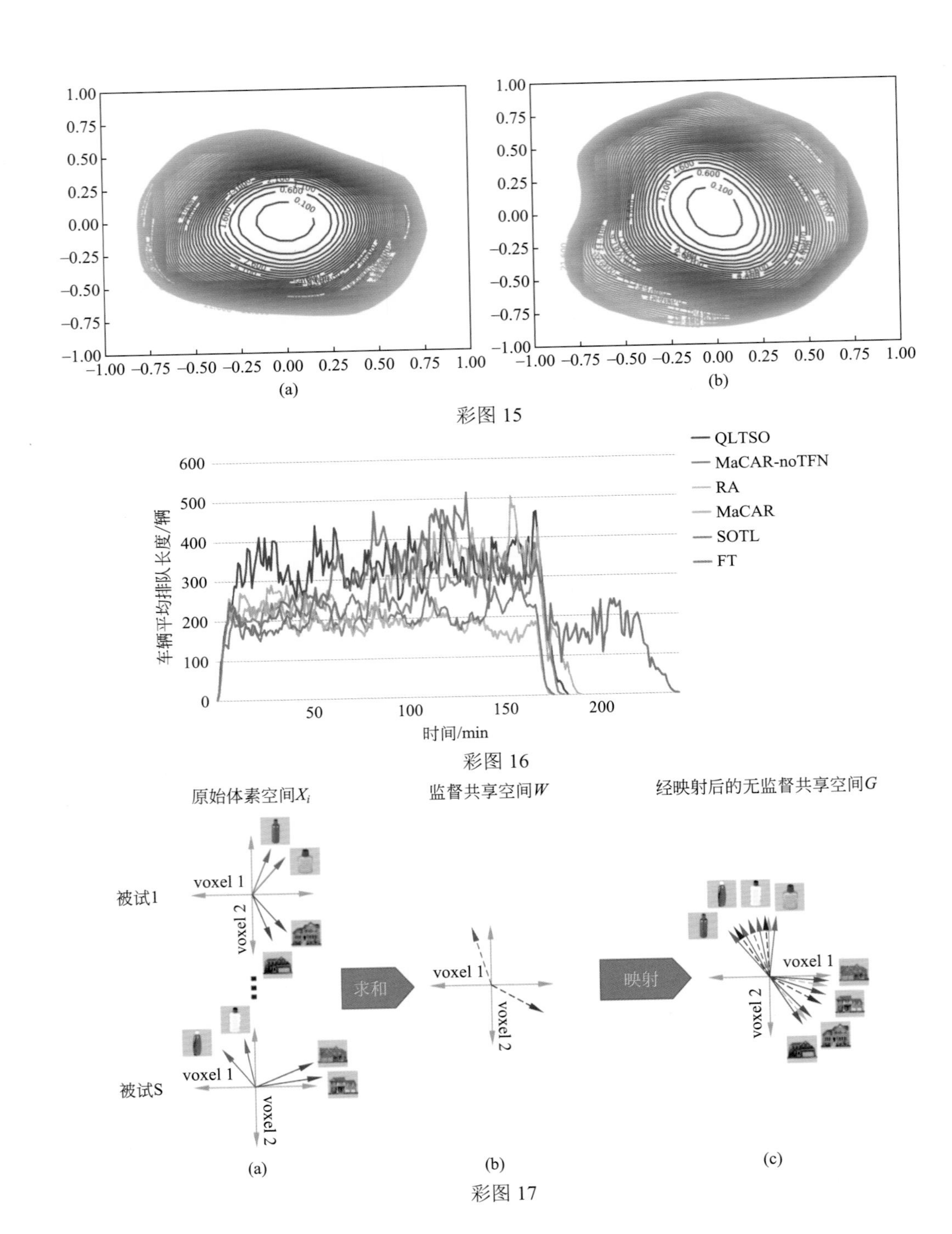

彩图 15

彩图 16

彩图 17

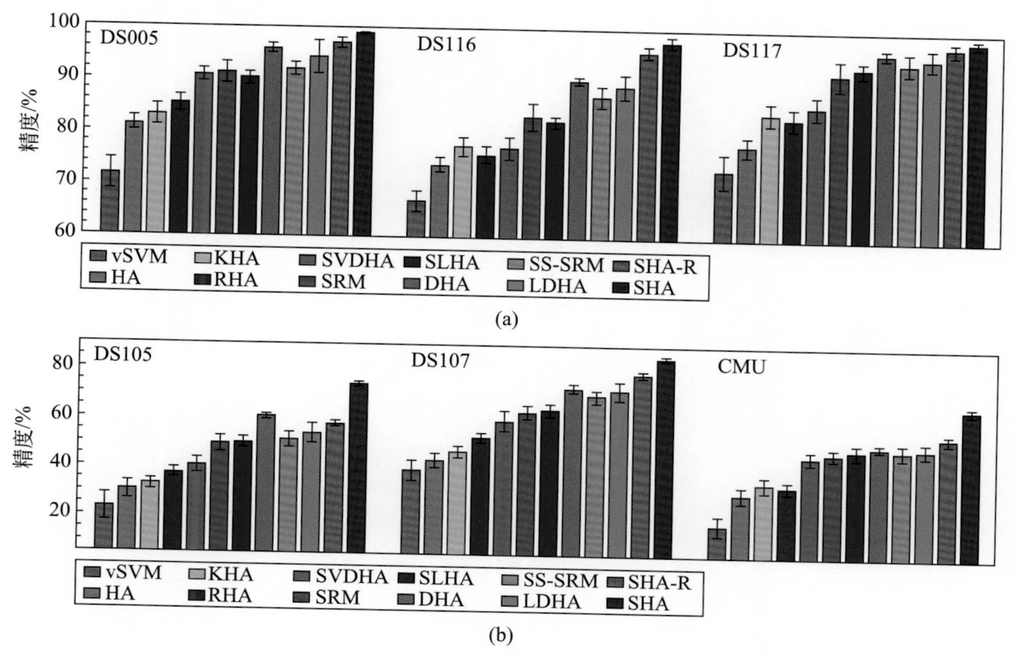

精度/%
DS005
DS116
DS117
vSVM
KHA
SVDHA
SLHA
SS-SRM
SHA-R
HA
RHA
SRM
DHA
LDHA
SHA
(a)
DS105
DS107
CMU
(b)

彩图 18